AF595460

Sensor Technology for Smart Homes

Sensor Technology for Smart Homes

Editors

Juan Ye
Michael O'Grady
Oresti Banos

MDPI • Basel • Beijing • Wuhan • Barcelona • Belgrade • Manchester • Tokyo • Cluj • Tianjin

Editors

Juan Ye
University of St Andrews
UK

Michael O'Grady
University College Dublin
Ireland

Oresti Banos
University of Granada (UGR)
Spain

Editorial Office
MDPI
St. Alban-Anlage 66
4052 Basel, Switzerland

This is a reprint of articles from the Special Issue published online in the open access journal *Sensors* (ISSN 1424-8220) (available at: https://www.mdpi.com/journal/sensors/special_issues/SmartHomes).

For citation purposes, cite each article independently as indicated on the article page online and as indicated below:

LastName, A.A.; LastName, B.B.; LastName, C.C. Article Title. *Journal Name* **Year**, *Volume Number*, Page Range.

ISBN 978-3-0365-0248-9 (Hbk)
ISBN 978-3-0365-0249-6 (PDF)

Contents

About the Editors

Juan Ye is a Senior Lecturer in the School of Computer Science at the University of St Andrews. She received a Bachelor's and Master's degree from Wuhan University, China, in 2002 and 2005, respectively, and a Ph.D. in Computer Science from University College Dublin, Ireland, in 2009. Her research area centers around sensor-based data analytics. The over-arching vision of her research is on how to make the best use of sensor data in understanding the environment and the people, and thus in making informed decisions. She has published three canonical reviews on human activity recognition in high-profile journals to identify the key research challenges and open questions. She has designed and developed major techniques advancing the area of human activity recognition with an emphasis on formal reasoning on relationships between context and human activities, designing a conceptual model to enable hierarchical reasoning, unsupervised learning of user behaviors, identifying multi-user concurrent activities, detecting abnormal and missing sensor data in real-world deployment, and discovering and recognizing unknown activities in long-term deployment. She has published papers with a H-index of 21 in top international conference proceedings and journals in pervasive computing. She has served as a General Chair of MobiHealth 2020, and as a Guest Editor for a Special Issue "Ambient Intelligence Environments" of Information.

Michael O'Grady has published in a wide range of international fora, contributing to over 176 peer-reviewed publications, resulting in a H-index of 28. He is a senior member of the ACM and IEEE. He has acted as an expert reviewer on European and international initiatives. His specific research interests include the applicability of intelligent systems in context and opportunistic service provision, as well as participatory science. From an educational perspective, he has a keen interest in Problem-Based Learning.

Oresti Banos (h-index 21) is a Tenured Professor of Computational Behaviour Modelling at the University of Granada (Spain, 2019-present). He is also a Senior Research Scientist affiliated with the Centre for Information and Communication Technologies of the University of Granada (CITIC-UGR). He is a Research Collaborator at Kyung Hee University (South Korea, 2016–present) and the University of Twente (Netherlands, 2018–present). He is a former Assistant Professor of Creative Technology and a Telemedicine Research Scientist at University of Twente (Netherlands, 2016–2018). Here, he worked with the Biomedical Signal and Systems Group (BSS), the Centre for Telematics and Information Technology (CTIT), the Research Centre for Biomedical Technology and Technical Medicine (MIRA), and the Centre for Monitoring and Coaching (CMC). He was a Postdoctoral Research Fellow at Kyung Hee University (South Korea, 2014–2016), a Predoctoral Research Fellow at the CITIC-UGR (Spain, 2010-2014) and a Visiting Scholar at the Technical University of Eindhoven (Netherlands, 2012), the Swiss Federal Institute of Technology Zurich (Switzerland, 2011), and the University of Alabama (USA, 2011). He holds three Master's degrees in telecommunications (2009), electrical (2011) and computer networking engineering (2011) and a PhD in computer science (2014), all with honours, from the University of Granada. Most recently, he has been Principal Investigator of the Dutch R&D project HoliBehave, Workpackage Lead of the EU-H2020 project COUCH and Technical Coordinator of the Korean project Mining Minds. He has participated in more than 20 Spanish, Korean, Dutch and FP7/H2020 European digital health and behavioural computing-related projects such as HPCBIO, AdaBIO, DSIPA-BIO, AmIVital, Symbionics, OPPORTUNITY, XOSOFT, eNHANCE or COUCH. He is author of more than 30 articles published in SCI journals and more than 50 papers in top-ranked international conferences. He is co-organiser of several digital health workshops and conferences (e.g. SUT4Coaching, Health-i-Coach, UCAmI) and he is also a member of the organising committee of a number of international conferences (e.g. PervasiveHealth, BodyNets or IWANN). He is also Associate Editor of various journals such as the Human-centric Computing and Information Sciences (Springer), International Journal of Environmental Research and Public Health (MDPI) and SENSORS (MPDI), and Guest Editor in top-tier journals such as the Journal of Distributed Sensor Networks (SAGE), Ambient Intelligence and Humanized Computing (Springer) and Computer (IEEE). His research works on the intersection of wearable, ubiquitous, and mobile computing with data mining and artificial intelligence for digital health and wellness applications. He is particularly interested in human-aware computing, behaviour and context modelling, intelligent coaching systems, and smart pervasive sensing.

Editorial

Sensor Technology for Smart Homes

Juan Ye [1,*], Michael O'Grady [2] and Oresti Banos [3]

1 School of Computer Science, University of St Andrews, St. Andrews KY16 9SX, UK
2 School of Computer Science and Informatics, University College Dublin, Dublin 4, Ireland; michael.j.ogrady@ucd.ie
3 Faculty ETSIIT, University of Granada, 18010 Granada, Spain; oresti@ugr.es
* Correspondence: jy31@st-andrews.ac.uk

Received: 7 December 2020; Accepted: 8 December 2020; Published: 9 December 2020

As advances in technology continue relentlessly, intriguing possibilities for smart home services have emerged. Nonetheless, significant obstacles remain before smart home and equivalent technologies reach a sufficiently high Technology Readiness Level (TRL) and thus become mainstream. The challenges that must be overcome are many; the inherent complexity of the smart home, though a constrained environment in many ways, poses many difficulties. Some difficulties are technical—the lack of robust and scalable technologies, for example. Human factors and even sociotechnical issues may drive other problems that arise. Holistic solutions for the smart home will only emerge from transdisciplinary research and collaborations.

At the time of writing, the second phase of the COVID-19 pandemic is peaking. A radical reassessment of what constitutes normalcy has resulted. Homes have increasingly become places of work and enterprise, posing many challenges for society, employers and families. Older adults and other vulnerable people have often been obliged to live in isolation for extended periods, often aggravating mental and physical health problems. Thus, going forward, it may be instructive to perhaps revisit the smart home both conceptually and phenomenologically, rather than as a manifestation of technology in the first instance.

Summary of this Special Issue

Resilient smart homes demand the availability of reliable, fault-tolerant sensor networks. Detecting sensor failure as early as possible is essential for ensuring trustworthiness and confidence. In [1], a mechanism for sensor failure detection and subsequent sensor isolation based on association rule mining is proposed for event-driven, ambient sensors. Promising results have emerged in the cases of unlabelled datasets and unknown sensor topologies.

In [2], the importance of occupancy management for smart homes, but without compromising privacy, is considered. An innovative method based on the prediction of CO_2 concentration is employed. An Artificial Neural Network, augmented with a wavelet transformation for additive noise cancelling, is harnessed. Initial results from this indirect monitoring of human daily living activities are impressive, showing significant potential for addressing the omnipresent ethical and privacy issues in smart home research.

A complementary approach, again using CO_2 as a predictor of occupancy, is described in [3]. In contrast, this approach focuses on occupancy detection and the number of occupants. A Fibre Bragg Grating (FBG) sensor is central to the proposed approach. Their small form factor makes their deployment easy and cost-effective, and thus an apt solution for smart homes.

Fall detection systems have been the subject of intense research over the last decade. The advantages of robust systems for the automated detection of falls is desirable for all the obvious reasons; however, as yet, none are on the market. The impossibility of generating real-world datasets from falls by older adults in real-world settings significantly exacerbates the difficulties in creating effective prediction systems. Thus, the eHomeSeniors Dataset [4] has been expressly constructed in collaboration with a physiotherapist, younger people, and performing artists, using privacy-preserving infrared sensors. This dataset is in the public domain.

Nonintrusive Appliance Load Monitoring (NIALM) offers a transparent approach to monitoring energy usage in the smart home at the device-level. Such functionality is indispensable for delivering services that advise on sustainable and cost-effective energy usage. Tracking appliance usage is also a useful proxy for occupancy and behaviour monitoring. Delivering robust, scalable NIALM systems for smart homes is an attractive proposition but one fraught with difficulty at present. In [5], the authors describe algorithms for signal processing within a NIALM context and present initial findings.

Potential occupants of the smart home will be the ultimate judges of the home's smartness. Multi-agent systems offer an intriguing solution for managing the various devices and catering to the occupant's needs. In [6], the authors describe a gamification approach, built on the multi-agent paradigm, for assigning and distributing tasks amongst occupants.

A key hindrance to the adaptation of smart home technologies concerns the need for a dedicated sensor infrastructure and the challenges that result from the installation and management of these infrastructures. Device-free activity-recognition systems offer a desirable alternative, minimizing the need for infrastructure, eliminating the need for Line-of-Sight (LoS) and maximining privacy issues. In [7], the authors explore the impact of the environment on the efficacy of Wi-Fi, resulting in insights as to how smart homes might be configured to harness Wi-Fi for device-free activity recognition.

One group of occupants for whom the promise of smart homes may be positively life-changing is people with severe disabilities. A novel assistive technology system based on eye-tracking is proposed in [8]. Here, usability assessments, by those both with and without disability, of the proposed system are positive.

To conclude: a scientific literature review of technologies used for activity recognition in smart homes and ambient assisted living scenarios is presented in [9].

Funding: This research received no external funding.

Acknowledgments: Thanks to all the authors who submitted their research for this Special Issue. The invaluable contribution of the international reviewers is gratefully acknowledged.

Conflicts of Interest: The authors declare no conflict of interest.

References

1. ElHady, N.E.; Jonas, S.; Provost, J.; Senner, V. Sensor Failure Detection in Ambient Assisted Living Using Association Rule Mining. *Sensors* **2020**, *20*, 6760. [CrossRef] [PubMed]
2. Vanus, J.; Fiedorova, K.; Kubicek, J.; Gorjani, O.M.; Augustynek, M. Wavelet-Based Filtration Procedure for Denoising the Predicted CO_2 Waveforms in Smart Home within the Internet of Things. *Sensors* **2020**, *20*, 620. [CrossRef] [PubMed]
3. Vanus, J.; Nedoma, J.; Fajkus, M.; Martinek, R. Design of a New Method for Detection of Occupancy in the Smart Home Using an FBG Sensor. *Sensors* **2020**, *20*, 398. [CrossRef] [PubMed]
4. Riquelme, F.; Espinoza, C.; Rodenas, T.; Minonzio, J.-G.; Taramasco, C. eHomeSeniors Dataset: An Infrared Thermal Sensor Dataset for Automatic Fall Detection Research. *Sensors* **2019**, *19*, 4565. [CrossRef] [PubMed]
5. Wójcik, A.; Łukaszewski, R.; Kowalik, R.; Winiecki, W. Nonintrusive Appliance Load Monitoring: An Overview, Laboratory Test Results and Research Directions. *Sensors* **2019**, *19*, 3621. [CrossRef]
6. Winnicka, A.; Kęsik, K.; Połap, D.; Woźniak, M.; Marszałek, Z. A Multi-Agent Gamification System for Managing Smart Homes. *Sensors* **2019**, *19*, 1249. [CrossRef]
7. Lee, H.; Ahn, C.R.; Choi, N.; Kim, T.; Lee, H. The Effects of Housing Environments on the Performance of Activity-Recognition Systems Using Wi-Fi Channel State Information: An Exploratory Study. *Sensors* **2019**, *19*, 983. [CrossRef] [PubMed]
8. Bissoli, A.; Lavino-Junior, D.; Sime, M.; Encarnação, L.; Bastos-Filho, T. A Human–Machine Interface Based on Eye Tracking for Controlling and Monitoring a Smart Home Using the Internet of Things. *Sensors* **2019**, *19*, 859. [CrossRef] [PubMed]

9. Sanchez-Comas, A.; Synnes, K.; Hallberg, J. Hardware for Recognition of Human Activities: A Review of Smart Home and AAL Related Technologies. *Sensors* **2020**, *20*, 4227. [CrossRef] [PubMed]

Publisher's Note: MDPI stays neutral with regard to jurisdictional claims in published maps and institutional affiliations.

Article

A Human–Machine Interface Based on Eye Tracking for Controlling and Monitoring a Smart Home Using the Internet of Things

Alexandre Bissoli [1,*], Daniel Lavino-Junior [2], Mariana Sime [3], Lucas Encarnação [1,2] and Teodiano Bastos-Filho [1,2]

1 Postgraduate Program in Electrical Engineering, Federal University of Espirito Santo (UFES), Vitoria 29075-910, Brazil; lucas@ele.ufes.br (L.E.); teodiano.bastos@ufes.br (T.B.-F.)
2 Electrical Engineering Department, Federal University of Espirito Santo (UFES), Vitoria 29075-910, Brazil; daniel_lavino@hotmail.com
3 Postgraduate Program in Biotechnology, Federal University of Espirito Santo (UFES), Vitoria 29047-105, Brazil; mariana.midori@gmail.com
* Correspondence: alexandre-bissoli@hotmail.com; Tel.: +55-27-98133-0070

Received: 2 January 2019; Accepted: 12 February 2019; Published: 19 February 2019

Abstract: People with severe disabilities may have difficulties when interacting with their home devices due to the limitations inherent to their disability. Simple home activities may even be impossible for this group of people. Although much work has been devoted to proposing new assistive technologies to improve the lives of people with disabilities, some studies have found that the abandonment of such technologies is quite high. This work presents a new assistive system based on eye tracking for controlling and monitoring a smart home, based on the Internet of Things, which was developed following concepts of user-centered design and usability. With this system, a person with severe disabilities was able to control everyday equipment in her residence, such as lamps, television, fan, and radio. In addition, her caregiver was able to monitor remotely, by Internet, her use of the system in real time. Additionally, the user interface developed here has some functionalities that allowed improving the usability of the system as a whole. The experiments were divided into two steps. In the first step, the assistive system was assembled in an actual home where tests were conducted with 29 participants without disabilities. In the second step, the system was tested with online monitoring for seven days by a person with severe disability (end-user), in her own home, not only to increase convenience and comfort, but also so that the system could be tested where it would in fact be used. At the end of both steps, all the participants answered the System Usability Scale (SUS) questionnaire, which showed that both the group of participants without disabilities and the person with severe disabilities evaluated the assistive system with mean scores of 89.9 and 92.5, respectively.

Keywords: human–machine interface (HMI); human–computer interaction (HCI); smart home; eye tracking; assistive technology; usability evaluation; user-centered design (UCD); home automation; Internet of Things (IoT)

1. Introduction

People with severe disabilities may have difficulties interacting with their home devices due to the limitations inherent to their disability. Simple activities such as turning on or off a lamp, fan, television, or any other equipment independently may even be impossible for this group of people. With technological advances in the field of sensors and actuators, in recent years some researchers have begun to transfer these technologies to improve the quality of life of people with disabilities, increasing their autonomy regarding the control of existing equipment in the environment [1–4].

Technologies dedicated to improving the lives of people with disabilities are known as assistive technologies. Assistive technology is an area of knowledge with an interdisciplinary characteristic which encompasses products, resources, methodologies, strategies, practices, and services that aim to promote the functionality related to the activity and participation of people with disabilities, inability, or reduced mobility, aiming at their autonomy, independence, quality of life, and social inclusion [5].

Although many works have been devoted to proposing new assistive technologies to improve the lives of people with disabilities [6–15], some studies have found that the abandonment of such technologies is quite high, reaching a rate of up to 30% [16–19]. The reasons for abandoning assistive technology are diverse, the most recurrent being that the user does not like the technology; the user is afraid to use the equipment; the user does not believe in the benefit of the device; the technology is not physically fit for the user; the price of the technology is very expensive; the user does not know how to use the equipment correctly; or the user disapproves of the equipment aesthetic factors [16–19].

Based on these facts, to avoid or at least reduce the abandonment of new technologies, in developing a new system, engineers should be concerned with developing a system that is useful to the user, i.e., that brings benefits; developing a system to suit the needs of the user; designing tests to validate the technology; evaluating the usability of the system; performing end-user testing; and testing the system outside the laboratory, i.e., testing the system where it will be actually used.

In order to increase the usability of an assistive system, it is also critical to consider the role of human–computer interaction (HCI). The concept of HCI refers to a discipline which studies information exchange between people and computers by using software. HCI mainly focuses on designing, assessing, and implementing interactive technological devices that cover the largest possible number of uses [20].

The ultimate goal of HCI is to make this interaction as efficient as possible, looking to minimize errors, increase satisfaction, lessen frustration, include users in development processes, work in multidisciplinary teams, and perform usability tests. In short, the goal is to make interaction between people and computers more productive [21].

New technologies have arisen with health-related developments which, by using HCI, meet the needs of different groups such as people with disabilities, the elderly, etc. [22,23]. Although these advances were unthinkable just a few years ago, they are gradually becoming a part of people's daily lives [24,25].

Human–computer interaction and the need for a suitable user interface has been an important issue in modern life. Nowadays, products and technologies used by society have created concerns about computer technology. For this reason, researchers and designers are interested in the assessment of human and machine behavior, where the machines varying according to the system functionality and the system or product requirements [26].

This work aims to assist people with physical disability to pursue daily living autonomously, taking into account concepts of user-centered design and usability, in order to avoid the abandonment of the proposed system. To this end, we present a new useful assistive system based on eye tracking for controlling and monitoring a smart home, based on the Internet of Things. With this assistive system, a person with severe disabilities was able to control everyday equipment in her residence, such as lamps, television, fan, and radio, and the caregiver was able to remotely monitor the use of the system by the user in real time. In addition, the user interface developed here has some functionalities that allowed improving the usability of the system as a whole.

The subsequent sections of this work are organized as follows. We firstly review the related work and cover some smart homes from around the world in Section 2. In Section 3, we introduce our assistive system and its detailed design. Tests protocols, experimental results, and evaluations are reported in Section 4. Finally, we draw conclusions from our work in Section 5.

2. Related Work

In this section, we introduce the state-of-the-art related works by dividing the literature into three parts: (i) user-centered design and usability; (ii) eye tracking; and (iii) smart homes.

2.1. User-Centered Design (UCD)

A User-Centered Design (UCD) approach can be used in any type of product from the perspective of HCI design. UCD, also called Human-Centered Design (HCD), is a method that defines the needs, desires, limitations, services, or processes that serve the end-user of a product/system at all stages of a project. In other words, UCD is a multistage troubleshooting process that follows all product development requirements. UCD tries to optimize how the user can use the product/system, what they want or what they need, instead of changing the user's behavior with the product/system [27].

The approach of UCD is to put human needs, resources, and behavior first, and then design technology to accommodate those needs, resources, and behaviors. It is necessary to understand the psychology and technology to start the design, which requires good communication, mainly between human and machine, indicating available options, the actual status, and the next step [28].

The term "interaction" from human–computer interaction (HCI) is a basis for designing or developing a user interface and an interaction between humans and machines. Preece et al. [29] define four basic activities of an interaction design: (i) identify needs and establish requirements; (ii) develop alternative projects; (iii) construct interactive versions of projects; and (iv) evaluate projects. They also describe three characteristics for interaction design: (i) focus on users; (ii) specific usability criteria; and (iii) iteration.

Regarding the user experience, Goodman et al. [30] claim that the process is not only to learn about the user experience with the technology, but also for designers to experience interacting in their own work. They report that user experience tests must be applied during the design, the approaches of which could be (i) reported approaches; (ii) anecdotal descriptions; and (iii) first-person research. In addition, Begum [31] presents the user interface (UI), proposing an extended UCD process that adds the "Understand" phase to the methods. The conventional steps of a UCD approach are (i) study, (ii) design, (iii) build, and (iv) evaluate; however, Begum [31] has extended it to (i) understand, (ii) study, (iii) design, (iv) construct, and (v) evaluate.

2.1.1. Usability

The definition of usability is when a product is used by specific users to achieve specific goals with effectiveness, efficiency, and satisfaction in a specific context of use [32]. Usability is more than just about whether users can perform tasks easily; it is also concerned with user satisfaction, where users will consider whether the product is engaging and aesthetically pleasing.

Usability testing is a technique in UCD which is used to evaluate a product by testing it with actual users. It allows developers to obtain direct feedback on how users interact with a product. Thus, through usability testing, it is possible to measure how well users perform against a reference and note if they meet predefined goals, also taking into account that users can do unexpected things during a test. Therefore, to create a design that works, it is helpful for developers to evaluate its Usability, i.e., to see what people do when they interact with a product [26,30].

Usability is then the outcome of a UCD process, which is a process that examines how and why a user will adopt a product and seeks to evaluate that use. That process is an iterative one and seeks to improve the design following each evaluation cycle continuously.

2.1.2. System Usability Scale (SUS)

The System Usability Scale (SUS) provides a reliable tool for measuring usability. It consists of a 10-item questionnaire with five response options which are scored by a 5-point Likert scale, ranging from "1—strongly disagree" to "5—strongly agree". Originally created by Brooke [33], it allows researchers, engineers, and designers to evaluate a wide variety of products and services, including hardware, software, mobile devices, websites, and applications.

The 10 statements on the SUS are as follows:

(1) I think that I would like to use this system frequently.
(2) I found the system unnecessarily complex.
(3) I thought the system was easy to use.
(4) I think that I would need the support of a technical person to be able to use this system.
(5) I found the various functions in this system were well integrated.
(6) I thought there was too much inconsistency in this system.
(7) I would imagine that most people would learn to use this system very quickly.
(8) I found the system very cumbersome to use.
(9) I felt very confident using the system.
(10) I needed to learn a lot of things before I could get going with this system.

After completion of the questionnaires by the interviewees, the SUS score is calculated as follows:

- For odd-numbered items: subtract 1 from the user response;
- For even-numbered items: subtract the user responses from 5;
- This scales all values from 0 to 4 (with 4 being the most positive response).
- Add the converted responses for each user and multiply that total by 2.5. This converts the range of possible values from 0 to 100 instead of from 0 to 40.

Although the scores are from 0 to 100, these are not percentages and should be considered only in terms of their percentile ranking. Based on the research of Brooke [33], an SUS score above 68 would be considered "above average", and anything below 68 is "below average". However, the best way to interpret the results involves normalizing the scores to produce a percentile ranking. This process is similar to grading on a curve based on the distribution of all scores. To get an A (the top 10% of scores), a product needs to score above an 80.3. This is also the score in which users are more likely to be recommending the product to a friend. Scoring at the mean score of 68 gives the product a C, and anything below 51 is an F (putting the product in the bottom 15%).

2.2. Eye Tracking

Eye tracking is a technique to measure either the point of gaze (where someone is gazing) or the motion of an eye relative to the head. Eye tracking technology has applications in industry and research in visual systems [34–37], psychology [38,39], assistive technologies [40–42], marketing [43], as an input device for human–computer interaction [44–47], and in product and website design [48].

Generally, eye tracking measures the eyeball position and determines the gaze direction of a person. The eye movements can be tracked using different methods which can be categorized into four categories: (i) infrared oculography (IROG); (ii) scleral search coil (SSC); (iii) electro-oculography (EOG); and (iv) video-oculography (VOG). SSC measures the movement of a coil attached to the eye [24,49]; VOG/IROG carries out optical tracking without direct contact to the eye [42,47]; and EOG measures the electric potentials using electrodes placed around the eyes [50]. Currently, most of the eye tracking research for HCI is based on VOG, as it has minimized the invasiveness to the user to some degree [40].

The eye is one of main human input media, and about 80 to 90 percent of the outside world information is obtained from the human eye [51]. For communication from user to computer, the eye movements can be regarded as a pivotal real-time input medium, which is especially important for people with severe motor disability, who have limited anatomical sites to use to control input devices [52].

The research into eye tracking techniques in HCI is mainly focused on incorporating eye movements into the communication with the computer in a convenient and natural way. The most intuitive solution for incorporating eye movements into HCI is the use of an eye tracker directly connected to a manual input source, such as a mouse. By installing an eye tracker and using its x, y coordinate output stream as a virtual mouse, the movement of the user's gaze directly causes the mouse cursor to move (eye mouse). In order to provide such appropriate interaction, several eye-tracking-based control systems have been developed, detailed as follows.

Chin et al. [53] proposed a cursor control system for computer users which integrated electromyogram signals from muscles on the face and point-of-gaze coordinates produced by an eye-gaze tracking system as inputs. Although it enabled a reliable click operation, it was slower than the control system that only used eye tracking, and its accuracy was low. Missimer and Betke [54] constructed a system that used the head position to control a mouse cursor and simulate left-click and right-click of the mouse by blinking the left or right eye. This system relied on the position of the user's head to control the mouse cursor position. However, irregular movement of the user's head affected the accuracy of the click function. Lupu et al. [41] proposed a communication system for people with disabilities which was based on a special device composed of a webcam mounted on the frame of a pair of glasses for image acquisition and processing. The device detected the eye movement, and the voluntary eye blinking was correlated with a pictogram or keyword selection, reflecting the patient's needs. The drawback of this system was that the image processing algorithm could not accurately detect the acquired image (low resolution) and was not robust to light intensity. Later, to improve the reliability of the communication system, they proposed an eye tracking mouse system using video glasses and a new robust eye tracking algorithm based on the adaptive binary segmentation threshold of the acquired images [42].

Lately, several similar systems have also been developed [55,56], and the main concept of these systems is to capture images from a camera, either mounted on headgear worn by the user or mounted remotely, and extract the information from different eye features to determine the point of the gaze. Since, at the time the research was performed, commercial eye trackers were prohibitively expensive to use in HCI, all the aforementioned eye tracking control systems were proposed with self-designed hardware and software. It was difficult for these systems to achieve widespread adoption, as the software and hardware designs were closed source.

2.3. Smart Homes

There are many motivations to design and develop applications in smart homes, the main ones being independent living [3,7,9–11,57]; wellbeing [4,6,8,12,58]; efficient use of electricity [59–72]; and safety and security [73–88].

The expression "smart home" is used for a home environment with advanced technology that enables control and monitoring for its occupants and boosts independent living through sensors and actuators to control the environment or through wellness forecasting based on behavioral pattern generation and detection. A variety of smart home systems for assisted living environments have been proposed and developed, but there are, in fact, few homes that apply smart technologies. One of the main reasons for this is the complexity and varied design requirements associated with different domains of the home, which are communication [89–95], control [96–110], monitoring [111–116], entertainment [117–120], and residential and living spaces [121,122].

As an important component of the Internet of Things (IoT), smart homes serve users effectively by connecting them with devices based on IoT. Smart home technology based on IoT has changed human life by providing connectivity to everyone regardless of time and place [123,124]. Home automation systems have become increasingly sophisticated in recent years, as these systems provide infrastructure and methods to exchange all types of appliance information and services [125]. A smart home is a domain of IoT, which is the network of physical devices that provides electronic, sensor, software, and network connectivity inside a home.

There are many smart home systems across the world, most notably in Asia, Europe, and North America. In Asia, it is important to highlight Welfare Techno Houses [126,127], Smart House

Ikeda [128], Robotics Room and Sensing Room [129], ActiveHome [130], ubiHome [131–134], Intelligent Sweet Home [135], UKARI Project and Ubiquitous Home [136,137], and Toyota Dream Home PAPI [138]. In Europe, there are comHOME [139], Gloucester Smart Home [140], CUSTODIAN Project [141], Siemens [142], myGEKKO [143], and MATCH [144]. In North America, there are Adaptive Smart House [145,146], Aware Home Research Initiative (AHRI) [147], MavHome Project [148,149], House_n (MIT House) [150,151], EasyLiving Project [152], Gator Tech Smart House [153], DOMUS Laboratory [154], Intelligent Home Project [155], CASAS [156,157], Smart Home Lab [158,159], AgingMO [160], and Home Monitoring at Rochester University [161].

3. Proposed Assistive System

The assistive system proposed here empowers people with severe physical disability and mitigates the limitations in everyday life with which they are confronted. The system aims at assisting people with physical disability to pursue daily living autonomously. In Figure 1, the local user is the person with disability who can control the equipment of his/her smart home through eye gaze using the device controller (gBox). At the same time, the caregiver (external user) can monitor the use of the system.

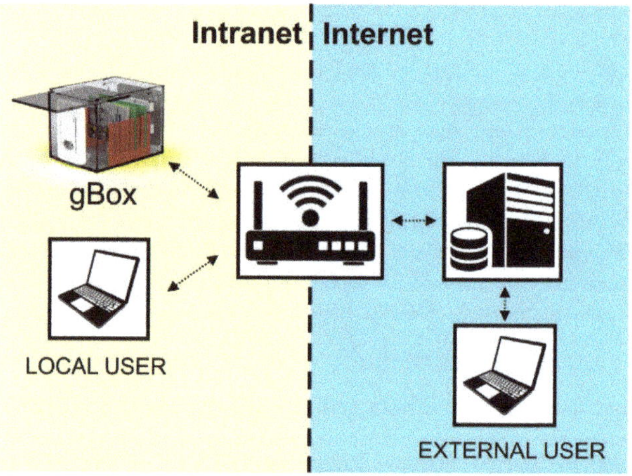

Figure 1. System overview.

The proposed eye-gaze-tracking-based control system is a software application using a low-cost eye tracker (e.g., The Eye Tribe 1.0 and Tobii Pro). The application detects the user's gaze with the "mouse cursor control" function provided by the eye tracker. The mouse cursor control allows users to redirect the mouse cursor to the gaze position. Therefore, the system realizes where the user is gazing according to the position of the mouse cursor. By gazing at the point for few seconds, the tool generates the corresponding event. This way, users can select and "click" the corresponding action. The eye tracker detects and tracks the coordinates of the user's eye gaze on the screen; this made it possible to create applications that can be controlled in this way.

The eye tracker software is based on an open Application Programming Interface (API) that allows applications (clients) to communicate with the eye tracker server to obtain eye gaze coordinates. The communication is based on messages sent asynchronously, via Socket, using the Transmission Control Protocol (TCP).

It should be noted that to use this assistive system, it is not necessary to install any software in addition to the Internet browser, as the web application was developed to run in Internet browsers.

It is only recommended to use the most up-to-date versions of well-known browsers, such as Google Chrome (preferable), Mozilla Firefox, or Internet Explorer.

3.1. System Architecture

Different systems for HCIs based on biological signals have been proposed with various techniques and applications [162]. Despite each work presenting unique properties, most of them fit into a common framework. Figure 2 shows the framework adopted in this system.

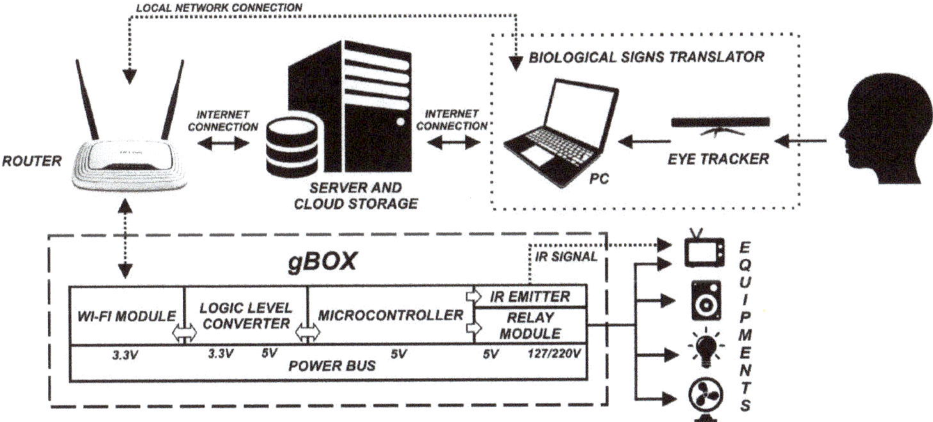

Figure 2. Functional model of the eye/gaze-tracking-based control system.

Observing the model, a loop structure can be recognized; it starts from the User, whose biological signals are the primary input, and ends with the environment that is affected by the actions of the system. Along this path, three main modules can be identified: the Biological Signal Translator module, the Server and Cloud module, and the Device Controller/Device module (gBox). Eventually, the loop is closed through user feedback of different kinds. The communication between the modules is bidirectional in order for a module to know the outcome of a command.

3.2. Connectivity

The web application was developed to work both online and offline. To open the online application and work in "online mode", the user must simply enter the Internet browser and the domain where it is hosted (https://ntagbox.000webhostapp.com). The online application can be hosted on any HTTP server; the domain used in this work is provided free of charge by "Hostinger", with limited, but sufficient, configurations. Because there is an external server, an Internet connection is necessary. In this case, the connection is via WebSocket (WS), which is the best option, as the connection between the application and the physical device is done over the Internet; in this way, the user commands are stored on the server instantly.

On the other hand, to use the offline application and work in the "offline mode", it is only necessary to have the site files in a folder on the computer, run the "intex.html" file, and the browser will run the application. In this case, the connection is via AJAX, which is used when there is no Internet access in the user's home (or if the Internet drops out temporarily); thus, the connection between the application and the physical device is made by the Intranet, and in this way, the user commands are stored temporarily on the computer until the connection via WS is established.

In order for the local and external connection to be implemented on the physical device, two ways of WEB communication were created (Figure 3): HTTP and WS. Once a command is launched from the application (APP), an internal mechanism identifies whether there is access to the external (Internet) server or not (local connection), and also identifies whether the physical device is properly connected to that server. The application has two communication clients related to the two communication channels. If the device is connected on the Internet, WS is used as the communication channel both in the application and in the device since it is capable of establishing an endless connection with the server, allowing the data to be sent bidirectionally and asynchronously. When there is no Internet connection and the application is in the same local network as the physical device, the communication channel used is HTTP, with an HTTP server on the physical device so that it has an IP that identifies it locally.

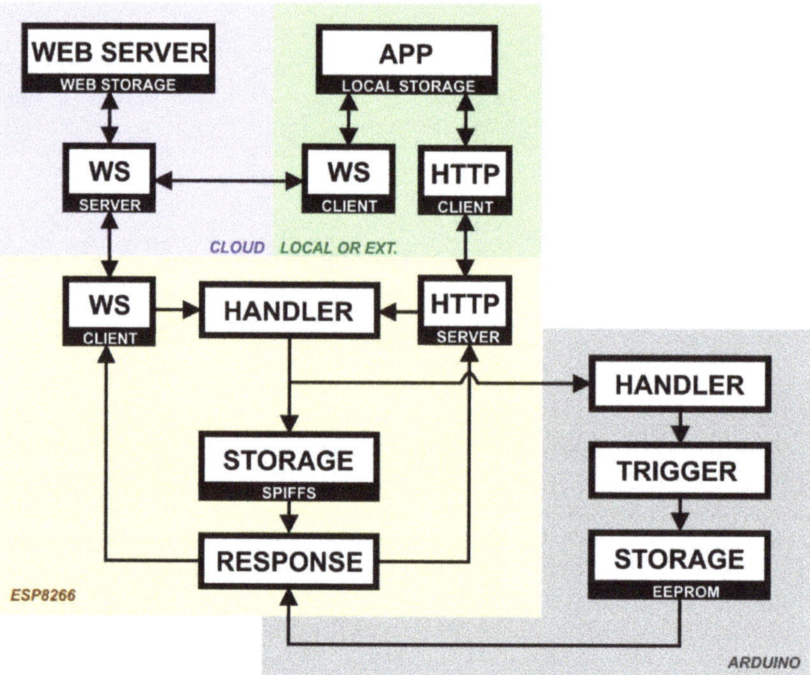

Figure 3. Connectivity of the assistive system.

In the physical device, the data packets can be received by the two communication channels. However, they are directed to a single channel—the HANDLER—that handles the information contained therein. The HANDLER has the function of authenticating the data received and identifying the command contained therein so that the TRIGGER can be activated. The TRIGGER is the set of triggers and sensors responsible for interacting with the user's devices. Some commands simply require changes in the internal variables of the system. For this purpose, it also has a small non-volatile SPIFFS memory module responsible for storing such variables, such as SSID and Wi-Fi password; user password; relay states; etc. It is also essential to keep the data stable if there is any power outage. If everything succeeds, the RESPONSE block returns the confirmation message to the application, returning to the input path of the packet.

If this input path is the WS client, the packet will be returned to the WS server in the cloud, which will store the command so that it is accessible via Internet, and finally send the confirmation to the WS client of the application. If this input path is the HTTP server, the device returns the response directly to the application, which places it in a rank to be sent to the server as soon as an Internet connection is established.

3.3. GlobalBox (gBox)

The GlobalBox (gBox) is the device controller module of the smart home. Figure 4 shows, in a simplified form, the main components of the gBox.

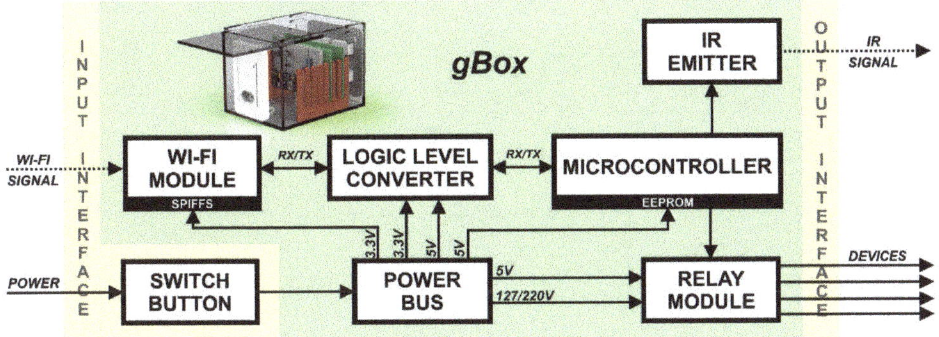

Figure 4. GlobalBox (gBox) of the assistive system.

Before starting the application, the switch button must be turned on. With the box powered up, it is possible to receive information through the Wi-Fi module. This information is the command sent by the user through the user interface running on the computer. The information received by the Wi-Fi module is processed in the microcontroller, and then the actions corresponding to the received commands are performed, being able to turn the relays of the equipment on or off or send specific commands by the infrared (IR) emitter to control the functions of the TV or radio.

3.4. Wireless Infrared Communication

Infrared (IR) signals are essential to control some residential devices, such as TVs and radios. The gBox has an internal library with a set of IR protocols (the most used) already implemented, which assists in the task of modulation and demodulation of IR signals. The hardware consists of an IR detector that receives signals at 38 kHz, an IR emitter, and the microcontroller, which is responsible for treating and storing the signal in memory for later use.

To store the code of any remote control, it is necessary to send the read command from the application so the demodulator will be activated; then, the caregiver can press the button (towards the IR detector) that he/she wants to record, which transforms the received IR signals into codes that can be stored in memory (Figure 5).

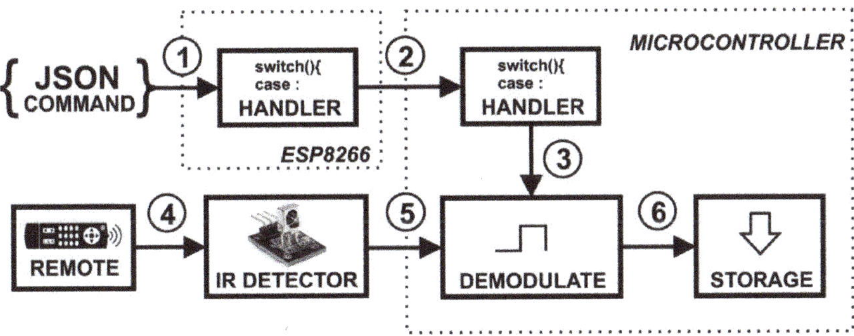

Figure 5. Detection and storage of an IR command.

To emit an IR signal, it is necessary to start the application whose command activates the signal modulator, which takes the codes stored in the memory of the microcontroller and transforms them into the original IR signal, sending it to the IR emitter module (Figure 6). The emitter module, when pointed towards the target device, acts in the same way as the remote control in its respective function.

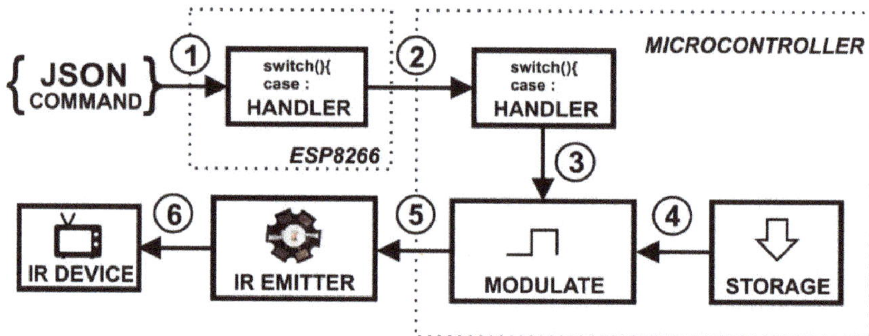

Figure 6. Circuit of IR emission commands previously stored.

3.5. User Interface

The most well-known strategy of eye tracking applications is using the gaze to perform tasks of pointing and selecting. However, the direct mapping of the gaze (more specifically, of fixations) to a command of system selection creates a problem, called the "Midas Touch", in which a selection can be activated in any screen position observed by the user, whether they intended to do it or not.

Thus, after filtering the eye tracker data, the Midas Touch problem must be avoided by implementing mechanisms for the user to indicate when he/she really desires to perform a selected command. The first approach to this problem is to implement a dwell time, in which the selection of one option is done only after a time interval.

Figure 7a shows the initial screen when the system is off. When the user turns the system on by selecting the "Start" icon, the main menu shown in Figure 7b appears. In Figure 7b, the user has three options: (i) "Close", to close the user interface of the system; (ii) "Start", to open the home devices control menu; and (iii) "Config", to open the configuration menu.

It is important for users to be able to turn the system on and off. That is why the interface presented in Figure 7b was included. If the user chooses "Close", the system is closed and returns to the initial interface.

When the user selects the "Config" option in Figure 7b, the configuration menu shown in Figure 7c appears. In the configuration menu, the user can choose the icon size and the dwell time. There are three options for the icon size: (i) "Small"; (ii) "Medium"; and (iii) "Large". There are four options for dwell time: (i) 0.5 s; (ii) 1.0 s; (iii) 2.0 s; and (iv) 3.0 s. After choosing any of the options, the interface automatically returns to the main menu with the new configuration saved.

When the user selects the "Start" option in Figure 7b, the control menu of the home devices shown in Figure 7d appears. In this menu, the user is presented with four options of devices to be turned on/off: a fan, TV, lamp, and radio. In addition, there is an option to close that menu to return to the main menu (Figure 7b) by selecting the center icon. The icon size on the interface and dwell time used are the ones that the user selected in the configuration menu.

When the device is turned on, the background of the icon becomes yellow, like "Lamp" and "Radio" are in Figure 7e. Fan and TV are turned off; thus, the background of the icons is white. After turning the desired device on or off, the user can give the command "Close" and turn the system off. This command closes the interface, but the system keeps the selected devices in their current state (on or off).

After selecting the TV icon, the interface displays the TV submenu shown in Figure 7f. In the TV submenu, the user can turn the TV on or off, change the channel "up" or "down", increase or decrease the volume, and close the TV submenu. Is this last option, the TV submenu is closed and the interface returns to the home devices control menu, but the system keeps the TV in its current state (on or off).

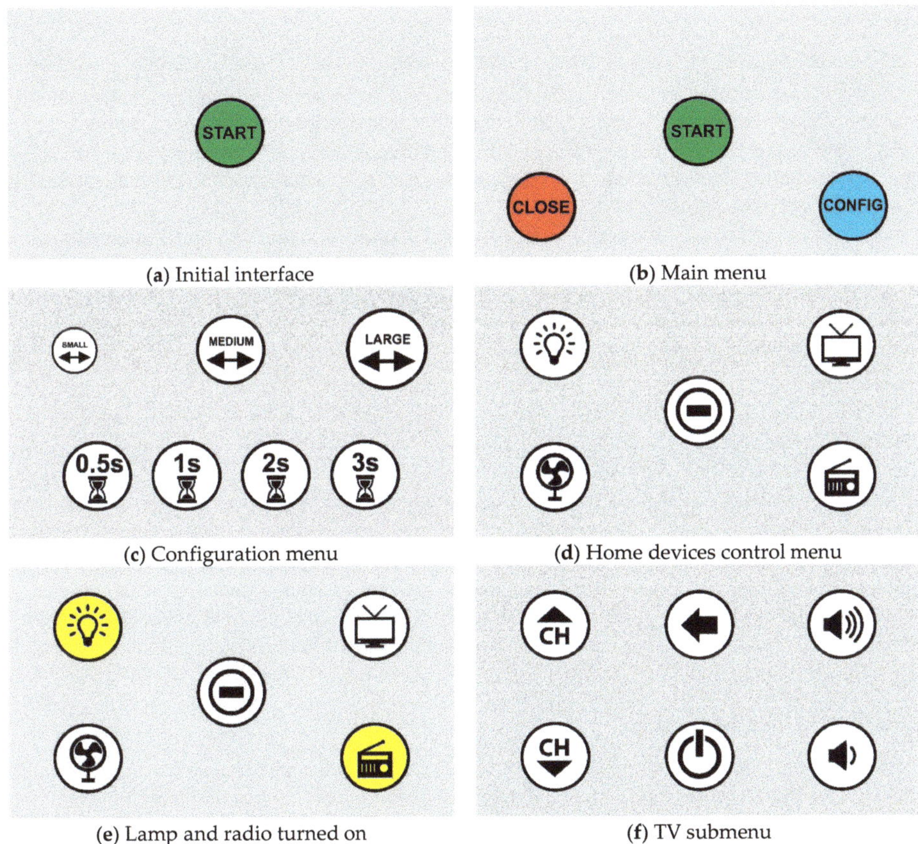

(**a**) Initial interface (**b**) Main menu

(**c**) Configuration menu (**d**) Home devices control menu

(**e**) Lamp and radio turned on (**f**) TV submenu

Figure 7. User interface.

3.6. Caregiver Interface

The gBox Central Management is accessible from the website (https://ntagbox.000webhostapp.com/). In the header of the caregiver interface are the top bar and titles. Before giving any command in the application, it is required to enter the password in the password field of the bar. After that, it is recommended to click "Update Status" so that the application synchronizes with the current states of the equipment. The "Start Application" button opens the user interface so the user can control the smart home with the eye tracker. On the right side of the bar, the connection status between the application and the physical devices is reported. The connection flag is independent of the access password to be updated. There are three possible connection status:

- Connected via WS. This is the best connection. It occurs when the connection between the application and the physical device is done by Internet; that way, user commands are stored on the server instantly.

- Connected via AJAX. This occurs when the connection between the application and the physical device is made by the Intranet, so the commands are stored temporarily on the user's computer until a connection via WS is established.
- Not Connected. This occurs when there is no connection between the application and the physical device. In this case, it is suggested to refresh the site and check the connections with the physical device.

The body of the application is composed of seven sections: (i) Last Commands; (ii) General Notifications; (iii) Infrared Remote Control Settings; (iv) Remote Actuation; (v) Data Acquisition; (vi) Change Password; and (vii) Change Wi-Fi Password, the details of which are as follows.

(i) "Last Commands": In this section, the commands successfully performed are presented, as well as the date and time they occurred. To appear in this list, it is necessary to update the states after establishing a WS connection.

(ii) "General Notifications": In this section, all the notifications from the application features are presented; for example, "updating status" or "the status are updated".

(iii) "Infrared Remote Control Settings": In this section, it is possible to update the IR commands by pressing the "READ" button and follow the instructions displayed in the general notifications section. In addition, hexadecimal codes and protocols of the buttons/commands of the IR control are presented.

(iv) "Remote Actuation": Remote actuation can be used to control the smart home application by using the caregiver application.

(v) "Data Acquisition": This section allows the download of the list of commands based on a specified time interval. The downloaded text file can be accessed in any text editor or spreadsheet analysis program.

(vi) "Change Password": This section allows the change of the user password. It is necessary that the password field of the upper bar be correctly filled with the old password.

(vii) "Change Wi-Fi Password": This functionality allows the change of the passwords of the SSID and of the Wi-Fi. It is necessary that the password field of the upper bar be correctly filled with the user password.

4. Tests, Results, and Discussion

In this section, we report the experiments, which were divided into two steps, and analyze and discuss the results. In the first step, the proposed assistive system was assembled in an actual home where tests were conducted with 29 participants (group of able-bodied participants). In the second step, the system was tested for seven days, with online monitoring, by a person with severe disability (end-user) in her own home, not only to increase her convenience and comfort, but also so that the system would be tested where it would in fact be used. The objective of this test was to explore the effectiveness of the assistive system, the ability of the participant to learn how to use the system, and the efficiency and the usability of the proposed user interface.

According to Resolution No. 466/12 of the National Health Council of Brazil, the research was approved by the Committee of Ethics in Human Beings Research of the Federal University of Espirito Santo (CEP/UFES) through opinion nº 2.020.868, of April 18, 2017.

4.1. Tests with a Group of Able-Bodied Participants

4.1.1. Pre-Test Preparation

Initially, we fully installed the system and tested all possible commands to verify that the system was working properly. Some errors were found and quickly corrected so that the system was considered to work perfectly before we started the tests with the participants.

Afterwards, a pilot test was conducted with one of the involved researchers to rehearse the procedure before conducting the study with the participants. The researcher completed all the data collection instruments. The problems encountered during the pilot test helped to identify changes before conducting the experiment with the participants.

4.1.2. Participants

To test the system, 29 healthy subjects (group of able-bodied participants) participated in the research with the assistive system in the smart home. The participants were 18 men and 11 women, all aged from 17 to 40 (average: 28) and height from 1.50 to 1.94 (average: 1.71). Some of the participants had had at least one experience with HCI through biological signals, but almost none of them had used eye tracking.

Of the total, 12 participants (#2, #8, #9, #11, #12, #14, #15, #16, #17, #22, #25, and #26) wear glasses; however, all of them performed the test without their glasses. In some cases, it was by preference of the participant himself/herself, but in most of the cases, their glasses had anti-reflective film which prevented, or at least disturbed, the infrared eye tracker passing through the lens of the glasses to obtain the correct position of the eyes of the participant. This is an important limitation of this study.

4.1.3. Experimental Sessions

The tests started by welcoming the participants and making them feel at ease. The participants were given an overview of the system and the test and were told that all their information will be kept private.

Each participant was seated on a couch in front of the laptop that contained the user interface, and the eye tracker was positioned properly pointing to his/her eyes (Figure 8).

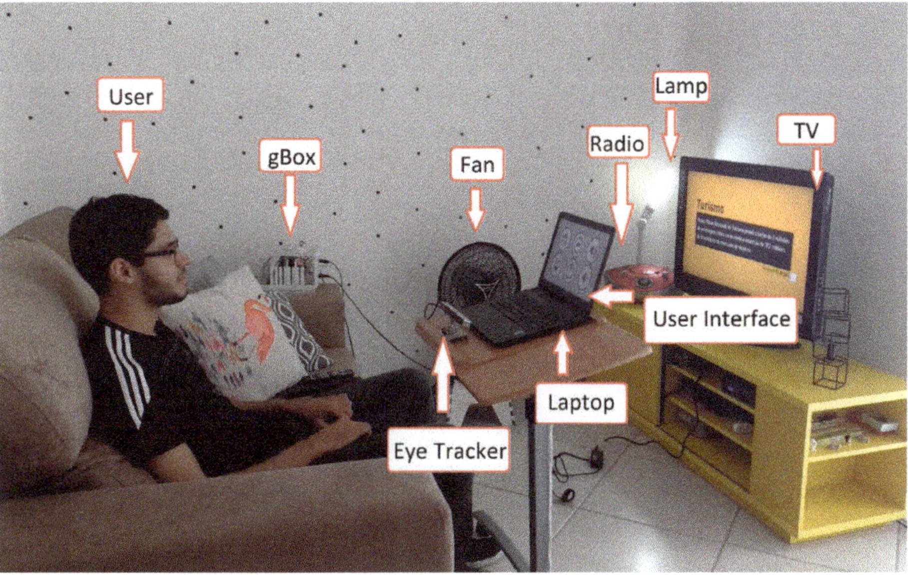

Figure 8. Able-bodied participant testing the system in the smart home.

After that, the participant performed the calibration stage of the eye tracker, following the manufacturer's guidelines. Each participant performed the test over about five to ten minutes. It was required of the participants to use the system long enough so they could test all the functionality options available.

It is important to mention that the user was positioned facing a glass door that provided access to a balcony, with high incidence of sunlight. Despite this, the sunlight did not cause a problem in performing the experiments, showing the robustness of the eye tracker.

It can be seen in Figure 8 that the eye tracker was positioned towards the user's eyes. In the course of the experiments, it is normal for a person to move his/her head a little, moving away from the location where the eye tracker was calibrated. We note that small vertical and horizontal position variations (around 5 to 10 cm) did not significantly interfere with the system performance. However, if the user's eyes are completely out of range of the eye tracker, then he/she will not be able to send commands to the system. In this case, it may be necessary to reposition the eye tracker and calibrate it again. Considering that this system was designed to be used by people with severe disabilities, it is not expected that they will have wide head movements.

After the end of the session, the participant answered the SUS questionnaire and was encouraged to make suggestions.

4.1.4. Results and Discussion

Of the able-bodied participants, 3 opted for the small interface icon size option, 22 opted for medium, and 4 opted for large. As predicted in our previous work [163], most users opted for the medium size option. However, it is important to note that seven participants (24% of the total) preferred another size, thus showing the advantage of having options available.

For dwell time, 6 participants opted for 0.5 s, 18 opted for 1.0 s, 5 opted for 2.0 s, and no one chose the option of 3.0 s. Again, as predicted in our previous work [163], most users opted for the option of 1.0 s. However, it is important to note that 11 participants (38% of the total) preferred another time interval, thus showing the advantage of having this functionality available.

In fact, only 14 participants opted for the combination of medium icon size and 1.0 s dwell time. In other words, 15 participants (52% of the total) preferred another size or other time interval, and this shows the importance of having options to choose from in order to increase the usability of the system.

Regarding the usability, three participants gave a maximum SUS score, and the lowest result was 75. Thus, the overall mean was 89.9, with a standard deviation of 7.1. It is worth mentioning that products evaluated above 80.3 are in the top 10% of the scores. In fact, according to Brooke [33] and Bangor [164], products evaluated in the 90 point range are considered exceptional, products evaluated in the range of 80 points are considered good, and products evaluated in the range of 70 points are acceptable. Figure 9 presents the SUS score of each item evaluated by the participants regarding the assistive system.

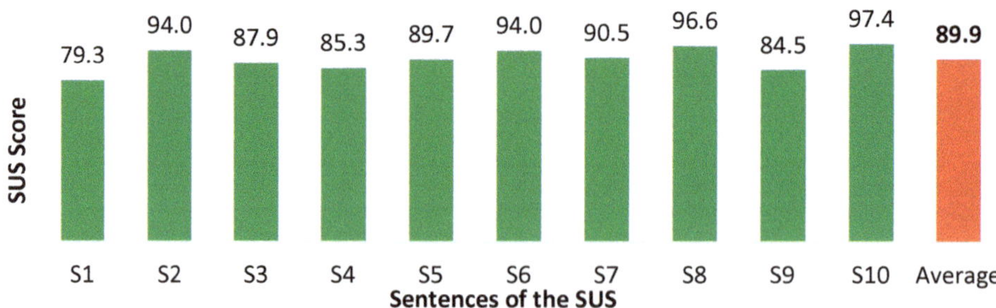

Figure 9. System Usability Scale (SUS) score of each of the ten items of the SUS.

The items regarding the available functionality, the complexity in using, and the confidence in functioning all received scores above 80. The lowest score obtained (79.3) was related to the sentence S1, which is about interest in using the system frequently. Many participants said that they would not have much interest in using this system, as the system was designed for a person with severe disability, which is not the case for the participant (able-bodied). This reinforced the need to test the system with people with severe disabilities.

4.2. Tests with a Person with Disabilities

4.2.1. Pre-Test Preparation

Initially, an interview was conducted by the occupational therapist of our research group to better understand the potential participant. At this point, relevant information was gathered about her disability and daily life, whether there was interest in participating in the study, in what activities she was most involved, and what tasks she would like to be able to do or have the assistance of the technology to execute.

After this first contact, the information was brought to the research group and the candidate was selected to participate in the experiments. A telephone appointment was made between the occupational therapist and the participant's husband at their home, where the system would be used.

4.2.2. Participant Background

The participant is female and 38 years old. She was diagnosed in June 2012 with Wernicke's Encephalopathy. Thus, the participant presents a lack of coordination of movements (ataxia) and extreme difficulty in balancing and walking. Her most difficult activities are those that require manual dexterity, such as typing on the computer, writing, using a mobile device, and handling the TV remote control. In addition, the participant has difficulty walking, considered practically impossible by her, or when necessary, causing enormous discomfort.

4.2.3. Experimental Sessions

To test the assistive system, it was firstly fully assembled and configured in the home of the end-user, who agreed to participate in the experiments (Figure 10). The participant was given an overview of the system and test. Before proceeding to the test, one of the researchers performed all possible commands to verify that the system was working properly.

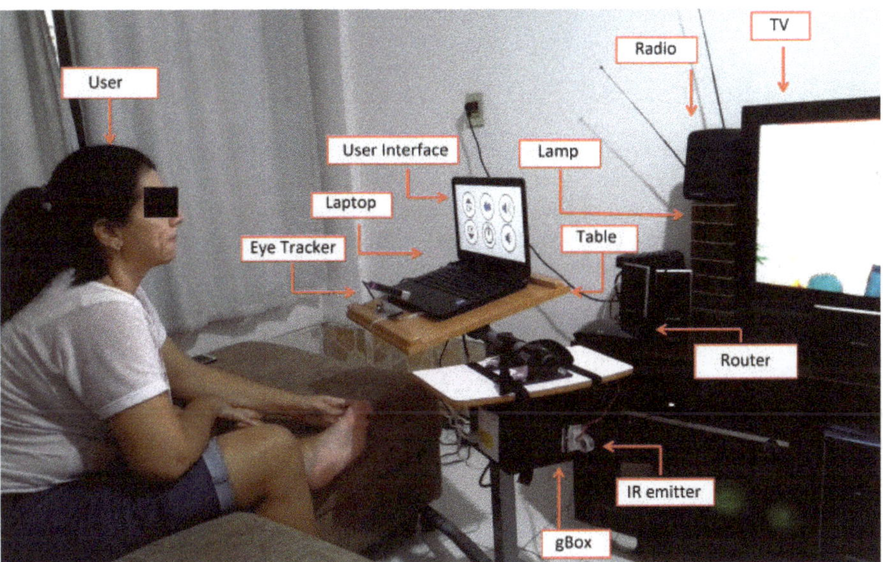

Figure 10. Participant with disabilities testing the system at her home.

The participant was seated on a couch in front of the laptop that contained the user interface, and the eye tracker was positioned properly, pointing towards her eyes (Figure 10). After that, the participant performed the calibration stage of the eye tracker, following the manufacturer's guidelines. The participant performed the test over seven days. It was required of the participant to use the system long enough so she could test all the functionality options available.

After the end of the session, the participant answered the SUS questionnaire and was also encouraged to make suggestions.

4.2.4. Results and Discussion

According to information obtained in the interview with the participant, Friday and Saturday were the best days of the week for her to receive the researchers in her home, so she opted to install the system on a Friday (09/14/2018) and uninstall it on a Saturday (10/06/2018). Before the experiments, the participant informed us that she was not able to use the equipment on Sundays and Mondays, as on Sunday she usually receives many relatives in her house, and on Monday she spends the whole day away from home. In addition, the participant informed that she would need to make a trip for personal reasons during the experiment (from 09/23/2018 until 10/01/2018). All these situations put forward by the participant were considered pertinent, as they actually depict the daily life of a person with disability, revealing how technology needs to adapt to the person's life. In addition, it was considered interesting to evaluate if the participant would use the system after she was away from home for a few days without using it. In many cases, assistive technology is abandoned, which did not happen with our system.

Table 1 shows the use of the assistive system by the participant, which shows the number of hours the system was used during the seven days of use.

Table 1. Summary of system usage information.

Date	Start	End	Duration	Commands
09/14/2018	14:20	18:10	03:50:00	163
09/28/2018	12:51	19:49	06:58:00	131
09/20/2018	13:58	15:55	01:57:00	36
09/22/2018	15:52	17:36	01:44:00	18
10/02/2018	17:38	18:01	00:23:00	31
10/04/2018	14:15	18:04	03:49:00	88
10/05/2018	14:35	15:56	01:21:00	75
Total			**20:02:00**	**542**

It is important to note that the system was not only used, but used for several hours over several days, which was considered better than expected. The tests previously performed with the group of able-bodied participants of only 5 to 10 minutes—although important for evaluating the system with several users—were much less representative than the present test with the person with disabilities, which had a total duration of more than 20 hours. Another fact that corroborates this is the number of commands performed by the system over the days, shown in Table 1. Note that the system received a total of 542 commands and, as reported by the user, worked perfectly.

To better understand how the system was used by the participant, Table 2 summarizes the complete information about the commands received by the system throughout the complete test. The system was always used between 2:00 p.m. and 10:00 p.m., and more than 80% of the commands were sent between 1:00 p.m. and 4:00 p.m., indicating a user routine.

Table 2. Hourly distribution of the commands throughout the days of use of the assistive system.

From	To	Day 1	Day 2	Day 3	Day 4	Day 5	Day 6	Day 7	Percentage
00	12	0	1	0	0	0	0	0	0%
12	13	0	1	0	0	0	0	0	0%
13	14	0	62	7	0	0	0	0	13%
14	15	94	7	22	0	0	75	31	42%
15	16	57	21	7	13	0	3	44	27%
16	17	0	6	0	1	0	0	0	1%
17	18	0	0	0	4	22	8	0	6%
18	19	12	3	0	0	5	0	0	4%
19	20	0	31	0	0	0	0	0	6%
20	21	0	0	0	0	0	0	0	0%
21	22	0	0	0	0	4	2	0	1%
22	00	0	0	0	0	0	0	0	0%
Total		**163**	**131**	**36**	**18**	**31**	**88**	**75**	**542**

Regarding the usability (SUS), the user gave the maximum score for all the questions regarding the willingness to use the system, available functions, ease and confidence in using it, etc. So, focusing on the only feature that the user rated low, according to her, she needed to learn many new things to be able to use the system and so she gave a low score on that item. She believes that after using the system more, she will not need to learn so much more additional information.

Despite this, the user evaluated the system with an average of 92.5, which is quite high, even higher than the previous tests with the group of able-bodied participants, in which the overall mean was 89.9. In fact, according to Brooke [33] and Bangor [164], products evaluated in the 90 point range are considered exceptional.

As recommended by Begum [31], in this work the methodology of UCD for the design of new products was used in order to put the needs and desires of the user first. This way, it was possible to understand, study, design, build, and evaluate the system from the user's point of view.

5. Conclusions

This work presented an assistive system, based on eye gaze tracking for controlling and monitoring a smart home using the Internet of Things, which was developed following concepts of user-centered design and usability. The proposed system allowed a user with disabilities to control everyday equipment in her residence (lamps, television, fan, and radio). In addition, the system could allow the caregiver to remotely monitor the use of the system by the user in real time. The user interface developed included some functionality to improve the usability of the system as a whole. The experiments were divided into two steps. In the first step, the assistive system was assembled in an actual home where tests were conducted with 29 participants (group of able-bodied participants). In the second step, the system was tested for seven days, with online monitoring, by a person with disability (end-user). The results of the SUS showed that the group of able-bodied participants and the end-user evaluated the assistive system with mean scores of 89.9 and 92.5, respectively, positioning the tool as exceptional.

Author Contributions: A.B. conducted the study and wrote this paper. A.B. and D.L.-J. developed the software and hardware. A.B., D.L.-J., and M.S. performed the tests with the participants. L.E. and T.B.-F. supervised and technically advised all the work and contributed to the editing of this manuscript.

Acknowledgments: The authors thank Google Inc. for the Google Research Awards for Latin America, the Federal University of Espirito Santo (UFES/Brazil), Fapes/Brazil, CNPq/Brazil, and CAPES/Brazil for their financial support and scholarships.

Conflicts of Interest: The authors declare no conflict of interest.

Abbreviations

The following abbreviations are used in this manuscript:

AJAX	Asynchronous JavaScript and XML
API	Application Programming Interface
CEP/UFES	Committee of Ethics in Human Beings Research of the Federal University of Espirito Santo
EEPROM	Electrically-Erasable Programmable Read-Only Memory
EOG	Electrooculography
HCI	Human-Computer Interaction
HMI	Human-Computer Interface
HTTP	Hypertext Transfer Protocol
IoT	Internet of Things
IP	Internet Protocol
IR	Infrared
IROG	Infrared Oculography
JSON	JavaScript Object Notation
SPIFFS	Serial Peripheral Interface Flash File System
SSC	Scleral Search Coil
SSID	Service Set IDentifier
SUS	System Usability Scale
TCP	Transmission Control Protocol
UCD	User-Centered Design
UI	User Interface
VOG	Video-Oculography
WS	WebSocket

References

1. Tang, L.Z.W.; Ang, K.S.; Amirul, M.; Yusoff, M.B.M.; Tng, C.K.; Alyas, M.D.B.M.; Lim, J.G.; Kyaw, P.K.; Folianto, F. Augmented reality control home (ARCH) for disabled and elderlies. In Proceedings of the 2015 IEEE Tenth International Conference on Intelligent Sensors, Sensor Networks and Information Processing (ISSNIP), Singapore, 7–9 April 2015; pp. 1–2.
2. Schwiegelshohn, F.; Wehner, P.; Rettkowski, J.; Gohringer, D.; Hubner, M.; Keramidas, G.; Antonopoulos, C.; Voros, N.S. A holistic approach for advancing robots in ambient assisted living environments. In Proceedings of the 2015 IEEE 13th International Conference on Embedded and Ubiquitous Computing, Porto, Portugal, 21–23 October 2015; pp. 140–147.
3. Konstantinidis, E.I.; Antoniou, P.E.; Bamparopoulos, G.; Bamidis, P.D. A lightweight framework for transparent cross platform communication of controller data in ambient assisted living environments. *Inf. Sci. (NY)* **2015**, *300*, 124–139. [CrossRef]
4. Boumpa, E.; Charalampou, I.; Gkogkidis, A.; Ntaliani, A.; Kokkinou, E.; Kakarountas, A. Assistive System for Elders Suffering of Dementia. In Proceedings of the 2018 IEEE 8th International Conference on Consumer Electronics, Berlin, Germany, 2–5 September 2018; pp. 1–4.
5. Brazil Assistive Technology. In *Proceedings of the National Undersecretary for the Promotion of the Rights of People with Disabilities*; Technical Assistance Committee: Geneva, Switzerland, 2009.
6. Elakkiya, J.; Gayathri, K.S. Progressive Assessment System for Dementia Care Through Smart Home. In Proceedings of the 2017 International Conference on Algorithms, Methodology, Models and Applications in Emerging Technologies (ICAMMAET), Chennai, India, 16–18 Febuary 2017; pp. 1–5.
7. Rafferty, J.; Nugent, C.D.; Liu, J.; Chen, L. From Activity Recognition to Intention Recognition for Assisted Living within Smart Homes. *IEEE Trans. Hum. -Mach. Syst.* **2017**, *47*, 368–379. [CrossRef]
8. Mizumoto, T.; Fornaser, A.; Suwa, H.; Yasumoto, K.; Cecco, M. De Kinect-based micro-behavior sensing system for learning the smart assistance with human subjects inside their homes. In Proceedings of the 2018 Workshop on Metrology for Industry 4.0 and IoT, Brescia, Italy, 16–18 April 2018; pp. 1–6.

9. Daher, M.; El Najjar, M.E.; Diab, A.; Khalil, M.; Charpillet, F. Multi-sensory Assistive Living System for Elderly In-home Staying. In Proceedings of the 2018 International Conference on Computer and Applications (ICCA), Beirut, Lebanon, 25–26 August 2012; pp. 168–171.
10. Ghayvat, H.; Mukhopadhyay, S.; Shenjie, B.; Chouhan, A.; Chen, W. Smart Home Based Ambient Assisted Living Recognition of Anomaly in the Activity of Daily Living for an Elderly Living Alone. In Proceedings of the 2018 IEEE International Instrumentation and Measurement Technology Conference (I2MTC), Houston, TX, USA, 14–17 May 2018; pp. 1–5.
11. Wan, J.; Li, M.; Grady, M.J.O.; Hare, G.M.P.O.; Gu, X.; Alawlaqi, M.A.A.H. Time-bounded Activity Recognition for Ambient Assisted Living. *IEEE Trans. Emerg. Top. Comput.* **2018**, 1–13. [CrossRef]
12. Kristály, D.M.; Moraru, S.-A.; Neamtiu, F.O.; Ungureanu, D.E. Assistive Monitoring System Inside a Smart House. In Proceedings of the 2018 International Symposium in Sensing and Instrumentation in IoT Era (ISSI), Shanghai, China, 6–7 September 2018; pp. 1–7.
13. Falcó, J.L.; Vaquerizo, E.; Artigas, J.I. A Multi-Collaborative Ambient Assisted Living Service Description Tool. *Sensors* **2014**, *14*, 9776–9812. [CrossRef]
14. Valadão, C.; Caldeira, E.; Bastos-filho, T.; Frizera-neto, A.; Carelli, R. A New Controller for a Smart Walker Based on Human-Robot Formation. *Sensors* **2016**, *16*, 1116. [CrossRef]
15. Kim, E.Y. Wheelchair Navigation System for Disabled and Elderly People. *Sensors* **2016**, *16*, 1806. [CrossRef]
16. Holloway, C.; Dawes, H. Disrupting the world of Disability: The Next Generation of Assistive Technologies and Rehabilitation Practices. *Healthc. Technol. Lett.* **2016**, *3*, 254–256. [CrossRef] [PubMed]
17. Cruz, D.M.C.D.; Emmel, M.L.G. Assistive Technology Public Policies in Brazil: A Study about Usability and Abandonment by People with Physical Disabilities. *Rev. Fac. St. Agostinho* **2015**, *12*, 79–106.
18. Da Costa, C.R.; Ferreira, F.M.R.M.; Bortolus, M.V.; Carvalho, M.G.R. Assistive technology devices: Factors related to abandonment. *Cad. Ter. Ocup. UFSCar* **2015**, *23*, 611–624.
19. Cruz, D.M.C.D.; Emmel, M.L.G. Use and abandonment of assistive technology for people with physical disabilities in Brazil. Available online: https://www.efdeportes.com/efd173/tecnologia-assistiva-com-deficiencia-fisica.htm (accessed on 19 February 2019).
20. Marcos, P.M.; Foley, J. HCI (human computer interaction): Concepto y desarrollo. *El Prof. La Inf.* **2001**, *10*, 4–16. [CrossRef]
21. Lamberti, F.; Sanna, A.; Carlevaris, G.; Demartini, C. Adding pluggable and personalized natural control capabilities to existing applications. *Sensors* **2015**, *15*, 2832–2859. [CrossRef] [PubMed]
22. Bisio, I.; Lavagetto, F.; Marchese, M.; Sciarrone, A. Smartphone-Centric Ambient Assisted Living Platform for Patients Suffering from Co-Morbidities Monitoring. *IEEE Commun. Mag.* **2015**, *53*, 34–41. [CrossRef]
23. Wang, K.; Shao, Y.; Shu, L.; Han, G.; Zhu, C. LDPA: A Local Data Processing Architecture in Ambient Assisted Living Communications. *IEEE Commun. Mag.* **2015**, *53*, 56–63. [CrossRef]
24. Lopez-Basterretxea, A.; Mendez-Zorrilla, A.; Garcia-Zapirain, B. Eye/head tracking technology to improve HCI with iPad applications. *Sensors* **2015**, *15*, 2244–2264. [CrossRef]
25. Butala, P.M.; Zhang, Y.; Thomas, D.C.; Wagenaar, R.C. Wireless System for Monitoring and Real-Time Classification of Functional Activity. In Proceedings of the 2012 Fourth International Conference Communication Systems Networks (COMSNETS 2012), Bangalore, India, 3–7 January 2012; pp. 1–5.
26. Ahamed, M.M.; Bakar, Z.B.A. Triangle Model Theory for Enhance the Usability by User Centered Design Process in Human Computer Interaction. *Int. J. Contemp. Comput. Res.* **2017**, *1*, 1–7.
27. Iivari, J.; Iivari, N. Varieties of user-centredness: An analysis of four systems development methods. *J. Inf. Syst.* **2011**, *21*, 125–153. [CrossRef]
28. Norman, D.A. *The Design of Everyday Things*; Basic Books: New York, NY, USA, 2002.
29. Preece, J.; Rogers, Y.; Sharp, H. *Interaction Design-Beyond Human-Computer Interaction*; John Wiley Sons: Hoboken, NJ, USA, 2002; pp. 168–186.
30. Goodman, E.; Stolterman, E.; Wakkary, R. Understanding Interaction Design Practices. In Proceedings of the SIGCHI Conference on Human Factors in Computing Systems, Vancouver, BC, Canada, 7–11 May 2011.
31. Begum, I. HCI and its Effective Use in Design and Development of Good User Interface. *Int. J. Res. Eng. Technol.* **2014**, *3*, 176–179.
32. *ISO 9241-210–Ergonomics of Human-System Interaction—Part 210: Human-Centred Design for Interactive Systems*; ISO: Geneva, Switzerland, 2010.

33. Brooke, J. *System Usability Scale (SUS): A Quick-and-Dirty Method of System Evaluation User Information*; Digital Equipment Co Ltd.: Reading, UK, 1986; pp. 1–7.
34. Khan, R.S.A.; Tien, G.; Atkins, M.S.; Zheng, B.; Panton, O.N.M.; Meneghetti, A.T. Analysis of eye gaze: Do novice surgeons look at the same location as expert surgeons during a laparoscopic operation. *Surg. Endosc.* **2012**, *26*, 3536–3540. [CrossRef]
35. Richstone, L.; Schwartz, M.J.; Seideman, C.; Cadeddu, J.; Marshall, S.; Kavoussi, L.R. Eye metrics as an objective assessment of surgical skill. *Ann. Surg.* **2010**, *252*, 177–182. [CrossRef]
36. Wilson, M.; McGrath, J.; Vine, S.; Brewer, J.; Defriend, D.; Masters, R. Psychomotor control in a virtual laparoscopic surgery training environment: Gaze control parameters differentiate novices from experts. *Surg. Endosc.* **2010**, *24*, 2458–2464. [CrossRef]
37. Wilson, M.R.; McGrath, J.S.; Vine, S.J.; Brewer, J.; Defriend, D.; Masters, R.S.W. Perceptual impairment and psychomotor control in virtual laparoscopic surgery. *Surg. Endosc.* **2011**, *27*, 2268–2274. [CrossRef] [PubMed]
38. Sun, Q.; Xia, J.; Nadarajah, N.; Falkmer, T.; Foster, J.; Lee, H. Assessing drivers' visual-motor coordination using eye tracking, GNSS and GIS: A spatial turn in driving psychology. *J. Spat. Sci.* **2016**. [CrossRef]
39. Moore, L.; Vine, S.J.; Cooke, A.M.; Ring, C. Quiet eye training expedites motor learning and aids performance under heightened anxiety: The roles of response programming and external attention. *Psychophysiology* **2012**. [CrossRef] [PubMed]
40. Eid, M.A.; Giakoumidis, N.; El-Saddik, A. A Novel Eye-Gaze-Controlled Wheelchair System for Navigating Unknown Environments: Case Study with a Person with ALS. *IEEE Access* **2016**, *4*, 558–573. [CrossRef]
41. Lupu, R.G.; Ungureanu, F. Mobile Embedded System for Human Computer Communication in Assistive Technology. In Proceedings of the 2012 IEEE 8th International Conference on Intelligent Computer Communication and Processing, Cluj-Napoca, Romania, 30 August–1 September 2012; pp. 209–212.
42. Lupu, R.G.; Ungureanu, F.; Siriteanu, V. Eye tracking mouse for human computer interaction. In Proceedings of the 2013 E-Health and Bioengineering Conference (EHB), Iasi, Romania, 21–23 November 2013; pp. 1–4.
43. Scott, N.; Green, C.; Fairley, S. Investigation of the use of eye tracking to examine tourism advertising effectiveness. *Curr. Issues Tour.* **2015**, *19*, 634–642. [CrossRef]
44. Cecotti, H. A Multimodal Gaze-Controlled Virtual Keyboard. *IEEE Trans. Hum.-Mach. Syst.* **2016**, *46*, 601–606. [CrossRef]
45. Lewis, T.; Pereira, T.; Almeida, D. Smart scrolling based on eye tracking. Design an eye tracking mouse. *Int. J. Comput. Appl.* **2013**, *80*, 34–37.
46. Nehete, M.; Lokhande, M.; Ahire, K. Design an Eye Tracking Mouse. *Int. J. Adv. Res. Comput. Commun. Eng.* **2013**, *2*, 1118–1121.
47. Frutos-Pascual, M.; Garcia-Zapirain, B. Assessing visual attention using eye tracking sensors in intelligent cognitive therapies based on serious games. *Sensors* **2015**, *15*, 11092–11117. [CrossRef]
48. Lee, S.; Yoo, J.; Han, G. Gaze-assisted user intention prediction for initial delay reduction in web video access. *Sensors* **2015**, *15*, 14679–14700. [CrossRef]
49. Takemura, K.; Takahashi, K.; Takamatsu, J. Estimating 3-D Point-of-Regard in a Real Environment Using a Head-Mounted Eye-Tracking System. *IEEE Trans. Hum.-Mach. Syst.* **2014**, *44*, 531–536. [CrossRef]
50. Manabe, H.; Fukumoto, M.; Yagi, T. Direct Gaze Estimation Based on Nonlinearity of EOG. *IEEE Trans. Biomed. Eng.* **2015**, *62*, 1553–1562. [CrossRef] [PubMed]
51. Holzman, P.S.; Proctor, L.R.; Hughes, D.W. Eye-tracking patterns in schizophrenia. *Science* **1973**, *181*, 179–181. [CrossRef] [PubMed]
52. Donaghy, C.; Thurtell, M.J.; Pioro, E.P.; Gibson, J.M.; Leigh, R.J. Eye movements in amyotrophic lateral sclerosis and its mimics: A review with illustrative cases. *J. Neurol. Neurosurg. Psychiatry* **2011**, *82*, 110–116. [CrossRef] [PubMed]
53. Chin, C.A.; Barreto, A.; Cremades, J.G.; Adjouadi, M. Integrated electromyogram and eye-gaze tracking cursor control system for computer users with motor disabilities. *J. Rehabil. Res. Dev.* **2008**, *45*, 161–174. [CrossRef] [PubMed]
54. Missimer, E.; Betke, M. Blink and wink detection for mouse pointer control. In Proceedings of the 3rd International Conference on Pervasive Technologies Related to Assistive Environments (PETRA '10), Samos, Greece, 23–25 June 2010.
55. Wankhede, S.; Chhabria, S. Controlling Mouse Cursor Using Eye Movement. *Int. J. Appl. Innov. Eng. Manag.* **2013**, *1*, 1–7.

56. Meghna, S.M.A.; Kachan, K.L.; Baviskar, A. Head tracking virtual mouse system based on ad boost face detection algorithm. *Int. J. Recent Innov. Trends Comput. Commun.* **2016**, *4*, 921–923.
57. Biswas, J.; Wai, A.A.P.; Tolstikov, A.; Kenneth, L.J.H.; Maniyeri, J.; Victor, F.S.F.; Lee, A.; Phua, C.; Jiaqi, Z.; Hoa, H.T.; et al. From context to micro-context–Issues and challenges in sensorizing smart spaces for assistive living. *Procedia Comput. Sci.* **2011**, *5*, 288–295. [CrossRef]
58. Visutsak, P.; Daoudi, M. The Smart Home for the Elderly: Perceptions, Technologies and Psychological Accessibilities. In Proceedings of the 2017 XXVI International Conference on Information, Communication and Automation Technologies (ICAT), Sarajevo, Bosnia-Herzegovina, 26–28 October 2017; pp. 1–6.
59. Beligianni, F.; Alamaniotis, M.; Fevgas, A.; Tsompanopoulou, P.; Bozanis, P.; Tsoukalas, L.H. An internet of things architecture for preserving privacy of energy consumption. In Proceedings of the Mediterranean Conference on Power Generation, Transmission, Distribution and Energy Conversion (MedPower 2016), Belgrade, Serbia, 6–9 November 2016; pp. 1–7.
60. Bouchet, O.; Javaudin, J.; Kortebi, A.; El-Abdellaouy, H.; Lebouc, M.; Fontaine, F.; Jaffré, P.; Celeda, P.; Mayer, C.; Guan, H. ACEMIND: The smart integrated home network. In Proceedings of the 2014 International Conference on Intelligent Environments, Shanghai, China, 30 June–4 July 2014; pp. 1–8.
61. Buhl, J.; Hasselkuß, M.; Suski, P.; Berg, H. Automating Behavior? An Experimental Living Lab Study on the Effect of Smart Home Systems and Traffic Light Feedback on Heating Energy Consumption. *Curr. J. Appl. Sci. Technol.* **2017**, *22*, 1–18. [CrossRef]
62. Lim, Y.; Lim, S.Y.; Nguyen, M.D.; Li, C.; Tan, Y. Bridging Between universAAL and ECHONET for Smart Home Environment. In Proceedings of the 2017 14th International Conference on Ubiquitous Robots and Ambient Intelligence (URAI), Jeju, South Korea, 28 June–1 July 2017; pp. 56–61.
63. Lin, C.M.; Chen, M.T. Design and Implementation of a Smart Home Energy Saving System with Active Loading Feature Identification and Power Management. In Proceedings of the 2017 IEEE 3rd International Future Energy Electronics Conference and ECCE Asia (IFEEC 2017–ECCE Asia), Kaohsiung, Taiwan, 3–7 June 2017; pp. 739–742.
64. Soe, W.T.; Mpawenimana, I.; Difazio, M.; Belleudy, C.; Ya, A.Z. Energy Management System and Interactive Functions of Smart Plug for Smart Home. *Int. J. Electr. Comput. Energ. Electron. Commun. Eng.* **2017**, *11*, 824–831.
65. Wu, X.; Hu, X.; Teng, Y.; Qian, S.; Cheng, R. Optimal integration of a hybrid solar-battery power source into smart home nanogrid with plug-in electric vehicle. *J. Power Sources* **2017**, *363*, 277–283. [CrossRef]
66. Melhem, F.Y.; Grunder, O.; Hammoudan, Z. Optimization and Energy Management in Smart Home considering Photovoltaic, Wind, and Battery Storage System with Integration of Electric Vehicles. *Can. J. Electr. Comput. Eng.* **2017**, *40*, 128–138.
67. Başol, G.; Güntürkün, R.; Başol, E. Smart Home Design and Application. *World Wide J. Multidiscip. Res. Dev.* **2017**, *3*, 53–58.
68. Han, J.; Choi, C.; Park, W.; Lee, I.; Kim, S.; Architecture, A.S. Smart Home Energy Management System Including Renewable Energy Based on ZigBee and PLC. *IEEE Trans. Consum. Electron.* **2014**, *60*, 198–202. [CrossRef]
69. Li, C.; Luo, F.; Chen, Y.; Xu, Z.; An, Y.; Li, X. Smart Home Energy Management with Vehicle-to-Home Technology. In Proceedings of the 2017 13th IEEE International Conference on Control & Automation (ICCA), Ohrid, Macedonia, 3–6 July 2017; pp. 136–142.
70. Kiat, L.Y.; Barsoum, N. Smart Home Meter Measurement and Appliance Control. *Int. J. Innov. Res. Dev.* **2017**, *6*, 64–70. [CrossRef]
71. Oliveira, E.L.; Alfaia, R.D.; Souto, A.V.F.; Silva, M.S.; Francês, C.R. SmartCoM: Smart Consumption Management Architecture for Providing a User-Friendly Smart Home based on Metering and Computational Intelligence. *J. Microw. Optoelectron. Electromagn. Appl.* **2017**, *16*, 732–751. [CrossRef]
72. Kibria, M.G.; Jarwar, M.A.; Ali, S.; Kumar, S.; Chong, I. Web Objects Based Energy Efficiency for Smart Home IoT Service Provisioning. In Proceedings of the 2017 Ninth International Conference on Ubiquitous and Future Networks (ICUFN), Milan, Italy, 4–7 July 2017; pp. 55–60.
73. Datta, S.K. Towards Securing Discovery Services in Internet of Things. In Proceedings of the 2016 IEEE International Conference on Consumer Electronics (ICCE), Las Vegas, NV, USA, 7–11 January 2016; pp. 506–507.

74. Huth, C.; Duplys, P.; Tim, G. Secure Software Update and IP Protection for Untrusted Devices in the Internet of Things Via Physically Unclonable Functions. In Proceedings of the 2016 IEEE International Conference on Pervasive Computing and Communication Workshops (PerCom Workshops), Sydney, Australia, 14–18 March 2016; pp. 1–6.
75. Rahman, R.A.; Shah, B. Security analysis of IoT protocols: A focus in CoAP. In Proceedings of the 2016 3rd MEC International Conference on Big Data and Smart City (ICBDSC), Muscat, Oman, 15–16 March 2016; pp. 1–7.
76. Rajiv, P.; Raj, R.; Chandra, M. Email based remote access and surveillance system for smart home infrastructure. *Perspect. Sci.* **2016**, *8*, 459–461. [CrossRef]
77. Wurm, J.; Hoang, K.; Arias, O.; Sadeghi, A.; Jin, Y. Security Analysis on Consumer and Industrial IoT Devices. In Proceedings of the 2016 21st Asia and South Pacific Design Automation Conference (ASP-DAC), Macau, China, 25–28 January 2016; pp. 519–524.
78. Arabo, A. Cyber Security Challenges within the Connected Home Ecosystem Futures. *Procedia Comput. Sci.* **2015**, *61*, 227–232. [CrossRef]
79. Golait, S.S. 3-Level Secure Kerberos Authentication for Smart Home Systems Using IoT. In Proceedings of the 2015 1st International Conference on Next Generation Computing Technologies (NGCT), Dehradun, India, 4–5 September 2015; pp. 262–268.
80. Han, J.H.; Kim, J. Security Considerations for Secure and Trustworthy Smart Home System in the IoT Environment. In Proceedings of the 2015 International Conference on Information and Communication Technology Convergence (ICTC), Jeju, South Korea, 28–30 October 2015; pp. 1116–1118.
81. Huth, C.; Zibuschka, J.; Duplys, P.; Tim, G. Securing Systems on the Internet of Things via Physical Properties of Devices and Communications. In Proceedings of the 2015 Annual IEEE Systems Conference (SysCon) Proceedings, Vancouver, BC, Canada, 13–16 April 2015; pp. 8–13.
82. Jacobsson, A.; Davidsson, P. Towards a Model of Privacy and Security for Smart Homes. In Proceedings of the 2015 IEEE 2nd World Forum Internet Things, Milan, Italy, 4–16 December 2015; pp. 727–732.
83. Peng, Z.; Kato, T.; Takahashi, H.; Kinoshita, T. Intelligent Home Security System Using Agent-based IoT Devices. In Proceedings of the 2015 IEEE 4th Global Conference on Consumer Electronics (GCCE), Osaka, Japan, 27–30 October 2015; pp. 313–314.
84. Peretti, G.; Lakkundit, V.; Zorzi, M. BlinkToSCoAP: An End-to-End Security Framework for the Internet of Things. In Proceedings of the 2015 7th International Conference on Communication Systems and Networks (COMSNETS), Bangalore, India, 6–10 January 2015; pp. 1–6.
85. Santoso, F.K.; Vun, N.C.H. Securing IoT for Smart Home System. In Proceedings of the 2015 International Symposium on Consumer Electronics (ISCE), Madrid, Spain, 24–26 June 2015; pp. 1–2.
86. Schiefer, M. Smart Home Definition and Security Threats. In Proceedings of the 2015 Ninth International Conference on IT Security Incident Management & IT Forensics, Magdeburg, Germany, 18–20 May 2015; pp. 114–118.
87. Alohali, B.; Merabti, M.; Kifayat, K. A Secure Scheme for a Smart House Based on Cloud of Things (CoT). In Proceedings of the 2014 6th Computer Science and Electronic Engineering Conference (CEEC), Colchester, UK, 25–26 September 2014; pp. 115–120.
88. Sivaraman, V.; Gharakheili, H.H.; Vishwanath, A.; Boreli, R.; Mehani, O. Network-Level Security and Privacy Control for Smart-Home IoT Devices. In Proceedings of the 2015 IEEE 11th International Conference on Wireless and Mobile Computing, Networking and Communications (WiMob), Abu Dhabi, United Arab, 19–21 October 2015; pp. 163–167.
89. Amadeo, M.; Briante, O.; Campolo, C.; Molinaro, A.; Ruggeri, G. Information-centric networking for M2M communications: Design and deployment. *Comput. Commun.* **2016**, *89–90*, 105–106. [CrossRef]
90. Li, H.; Seed, D.; Flynn, B.; Mladin, C.; Di Girolamo, R. Enabling Semantics in an M2M/IoT Service Delivery Platform. In Proceedings of the 2016 IEEE Tenth International Conference on Semantic Computing (ICSC), Laguna Hills, CA, USA, 4–6 Febuary 2016; pp. 206–213.
91. Rizopoulos, C. Implications of Theories of Communication and Spatial Behavior for the Design of Interactive Environments. In Proceedings of the 2011 Seventh International Conference on Intelligent Environments, Nottingham, UK, 25–28 July 2011; pp. 286–293.

92. Waltari, O.; Kangasharju, J. Content-Centric Networking in the Internet of Things. In Proceedings of the 2016 13th IEEE Annual Consumer Communications & Networking Conference (CCNC), Las Vegas, NV, USA, 9–12 January 2016; pp. 73–78.
93. Seo, D.W.; Kim, H.; Kim, J.S.; Lee, J.Y. Hybrid reality-based user experience and evaluation of a context-aware smart home. *Comput. Ind.* **2016**, *76*, 11–23. [CrossRef]
94. Amadeo, M.; Campolo, C.; Iera, A.; Molinaro, A. Information Centric Networking in IoT scenarios: The Case of a Smart Home. In Proceedings of the 2015 IEEE International Conference on Communications (ICC), London, UK, 8–12 June 2015; pp. 648–653.
95. Elkhodr, M.; Shahrestani, S.; Cheung, H. A Smart Home Application based on the Internet of Things Management Platform. In Proceedings of the 2015 IEEE International Conference on Data Science and Data Intensive Systems, Sydney, Australia, 11–13 December 2015; pp. 491–496.
96. Bhide, V.H.; Wagh, S. I-learning IoT: An intelligent self learning system for home automation using IoT. In Proceedings of the 2015 International Conference on Communications and Signal Processing (ICCSP), Melmaruvathur, India, 2–4 April 2015; pp. 1763–1767.
97. Bian, J.; Fan, D.; Zhang, J. The new intelligent home control system based on the dynamic and intelligent gateway. In Proceedings of the 2011 4th IEEE International Conference on Broadband Network and Multimedia Technology, Shenzhen, China, 28–30 October 2011; pp. 526–530.
98. Cheuque, C.; Baeza, F.; Marquez, G.; Calderon, J. Towards to responsive web services for smart home LED control with Raspberry Pi. A first approach. In Proceedings of the 2015 34th International Conference of the Chilean Computer Science Society (SCCC), Santiago, Chile, 9–13 November 2015.
99. Hasibuan, A.; Mustadi, M.; Syamsudin, D.I.E.Y.; Rosidi, I.M.A. Design and Implementation of Modular Home Automation Based on Wireless Network, REST API and WebSocket. In Proceedings of the Internastional Symposium on Intelligent Signal Processing and Communication Systems (ISPACS), Nusa Dua, Indonesia, 9–12 November 2015; pp. 9–12.
100. Hernandez, M.E.P.; Reiff-Marganiec, S. Autonomous and self controlling smart objects for the future Internet. In Proceedings of the 2015 3rd International Conference on Future Internet of Things and Cloud, Rome, Italy, 24–26 August 2015; pp. 301–308.
101. Jacobsson, A.; Boldt, M.; Carlsson, B. A risk analysis of a smart home automation system. *Futur. Gener. Comput. Syst.* **2016**, *56*, 719–733. [CrossRef]
102. Jacobsson, A.; Boldt, M.; Carlsson, B. On the risk exposure of smart home automation systems. In Proceedings of the 2014 International Conference on Future Internet of Things and Cloud, Barcelona, Spain, 27–29 August 2014; pp. 183–190.
103. Lazarevic, I.; Sekulic, M.; Savic, M.S.; Mihic, V. Modular home automation software with uniform cross component interaction based on services. In Proceedings of the 2015 IEEE 5th International Conference on Consumer Electronics Berlin (ICCE-Berlin), Berlin, Germany, 6–9 September 2015.
104. Lee, K.M.; Teng, W.G.; Hou, T.W. Point-n-Press: An Intelligent Universal Remote Control System for Home Appliances. *IEEE Trans. Autom. Sci. Eng.* **2016**, *13*, 1308–1317. [CrossRef]
105. Yeazell, S.C. Teaching Supplemental Jurisdiction. *Indiana Law J.* **1998**, *74*, 222–249.
106. Miclaus, A.; Riedel, T.; Beigl, M. Computing Corner. *Teach. Stat.* **1990**, *12*, 25–26.
107. Mittal, Y.; Toshniwal, P.; Sharma, S.; Singhal, D.; Gupta, R.; Mittal, V.K. A voice-controlled multi-functional Smart Home Automation System. In Proceedings of the 2015 Annual IEEE India Conference (INDICON), New Delhi, India, 17–20 December 2015; pp. 1–6.
108. Moravcevic, V.; Tucic, M.; Pavlovic, R.; Majdak, A. An approach for uniform representation and control of ZigBee devices in home automation software. In Proceedings of the 2015 IEEE 5th International Conference on Consumer Electronics - Berlin (ICCE-Berlin), Berlin, Germany, 6–9 September 2015; pp. 237–239.
109. Papp, I.; Velikic, G.; Lukac, N.; Horvat, I. Uniform representation and control of Bluetooth Low Energy devices in home automation software. In Proceedings of the 2015 IEEE 5th International Conference on Consumer Electronics - Berlin (ICCE-Berlin), Berlin, Germany, 6–9 September 2015; pp. 366–368.
110. Ryan, J.L. Home automation. *IEE Rev.* **1988**, *34*, 355. [CrossRef]
111. Lee, Y.-T.; Hsiao, W.-H.; Huang, C.-M.; Chou, S.-C.T. An Integrated Cloud-Based Smart Home Management System with Community Hierarchy. *IEEE Trans. Consum. Electron.* **2016**, *62*, 1–9. [CrossRef]
112. Kanaris, L.; Kokkinis, A.; Fortino, G.; Liotta, A.; Stavrou, S. Sample Size Determination Algorithm for fingerprint-based indoor localization systems. *Comput. Netw.* **2016**, *101*, 169–177. [CrossRef]

113. Mano, L.Y.; Faiçal, B.S.; Nakamura, L.H.V.; Gomes, P.H.; Libralon, G.L.; Meneguete, R.I.; Filho, G.P.R.; Giancristofaro, G.T.; Pessin, G.; Krishnamachari, B.; et al. Exploiting IoT technologies for enhancing Health Smart Homes through patient identification and emotion recognition. *Comput. Commun.* **2016**, *89–90*, 178–190. [CrossRef]
114. Zanjal, S.V.; Talmale, G.R. Medicine Reminder and Monitoring System for Secure Health Using IOT. *Procedia Comput. Sci.* **2016**, *78*, 471–476. [CrossRef]
115. Bhole, M.; Phull, K.; Jose, A.; Lakkundi, V. Delivering Analytics Services for Smart Homes. In Proceedings of the 2015 IEEE Conference on Wireless Sensors (ICWiSe), Melaka, Malaysia, 24–26 August 2015; pp. 28–33.
116. Thomas, S.; Bourobou, M.; Yoo, Y. User Activity Recognition in Smart Homes Using Pattern Clustering Applied to Temporal ANN Algorithm. *Sensors* **2015**, *15*, 11953–11971.
117. Scholz, M.; Flehmig, G.; Schmidtke, H.R.; Scholz, G.H. Powering Smart Home intelligence using existing entertainment systems. In Proceedings of the 2011 Seventh International Conference on Intelligent Environments, Nottingham, UK, 25–28 July 2011; Volume 970, pp. 230–237.
118. Jiang, S.; Peng, J.; Lu, Z.; Jiao, J. 802.11ad Key Performance Analysis and Its Application in Home Wireless Entertainment. In Proceedings of the 2014 IEEE 17th International Conference on Computational Science and Engineering, Chengdu, China, 19–21 December 2014; Volume 5, pp. 1595–1598.
119. Technologies, I. Analyzing Social Networks Activities to Deploy Entertainment Services in HRI-based Smart Environments. In Proceedings of the 2017 IEEE International Conference on Fuzzy Systems (FUZZ-IEEE), Naples, Italy, 9–12 July 2017; pp. 1–6.
120. Hossain, M.A.; Alamri, A.; Parra, J. Context-Aware Elderly Entertainment Support System in Assisted Living Environment. In Proceedings of the 2013 IEEE International Conference on Multimedia and Expo Workshops (ICMEW), San Jose, CA, USA, 15–19 July 2013; pp. 1–6.
121. Dooley, J.; Henson, M.; Callaghan, V.; Hagras, H.; Al-Ghazzawi, D.; Malibari, A.; Al-Haddad, M.; Al-ghamdi, A.A. A Formal Model For Space Based Ubiquitous Computing. In Proceedings of the 2011 Seventh International Conference on Intelligent Environments, Nottingham, UK, 25–28 July 2011; pp. 294–299.
122. De Morais, W.O.; Wickström, N. A "Smart Bedroom" as an Active Database System. In Proceedings of the 2013 9th International Conference on Intelligent Environments, Athens, Greece, 16–17 July 2013; pp. 250–253.
123. Gaikwad, P.P.; Gabhane, J.P.; Golait, S.S. Survey based on Smart Homes System Using Internet-of-Things. In Proceedings of the 2015 International Conference on Computation of Power, Energy, Information and Communication (ICCPEIC), Chennai, India, 22–23 April 2015; pp. 330–335.
124. Samuel, S.S.I. A Review of Connectivity Challenges in IoT-Smart Home. In Proceedings of the 2016 3rd MEC International Conference on Big Data and Smart City (ICBDSC), Muscat, Oman, 15–16 March 2016; pp. 1–4.
125. Kim, Y.; Lee, S.; Jeon, Y.; Chong, I.; Lee, S.H. Orchestration in distributed web-of-objects for creation of user-centered iot service capability. In Proceedings of the 2013 Fifth International Conference on Ubiquitous and Future Networks (ICUFN), Da Nang, Vietnam, 2–5 July 2013.
126. Tamura, T.; Kawarada, A.; Nambu, M.; Tsukada, A.; Sasaki, K.; Yamakoshi, K. E-Healthcare at an Experimental Welfare Techno House in Japan. *Open Med. Inform. J.* **2007**, 1–7. [CrossRef] [PubMed]
127. Mohktar, M.S.; Sukor, J.A.; Redmond, S.J.; Basilakis, J.; Lovell, N.H. Effect of Home Telehealth Data Quality on Decision Support System Performance. *Proc. Comput. Sci.* **2015**, *64*, 352–359. [CrossRef]
128. Matsuoka, K. Aware Home Understanding Life Activities. In Proceedings of the 2nd International Conference Smart Homes Health Telematics (ICOST 2004), Ames, IA, USA, 28 June–2 July 2004; Volume 14, pp. 186–193.
129. Mori, T.; Noguchi, H.; Takada, A.; Sato, T. Sensing room environment: Distributed sensor space for measurement of human dialy behavior. *Ransaction Soc. Instrum. Control Eng.* **2006**, *1*, 97–103.
130. Lee, H.; Kim, Y.-T.; Jung, J.-W.; Park, K.-H.; Kim, D.-J.; Bang, B.; Bien, Z.Z. A 24-hour health monitoring system in a smart house. *Gerontechnol. J.* **2008**, *7*, 22–35. [CrossRef]
131. Ubihome | Mind the Gap. Available online: https://mindthegap.agency/client/ubihome (accessed on 30 November 2018).
132. ubiHome. Available online: http://ubihome.me/Home (accessed on 30 November 2018).
133. Oh, Y.; Lee, S.; Woo, W. User-Centric Integration of Contexts for a Unified Context-Aware Application Model. CEUR Workshop Proceeding. 2005, pp. 9–16. Available online: https://www.semanticscholar.org/paper/User-centric-Integration-of-Contexts-for-A-Unified-Oh-Lee/0b36ca616e47b66487372625df44aeb1d919fe48 (accessed on 29 November 2018).

134. Oh, Y.; Woo, W. A Unified Application Service Model for ubiHome by Exploiting Intelligent Context-Awareness. In *International Symposium on Ubiquitious Computing Systems*; Springer: Berlin, Germnay, 2004; pp. 192–202.
135. Bien, Z.Z.; Park, K.-H.; Bang, W.-C.; Stefanov, D.H. LARES: An Intelligent Sweet Home for Assisting the Elderly and the Handicapped. 1st International Conference Smart Homes Health Telematics, Assistive Technology, 2003. pp. 151–158. Available online: https://edurev.in/studytube/LARES-An-Intelligent-Sweet-Home-for-Assisting-the-/bf43423b-0daf-42b3-bf3f-54c8fa4e2bbd_p (accessed on 29 November 2018).
136. Minoh, M. Experiences in UKARI Project. *J. Natl. Inst. Inf. Commun. Technol.* **2007**, *54*, 147–154.
137. Tetsuya, F.; Hirotada, U.; Michihiko, M. A Looking-for-Objects Service in Ubiquious Home. *J. Natl. Inst. Inf. Commun. Technol.* **2007**, *54*, 175–181.
138. Toyota Dream House PAPI. Available online: http://tronweb.super-nova.co.jp/toyotadreamhousepapi.html (accessed on 30 November 2018).
139. Junestrand, S.; Keijer, U.; Tollmar, K. Private and Public Digital Domestic Spaces. *Int. J. Hum. Comput. Stud.* **2001**, *54*, 753–778. [CrossRef]
140. Orpwood, R.; Gibbs, C.; Adlam, T.; Faulkner, R.; Meegahawatte, D. The Gloucester Smart House for People with Dementia—User-Interface Aspects. In *Designing a More Inclusive World*; Springer: Berlin, Germany, 2004; pp. 237–245.
141. Davis, G.; Wiratunga, N.; Taylor, B.; Craw, S. Matching smarthouse technology to needs of the elderly and disabled. In Proceedings of the Workshop Proceedings of the 5th International Conference on Case-Based Reasoning, Trondheim, Norway, 23–26 June 2003; pp. 29–36.
142. Mehrotra, S.; Dhande, R. Smart cities and smart homes: From realization to reality. In Proceedings of the 2015 International Conference on Green Computing and Internet of Things (ICGCIoT), Noida, India, 8–10 October 2015; pp. 1236–1239.
143. myGEKKO. Available online: https://www2.my-gekko.com/en/ (accessed on 30 November 2018).
144. MATCH–Mobilising Advanced Technologies for Care at Home. Available online: http://www.cs.stir.ac.uk/~{}kjt/research/match/main/main.html (accessed on 30 November 2018).
145. The Adaptive House Boulder, Colorado. Available online: http://www.cs.colorado.edu/~{}mozer/index.php?dir=/Research/Projects/Adaptivehouse/ (accessed on 29 November 2018).
146. Lindsey, R.; Daluiski, A.; Chopra, S.; Lachapelle, A.; Mozer, M.; Sicular, S.; Hanel, D.; Gardner, M.; Gupta, A.; Hotchkiss, R.; et al. Deep neural network improves fracture detection by clinicians. *Proc. Natl. Acad. Sci. USA* **2018**, *115*, 1–6. [CrossRef] [PubMed]
147. Aware Home Research Initiative (AHRI). Available online: http://awarehome.imtc.gatech.edu/ (accessed on 29 November 2018).
148. MavHome: Managing an Adaptive Versatile Home. Available online: http://ailab.wsu.edu/mavhome/ (accessed on 30 November 2018).
149. Cook, D.J.; Youngblood, M.; Heierman, E.O.; Gopalratnam, K.; Rao, S.; Litvin, A.; Khawaja, F. MavHome: An agent-based smart home. In Proceedings of the First IEEE International Conference on Pervasive Computing and Communications, 2003. (PerCom 2003), Fort Worth, TX, USA, 26 March 2003; pp. 521–524.
150. House_n Materials and Media. Available online: http://web.mit.edu/cron/group/house_n/publications.html (accessed on 29 November 2018).
151. Intille, S.S. The Goal: Smart People, Not Smart Homes. In Proceedings of the International Conference on Smart Homes and Health Telematics, Belfast, Northern Ireland, 26–28 June 2006; pp. 1–4.
152. Shafer, S.; Krumm, J.; Brumitt, B.; Meyers, B.; Czerwinski, M.; Robbins, D. The New EasyLiving Project at Microsoft Research. In Proceedings of the Joint DARPA/NIST Smart Spaces Workshop, Gaithersburg, MD, USA, 30–31 Jully 1998; Volume 5.
153. Helal, A.; Mann, W. Gator Tech Smart House: A Programmable Pervasive Space. *IEEE Comput. Mag.* **2005**, 64–74. [CrossRef]
154. Pigot, H.; Lefebvre, B. The Role of Intelligent Habitats in Upholding Elders in Residence. *WIT Trans. Biomed. Health* **2003**. [CrossRef]
155. Lesser, V.; Atighetchi, M.; Benyo, B.; Horling, B.; Raja, A.; Vincent, R.; Wagner, T.; Xuan, P.; Zhang, S. *A Multi-Agent System for Intelligent Environment Control*; UMass Computer Science Technical Report 1998-40; University of Massachusetts: Amherst, MA, USA, 1999.

156. CASAS–Center for Advanced Studies in Adaptive Systems. Available online: http://casas.wsu.edu/ (accessed on 29 November 2018).
157. Ghods, A.; Caffrey, K.; Lin, B.; Fraga, K.; Fritz, R.; Schmitter-Edgecombe, M.; Hundhausen, C.; Cook, D.J. Iterative Design of Visual Analytics for a Clinician-in-the-loop Smart Home. *IEEE J. Biomed. Heal. Inf.* **2018**, 2168–2194. [CrossRef] [PubMed]
158. Yang, H.I.; Babbitt, R.; Wong, J.; Chang, C.K. A framework for service morphing and heterogeneous service discovery in smart environments. In *International Conference on Smart Homes and Health Telematics*; Springer: Berlin, Germany, 2012; pp. 9–17.
159. Heinz, M.; Martin, P.; Margrett, J.A.; Yearns, M.; Franke, W.; Yang, H.I. Perceptions of technology among older adults. *J. Gerontol. Nurs.* **2013**. [CrossRef] [PubMed]
160. Wang, F.; Skubic, M.; Rantz, M.; Cuddihy, P.E. Quantitative gait measurement with pulse-doppler radar for passive in-home gait assessment. *IEEE Trans. Biomed. Eng.* **2014**, *61*, 2434–2443. [CrossRef] [PubMed]
161. Rochester. Available online: https://www.rochester.edu/pr/Review/V64N3/feature2.html (accessed on 30 November 2018).
162. Du, K.K.; Wang, Z.L.; Hong, M. Human machine interactive system on smart home of IoT. *J. China Univ. Posts Telecommun.* **2013**, *20*, 96–99. [CrossRef]
163. Bissoli, A.L.C. *Multimodal Solution for Interaction with Assistance and Communication Devices*; Federal University of Espirito Santo (UFES): Vitoria, Espirito Santo, Brazil, 2016.
164. Bangor, A.; Philip, K.; James, M. Determining what individual SUS scores mean: Adding an adjective rating scale. *J. Usability Stud.* **2009**, *4*, 114–123.

Article

The Effects of Housing Environments on the Performance of Activity-Recognition Systems Using Wi-Fi Channel State Information: An Exploratory Study

Hoonyong Lee [1], Changbum R. Ahn [2,*], Nakjung Choi [3], Toseung Kim [4] and Hyunsoo Lee [4]

1. Department of Architecture, College of Architecture, Texas A&M University, College Station, TX 77843-3137, USA; onarcher@tamu.edu
2. Department of Construction Science, College of Architecture, Texas A&M University, College Station, TX 77843-3137, USA
3. Nokia Bell Labs, Murray Hill, NJ 07974-0636, USA; nakjung.choi@nokia-bell-labs.com
4. Department of Architecture and Architectural Engineering, Seoul National University, Seoul 08826, Korea; toeskim92@gmail.com (T.K.); hyunslee@snu.ac.kr (H.L.)

* Correspondence: ryanahn@tamu.edu

Received: 18 January 2019; Accepted: 21 February 2019; Published: 26 February 2019

Abstract: Recently, device-free human activity–monitoring systems using commercial Wi-Fi devices have demonstrated a great potential to support smart home environments. These systems exploit Channel State Information (CSI), which represents how human activities–based environmental changes affect the Wi-Fi signals propagating through physical space. However, given that Wi-Fi signals either penetrate through an obstacle or are reflected by the obstacle, there is a high chance that the housing environment would have a great impact on the performance of a CSI-based activity-recognition system. In this context, this paper examines whether and to what extent housing environment affects the performance of the CSI-based activity recognition systems. Activities in daily living (ADL)–recognition systems were implemented in two typical housing environments representative of the United States and South Korea: a wood-frame apartment (Unit A) and a reinforced concrete-frame apartment (Unit B), respectively. The experimental results show that housing environments, combined with various environmental factors (i.e., structural building materials, surrounding Wi-Fi interference, housing layout, and population density), generate a significant difference in the accuracy of the applied CSI-based ADL-recognition systems. This outcome provides insights into how such ADL systems should be configured for various home environments.

Keywords: smart home; occupant activity recognition; channel state information (CSI); Wi-Fi; housing environment

---

1. Introduction

In order to advance smart home environments that can deliver elderly healthcare, energy savings, and home security, systems must first be able to accurately recognize human activity in daily living (ADL) [1,2]. Traditionally, ADL-recognition systems rely on dedicated sensors such as cameras, motion sensors, or other special sensors (e.g., inertial measurement units). However, these device-based ADL-recognition solutions have limitations in their use, especially since these systems require significant infrastructural installations in the environment—for example, some such systems require cameras or motion sensors to be attached to walls or doors to detect activity. Problematically, some of these technologies face inherent limitations—e.g., cameras raise privacy issues and require line-of-sight for human movements—which makes installation concerns and considerations more pressing. Alternatively, wearable, sensor-based approaches that require users to wear the sensors to detect activities may offset the infrastructural concerns, but these wearable approaches demand users' diligent applications of the devices, a fact that can challenge the effectiveness of the technology [1–3]. For these reasons, ADL-recognition still represents a puzzle to many smart-environment developers.

In recent years, device-free activity-recognition approaches have been the focus of ADL-recognition systems [2], with Wi-Fi signal systems presenting an especially attractive option. Due to its ubiquitous presence in home environments, Wi-Fi signals have already been employed for human-activity recognition without additional devices. Such Wi-Fi signal-based approaches do not require a dense placement of sensors to generate detecting areas of interest. These systems consist of a Wi-Fi transmitter (an Access Point, AP) and one or several Wi-Fi devices (Receivers) located in different places within the environment. Also, these Wi-Fi signal-based ADL-recognition systems do not require human activity in the Line-of-Sight (LOS) or face privacy problems [4], making them a reasonable alternative to device-based ADL-recognition systems.

Wi-Fi signal-based ADL-recognition approaches exploit fine-grained Wi-Fi signal signatures. Specifically, Wi-Fi signals propagate from a transmitter to receivers, so human activity may affect the signals' propagation paths, which in turn cause the signals to be changed at receivers. For example, Received Signal Strength (RSS) has been used for fingerprint-based localization systems [5–12]; when a subject is located between the AP and the receiver, the RSS is changed due to signal attenuation. Exploiting this RSS, coarse human activity (e.g., vacant home, occupied home, human movement, walking activity) has been detected with an average 90% accuracy [13]. However, the option to exploit RSS for ADL-recognition is only available for coarse-level activity detection because RSS captures total power and exhibits signal variance as a single amplitude, rendering the approach ineffective in a static and/or complex environment [14,15].

Unlike RSS which has a superimposition layer of multipath signals, Channel State Information (CSI) exploits channel information between the AP and the receiver at the individual subcarrier level. Specifically, CSI represents a channel response to the physical environmental changes, which depicts the multipath effects of signals. Consequently, CSI is more stable and robust than RSS, and CSI data can be captured from commodity Wi-Fi devices using a Linux CSI 802.11n tool [16]. Recently, CSI-based ADL-recognition approaches have succeeded at detecting human activities at different levels of granularity ranging from coarse to fine, as evidenced by the detection successes of technologies such as Wi-Sleep [17], Wi-Chase [15], E-eyes [18], and RT-Fall (Real-Time Fall) [3]. These approaches can not only measure various daily activities, but also the fine movement of the chest during breathing in real time.

Despite these advantages, these CSI-based ADL-recognition systems environments still need further verification in terms of their performance in various real-world housing environments; Most previous studies [15,19–22] were conducted in controlled laboratory settings and several few attempts in real-world environments were mainly conducted in a representative housing environment in the United States [3,18]. The Housing environment greatly varies by its location and housing type, and the difference of the housing environment may create a significant impact on the performance of CSI-based ADL-recognition. For example, Wi-Fi signals have different propagation loss as Wi-Fi signals penetrate different building materials, such as wood, glass, and concrete [23,24]. In addition to building materials, other factors such as unit layouts and distance to neighbor units would affect the performance of CSI-based ADL-recognition. However, it has not been investigated whether and to what extent existing CSI-based ADL recognition systems can be resistant to different housing environments. To this end, this paper examines the effects of housing environments on the performance of CSI-based ADL-recognition systems. In particular, this study selected two units that provide a great difference in their housing environments, including structural and finish materials, and unit density, and investigated whether such a difference in housing environment creates a noticeable difference in the performance of CSI-based ADL-recognition, regardless of the algorithms used for ADL recognition.

2. Related Work

Recently, exploiting Wi-Fi signals for activity detection has risen in popularity due to the availability and ubiquitous distribution of these wireless networks and their corresponding commercial devices. Human-body movements cause Wi-Fi signals to change when the signals are reflected from the body, which results in variations at the Wi-Fi receiver and, consequently, opportunities to estimate human activities by analyzing the signal variance. This approach further benefits from the fact that Wi-Fi signal–based ADL-recognition systems do not require LOS for the activities since Wi-Fi signals propagate through walls. Thus, systems such as Wi-Vi [25] and WiSee [26] have successfully detected the existence of humans and have differentiated several human gestures—such as punch, kick, and push—even on the other side of a wall.

While such RSS approaches have been successfully used for gross-level ADL recognition, CSI manifests even greater sensitivity to fine-activity differentiation. CSI is more stable and robust than RSS [20], and the approach can be accessed using several off-the-shelf Wi-Fi devices (e.g., Intel Wi-Fi Link 5300 NICs and Atheros AR9580 NICs). Furthermore, CSI can be successfully extracted from Wi-Fi signals using a readily available CSI 802.11n tool [16].

To-date, CSI-based ADL-recognition systems have been applied to various activity-detection tasks. Firstly, CSI has been used for indoor localization. Location detection indoors is required in various settings, such as in hospitals (patient tracking) and disaster areas (personnel locating). Wu et al. [27] compared the accuracies between an RSS-based indoor localization solution and a CSI-based indoor localization solution using probabilistic approaches (which provide more accuracy than deterministic approaches [28]); deep learning techniques were employed to reduce the location error of the CSI-based indoor location systems [20]. In order to reduce complexity, time or system processes, Wu et al. [27] divided their system into two states: an offline process and an online process. The offline process served as the training stage for database construction, which in turn trained the CSI fingerprinting function; then, the online process recorded real-time data and tested the CSI-based approach by exploiting the database. Their results outperformed existing indoor localization systems.

Another CSI-based approach, Wi-Chase [15], recognized coarse activities—such as walking, running, and moving hands—by exploiting all CSI-subcarrier data. This approach used two machine-learning algorithms, k-Nearest Neighbor (kNN) and Support Vector Machine (SVM), and obtained the highest accuracy for hands moving because hands moving has similar repeating patterns in a fixed position, unlike locomotive activities. The Wi-Chase study also showed that when more subcarriers were used with multiple AP and receiver links, the performance of the system improved.

For more diverse in-place activities and for walking-direction recognition, Wang et al. [18] proposed the E-eyes algorithm. Unlike Wi-Chase, E-eyes algorithm selected known activity data and measured the similarity between the known activity data and the unknown activity data to recognize the unknown activity. Specifically, the E-eyes algorithm first differentiated between walking activity and in-place activity using CSI variance, since walking activity causes higher variance in CSI than in-place activity. Then, in-place activities were estimated based on their similarity with known activities using Earth Mover's Distance (EMD), and walking directions were detected by Dynamic Time Warping (DTW). The results of the Wang et al. [18] study showed that higher packet-transmission rates yielded higher accuracy in activity recognition.

Building upon these successes, researchers have applied CSI-based ADL-recognition solutions to various smart home healthcare systems. Wang et al. [3] proposed RT-Fall algorithms to detect falls, since fall detection is essential for elderly healthcare in a smart home. The process of detecting falls using CSI requires high packet-transmission rates and small window sizes because falling occurs in a very short time [3]. In general, the frequencies of fall and fall-like activities lie between 5 and 10 Hz, while in-place activities lie in the lower frequency range, from 0 to 4 Hz. RT-Fall analyzed the frequency of CSI to recognize fall and fall-like activities such as sitting down and lying down. They recognized falls in real time with approximately 90% accuracy. Borhani and Pätzold [29] developed a simulation model for the Wi-Fi-based fall detection system using a stochastic 3D trajectory model.

In the simulation, human body, such as the head, arms, and legs, are molded as moving scatters, and fixed scatters represents static objects (e.g., the walls, appliances, or furniture). The simulation model detects when a fall occurs during random walking by analyzing the time-variant Doppler effect caused by an occupant's activity.

In keeping with the healthcare applications, CSI has been exploited to detect vital signs such as respiration and heartbeat rates. PhaseBeat [30] used CSI phase data to detect vital signs because phase data are more stable than amplitude and manifest periodicity. In order to detecting heart rates, the researchers used a directional antenna at the AP to improve the reflected signal power—heart movements are too weak to cause variance in the reflected signals. Similarly, Liu et al. [17] detected breathing rates using CSI. In their study, an AP and receiver were placed at two sides of the subject for better signal quality. They noticed that sleeping positions affected the performance of respiration detection: If a person is in 'Fetus,' 'Log,' or 'Yearner' sleeping positions, the back of the subject blocks the Wi-Fi signals' paths. Thus, the researchers determined that users should change the location of the AP-Receiver pair to detect chest movement.

CSI-based ADL-recognition systems have also been used in place of human-device interactions. Various smart home gadgets control such home appliances as TVs, laptops, and mobile phones. Alternatively, Nandakumar et al. [31] used CSI to control these home appliances by detecting human gestures; they obtained an average 91% and 89% accuracies when the receiver was located in LOS and in a bag, respectively. In another study, Ali et al. [4] focused on keystroke recognition using CSI. When a person types a specific key, his hands and fingers move in a unique pattern; however, the movements of hands and fingers are micro-movements and some unique patterns for different keys are almost identical—for example, 'F' and 'G' keys are closely placed and may easily be confused. In order to solve such nuance, the researchers extracted features from the shapes of keystroke waveforms instead of from the CSI values themselves, since the CSI values of many keys have similar features—such as maximum value, mean value, or root mean square deviation—but have different waveforms, and the shapes contain both a time and a frequency domain. Their WiKey algorithm obtained an approximately 94% keystroke-recognition accuracy.

These successfully developed and verified CSI-based ADL-recognition systems benefit from CSI's stability and accessibility. However, Wi-Fi signals are affected by various environmental factors, which can in turn have impact on the performance of CSI-based ADL-recognition systems. Thus, in order to advance the opportunities for applying CSI to human ADL recognition in smart homes, studies must verify that these systems' performances will yield consistent accuracy in different housing environments.

3. Background

As Wi-Fi signals propagate in physical space, the signals reach receivers (Wi-Fi devices) through various routes, a concept known as multipath. Figure 1 shows the multipath of signal propagation: The received signal is composed of signals arriving over many different paths, all which can be affected by environmental factors [3]. The environmental factors therefore combine with scattering, fading, and power decay over distances, and these environmental effects on the signals can be manifested in the CSI [18]. For example, while CSI remains stable in a static home environment, if a person performs activities, Wi-Fi signals scatter in response to the body's movement, thereby causing bistatic Doppler shift at the receiver. The bistatic Doppler frequency depends on the occupant's moving speed, the Wi-Fi frequency band, and the relative position between the occupant and the Wi-Fi transceivers [32].

Mathematically, the CSI matrix, H_i, is related to the transmitted signal vector X_i and the received signal vector Y_i, as shown in Equation (1) [15].

$$Y_i = H_i X_i + N_i \tag{1}$$

where i is the data packet, $i \in [1, N]$; N is the number of received packets; Y_i is the received signal vector; H_i is the CSI matrix; X_i is the signal vector; and N_i is the noise vector.

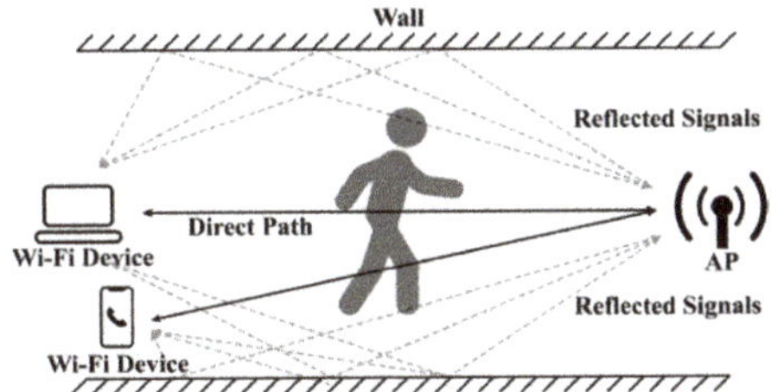

Figure 1. Multipath propagation of Wi-Fi signal indoors.

CSI has several subcarriers that are divided by Orthogonal Frequency Division Multiplexing (OFDM) [2]. Thus, the CSI matrix for a packet, H, has 30 subcarriers, each with three transmission antennas and three receiver antennas. Hence, the total number of information pathways for a sending packet is 270 CSI amplitude and another CSI 270 phase. Unlike RSS, which only has one path per packet, CSI exploits multiple subcarriers that travel along different fading or scattering multipaths [27] in order to better denote data across dimensions of time and space.

Figure 2 shows the raw CSI amplitude data of 30 subcarriers for a walking activity. The different colors indicate 30 subcarriers that have different amplitude values but show a similar tendency. When a subject is standing, the CSI is relatively stable. However, the amplitudes fluctuate as the subject starts to walk. The black area in Figure 2—which indicates when the walking activity occurred—has relatively high amplitude variance. CSI-based ADL-recognition approaches exploit such CSI pattern changes [18] to identify human activity.

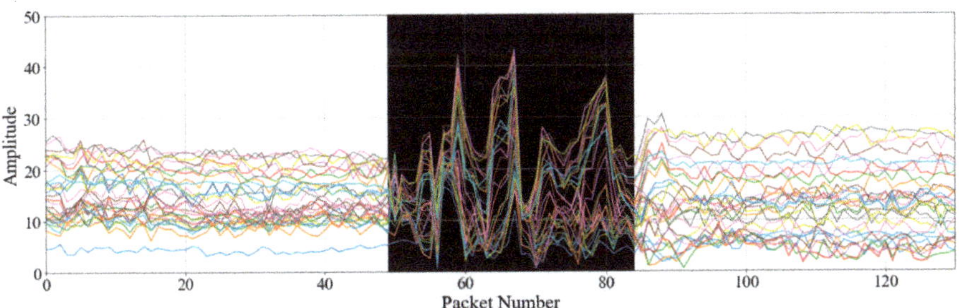

Figure 2. Raw CSI amplitude data of 30 subcarriers captured during walking. Colors indicate different subcarriers, and the black region indicates the walking activity.

4. Methodology

In order to examine whether and how CSI-based ADL-recognition systems can be resistant to varying housing environment, two different housing environments were selected: A wood-frame, low-rise apartment in the United States (Unit A), and a reinforced concrete-frame, high-rise apartment in South Korea (Unit B). These two housing environments show clear differences in construction materials and population densities. The exterior walls in Unit A were composed of wood framing and insulation, and wood sheathing, and its interior walls were also composed of wood framing and drywalls with painting finishes. Unit B was built with reinforced concrete structure and its interior walls were masonry and drywall, and wallpapers were used for the interior wall finishing. While Unit A is also a part of low-rise multi-family housing, the population density of the high-rise apartment complex that Unit B belongs to is much higher. The layouts of two units present clear differences, as shown in Figure 3. However, the sizes of two units in terms of floor space are quite similar, a Wi-Fi router and a receiver were located similarly in each unit (in living room), and the distance between the router and the receiver was also quite similar. Because the two units were actual living spaces, there were miscellaneous household items. The type, number, and location of household items were different in the two units. Both Unit A and Unit B contains large household items: a refrigerator in the kitchen, a dining table in the dining room, and a desk in the living room. However, there were more small and medium sized household items (e.g., furniture, appliances) in Unit B than in Unit A. Four different activities, including walking, eating, typing, and no-activity, were recognized by the CSI-based systems. Two algorithms were used in ADL recognition in order to analyze the housing environment effect independent from algorithms. The ADL-recognition using these algorithms followed the four steps: (1) Data collection, (2) Data preprocessing, (3) Activity segmentation, and (4) Activity classification. This section discusses the experimental setup, the two activity detection approaches, and how they were compared.

4.1. Data Collection

Two subjects were recruited for the experiments at each Unit; Recruiting for the experiment at each Unit was conducted separately due to the geographical distance. All the four subjects participating in both Units were male having similar physical characteristics; Their heights ranged from 175 cm to 180 cm, while their weights range from 68 kg to 70 kg. During the experiments, the subjects were instructed to perform the identical activities, and tests were performed by one subject at a time. Each subject walked in ten rounds, as indicated by the arrows, and performed eating and typing in ten rounds in the dining room and living room, respectively. Each round required 10 s of activity and 20 s of no-activity. The 20 s interval between activities clearly differentiated the multiple activity rounds. Figure 3 shows the walking trajectories in the test beds.

During the test, the activities were recorded using a camera and were labelled with time stamps for the activities to establish the ground truth. As shown in Figure 3, one AP and one receiver were used for this test: An AC1750 MU-MIMO gigabit router (Linksys, Irvine, CA, USA) was used for the AP and a Lenovo T400 laptop with Intel 5300 NIC was used for the receiver. The router provided a 3×3 multi-input and multi-output (MIMO) system using three built-in antennas. The router was configured to support the 802.11n AP mode at 5 GHz frequency. Internet control message protocol (ICMP) packets were transmitted at a sampling rate of 10 Hz (10 packets per second). Then, using a Linux CSI 802.11n tool [16], the CSI data were captured and extracted for 30 subcarriers for the first AP-Receiver antenna pair. The CSI amplitude data were then used to perform data processing for human-activity recognition.

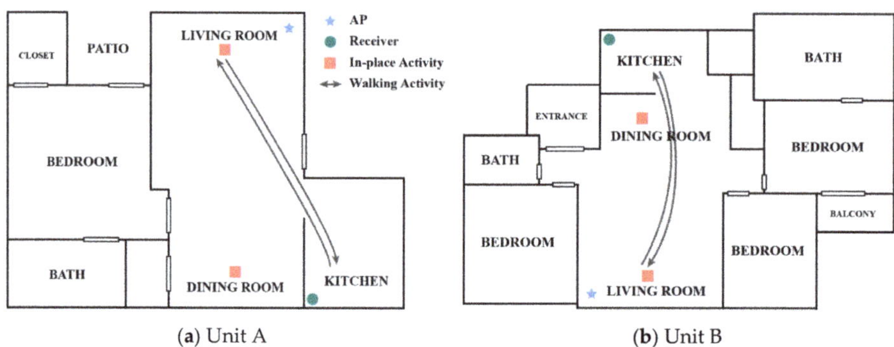

(a) Unit A (b) Unit B

Figure 3. Floor plans for experimental test beds.

4.2. Data Preprocessing

The raw CSI data contains high frequency noise, outliers, and artifacts introduced by rate adaptation rather than by human activities—high frequency noise manifests when the radios are switched to different modulation and coding schemes. Thus, a second-order, low-pass Butterworth filter was used to remove high frequency noise from the raw CSI data. The filter was configured to keep the sampling rate at 10 packets/s, and the cut-off frequency was set at 1 Hz. Although the walking activity causes higher Doppler shift than 1 Hz, the phase shift caused by occupant's activity would be rotated due to hardware imperfection. Thus, the 1 Hz filtered CSI still contained the variation caused by walking activity. The variation in the filtered CSI was distinguished from other activities [15]. Figure 4 shows the raw and filtered amplitude data of one subcarrier. The filtered data became smooth after the high-frequency noise reduction.

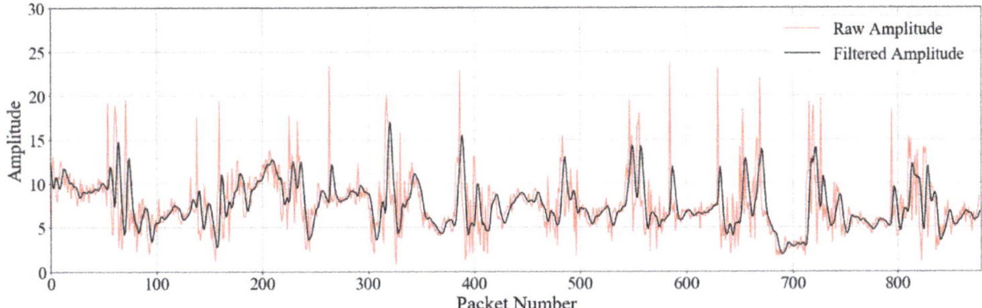

Figure 4. Comparison of raw and filtered amplitude data for one subcarrier.

4.3. Activity Segmentation

As shown in Figure 2, human activities cause signal changes, and the CSI variance has relatively high values when activities occur. If the CSI variances are greater than a threshold determined empirically, the CSI data may be considered to contain certain activities. While this approach somewhat successfully segments activities, when a human activity has small body movements or when some noise remains in the data after filtering has occurred, this approach will not always appropriately separate all activities, and the segmented data may additionally continue to contain unrelated data packets.

As developing high-efficiency activity-segmentation algorithms is beyond the scope of this study, to increase recognition accuracy, the walking and in-place activities were manually segmented by referring to ground truth. A total of 400 data samples were segmented.

4.4. Activity Classification

Activity-classification approaches exploit various activity-recognition models to classify the segmented activity data. In this study, two different algorithms were used for activity classification: SVM and EMD reconstructed from literature [1–3,15,18,30]. An SVM model requires users to input features and their labels for training the model. In this case, the labels were extracted from the ground truth recorded via cameras, and the features were extracted from the input data—i.e., the segmented activity data. In the segmented activity data, the CSI data of the first data packet was represented as $H_a(s)$—with an $S_b \times 1$ dimensional vector—where S_b indicates the 30 subcarriers, and s indicates the order of the N_a successive data packets. The six characteristic features present in all subcarriers of the CSI amplitudes that were used for learning are: (1) the average of $H_k(s)$, (2) the standard deviation of $H_k(s)$, (3) the 25th percentile of $H_k(s)$, (4) the 75th percentile of $H_k(s)$, (5) the maximum of $H_k(s)$, and (6) the median absolute deviation of $H_k(s)$, where $\forall k \in [1, S_b]$ and $\forall s \in [1, N_a]$. 60 percent of the total segmented data were used for training, and the remaining 40 percent were used for testing. The SVM showed the anticipated labels for the testing data, so the labels resulting from the SVM were compared to the labels from the ground truth to determine model accuracy.

Unlike an SVM, an EMD finds the minimum cost of matching one distribution into another and thereby represents to what extent two distributions are similar to each other. Notably, under EMD, the same activity will have similar distributions across the CSI amplitudes, whereas different activities will have distinctive distributions. Figure 5 shows a histogram of the CSI distributions across all 30 amplitude subcarriers; here, "Bin" refers to the range of amplitudes, and "Amplitude Count" refers to the number of times the corresponding amplitudes appear in each Bin. In order to recognize activities, a known CSI-amplitude distribution for each activity needs to be selected; in this study, we selected three known distributions for walking, eating, and typing. Then, the EMD algorithm [33] was employed to calculate the EMD between the known distribution for the labeled activities and the unknown distribution for the unlabeled activities. If the EMD of an unknown activity manifested a minimal distance to one of the three known activity distributions, the unknown activity received a label for the corresponding activity.

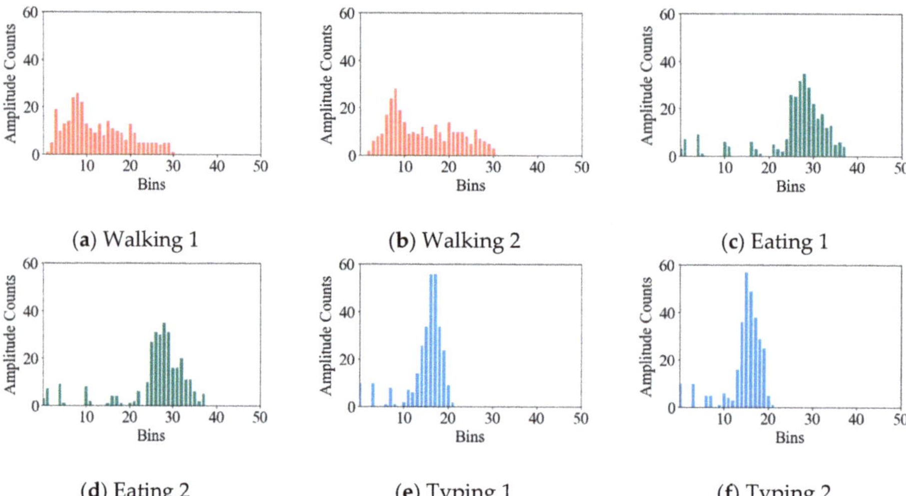

Figure 5. Amplitude counts in the amplitude bins for walking and in-place activities.

5. Results

In this study, to determine the impact of environmental factors on the accuracy of two different CSI-based ADL-recognition systems, two subjects performed a walking activity and two in-place activities in different housing environments. To recognize activities, the authors used two different activity-recognizing approaches, an SVM-based model and an EMD-based model. The activity-recognition accuracy of the SVM-based model was 94.38% for Unit A and 87.50% for Unit B, whereas the EMD-based model showed 68.75% and 60.25% accuracy levels for Unit A and Unit B, respectively. Tables 1 and 2 summarize the results from the SVM-based model and the EMD-based model, respectively. Both of the two ADL-detection algorithms show lower accuracies for Unit B than Unit A, which implies a mediating effect of environmental factors on the accuracy of the ADL-recognition systems.

Table 1. Confusion matrix for Unit A and Unit B using SVM-based model (Data shown here represent 40% of garnered data).

	Unit A				Unit B			
	Walking	**Eating**	**Typing**	**No Activity**	**Walking**	**Eating**	**Typing**	**No Activity**
Walking	38	0	0	0	38	0	0	0
Eating	3	35	0	0	0	38	0	0
Typing	2	0	31	0	0	1	30	2
No Activity	4	0	0	47	2	0	15	34

Table 2. Confusion matrix for Unit A and Unit B using EMD-based model.

	Unit A				Unit B			
	Walking	**Eating**	**Typing**	**No Activity**	**Walking**	**Eating**	**Typing**	**No Activity**
Walking	96	4	0	0	92	0	0	8
Eating	0	73	2	15	0	28	21	41
Typing	0	21	46	23	0	28	62	0
No Activity	0	26	34	60	0	24	37	59

Under the SVM-based model, features used for activity classification related to the amplitude of signal variance. Thus, even though an activity signal forms a specific waveform, its amplitude appeared to be low, so the SVM classifier considered the activity as no activity. The confusion matrix for Unit B showed 15 'no-activity' data were predicted as examples of 'typing activity,' which mainly reduced the model's accuracy. 'Typing activity' is one of the smallest movements in daily life, since when people type, they only move their fingers and arms a small amount. Thus, the signal variance of the 'typing activity' has a low amplitude, and if the signal power is reduced by environmental factors, the 'typing activity' signal may be too weak to differentiate from a 'no activity' signal. Accordingly, we expected—and observed—that Unit B environment would have a greater impact on signal propagation than Unit A environment.

Under the EMD-based model, the distribution of the signals' amplitudes has more impact on the activity classification than the amplitudes' variance within the signal, since the EMD-based model classifies unknown activities based on the distributions' similarity to known distributions. As shown in Figure 5, the amplitude of the 'walking' activity is evenly distributed over the bins, but the histograms of 'eating' and 'typing' activities show concentrated amplitude distributions in specific bin ranges. Thus, if the peak point of the amplitude distribution was shifted by changing the overall signal strength, the 'eating' and 'typing' activities could have similar amplitude distributions, which would mean that the EMD-based approach would consider those two different activities as the same activity. As shown in Table 2, the 'walking' activity in both Unit A and Unit B is well recognized, but the EMD-based approach achieved a low accuracy when predicting 'eating' and 'typing' activities. Table 2 also shows that many identified 'no activity' data were predicted to be 'eating' and 'typing' activities because the 'no activity' data manifested a stable signal wherein the signal strength was concentrated in a certain amplitude range. Thus, the histogram of the 'no activity' data appeared to be similar to those of 'eating' and 'typing' activities under the EDM approach.

The results also show that the EMD-based model has lower average accuracies compared to the SVM-based model. The SVM-based approach exploits 60% of the total 400 activities data to train the model and classified the remaining 40% activity data—160 samples. On the other hand, the EMD-based approach does not need training data, and consequently evaluates activity data for the full 400 samples. Thus, the actual number of recognized activities evaluated by the EMD-based model is greater than the number recognized by SVM-based model, which means the EMD-based model has more opportunities to make false predictions than the SVM-based model. Also, the SVM-based model used 60% of the total data as reference data, whereas the EMD-based model exploits only one reference datum for each activity to calculate the similarity of the known activity against the unknown activity data. Consequently and foreseeably, the EMD-based model does not cope with the variability of unknown activity data as well as the SVM-based model does.

To investigate the effect of time-dependency or subject-dependency of the result, we split data into each subject data and classified activities for each subject data; Each subject performed the experiment in different days. The results in different data segmentation showed that the accuracy of Unit A was generally higher than Unit B and the accuracy of SVM was higher than EMD, meaning that the time dependency and subject dependency did not have much impact on this result.

It is important to note that rather than identify effective ADL-recognition methods, this study compares the effect different environmental factors have on the performance of CSI-based ADL-recognition systems. Even though the SVM-based model and EMD-based model both show different accuracy levels for activity recognition, the results clearly indicate that both CSI-based ADL-recognition systems show less accuracy for the Unit B environment, meaning that the Unit B environment mediates the accuracy of the CSI-based ADL-recognition systems more than the Unit A environment does. We further discuss this important finding about the environmental effects mediating performance of the CSI-based ADL-recognition systems in the next section.

6. Discussion

6.1. Housing Environmental Factors

The results indicate that both of the tested CSI-based ADL-recognition models are quite sensitive to the difference of housing environment and yield significantly different accuracies between two units. Here, we mainly discuss the potential factors that affect the performance of the CSI-based ADL-recognition systems. While these factors may appear to be limitations to the present study, they serve—in fact—as a further defense of the justification of this study, since smart-home engineers must mediate these factors when designing tools within different environments.

6.1.1. Building Materials and Household Items

One potential factor contributing the performance of ADL recognition is the difference of building materials. The main building materials of Unit B were reinforced concrete and concrete masonry, which is one of the hardest building materials for wireless signals to penetrate [23], while Unit A was built with predominantly plywood or drywall, which represent building materials that yield less signal loss. Such Wi-Fi resistant properties of concrete materials could potentially benefit the CSI-based ADL-recognition systems by blocking interferences from surrounding Wi-Fi networks and noises created from dynamic object movements outside of the unit, but it could inhibit the performance of CSI-based ADL-recognition systems, as the amplitude of transmitted signals becomes reduced during the signals' travel indoors. This signal loss means that the signal variance caused by activities decreases, which challenges processes for differentiating noise, specific activities, and no activity. At least in our experiment, the disadvantage of having Wi-Fi resistant interior walls would have outweighed the advantage of having Wi-Fi resistant exterior walls and would have potentially contributed to the lower performance in Unit B.

The household item (e.g., furniture, appliance) is another potential factor for Wi-Fi signal propagation. Depending on the materials of household items, Wi-Fi signal is reflected from the object more or a lot of energy is absorbed. Unit B had more small and medium sized household items, which were not all made of metal, the more reflective objects compared to the wooden or plastic materials. Thus, the small and medium sized objects absorbed more energy of the Wi-Fi signal in Unit B than Unit A, which resulted in a reduced amplitude of the CSI. The decreased variation in the CSI, then, became hard to distinguish from no-activity or other activities in Unit B.

6.1.2. Population Density with Surrounding Wi-Fi Interference

The other contributing factor is the difference of population densities. Just as people can access home networks from outside their walls, people outside of a house can affect the Wi-Fi network environment inside. Thus, neighbors' activities or outdoor movements will conceivably affect the performance of CSI-based ADL-recognition systems. Unit B had more nearby units than Unit A. Significantly, for Unit B, neighboring households shared walls on both sides of the apartment, whereas Unit A only had neighbors five meters away on one side of the testbed. Also, the elevator located next to Unit B continuously ran, which influenced the signals within Unit B; Unit A had no elevator. Thus, more movements unrelated to the test activities took place in proximity to Unit B, which could affect the received signals used for activity detection. Unit B was in a higher population apartment complex than Unit A testbed, meaning that there were more Wi-Fi networks around to influence the test. Thus, the surrounding Wi-Fi signals had a higher chance to interfere with the testing Wi-Fi signals in Unit B than Unit A. However, as mentioned in the previous section, such interference from surrounding Wi-Fi signals might have been alleviated by the building materials of Unit B, although Unit B had many openings.

6.2. Potential Strategies to Address Housing Environmental Factors

In this section, we will discuss possible solutions for coping with the housing environmental effects influencing the accuracy of the CSI-based ADL-recognition system, including (1) using multiple receivers; (2) filtering external noise; and (3) using Wi-fi-friendly construction materials.

6.2.1. Using Multiple Receivers

While our experimental setting used only one receiver, using additional receivers would improve the recognition accuracies of CSI-based systems by augmenting the variance of the received signals. In larger houses or those with multiple rooms, the strength of the signals arriving at the receivers becomes weaker, as the signals have to penetrate through multiple walls [10,23,24]. Under such circumstances, the reduced amplitude of the signal both becomes hard to distinguish from noise [3] and directly impacts the performance of the CSI-based ADL-recognition systems. For example, a receiver located in the living room may only capture reduced signal variance if an occupant performs an activity in the bedroom since the signal affected by the occupant must move through the environment—and walls—to arrive at the receiver in the living room; if the transmitter is not within the bedroom, the signal will lose even more energy as it travels from the transmitter to the bedroom and back to the receiver in the living room. However, if the environment includes an additional receiver in the bedroom, the affected signal will reach the receiver in the bedroom, thus reducing the additional energy loss. Having multiple receivers located in different rooms would therefore resolve the reduced-amplitude issue and improve accuracy.

6.2.2. Filtering External Noise

In order to address the effects resulting from a neighbor's activity, advanced filtering methods must be proposed and used. While such filtering would represent a great challenge, developers may conceivably be able to distinguish the signal changes due to a neighbor's activities in another unit, especially as such signals would have to penetrate multiple walls (possibly including exterior walls) and a neighbor's activities may manifest different patterns in the frequency domain (e.g., different walking patterns). Lee et al. [34] demonstrated this possibility of differentiating simultaneous activities of multiple occupants using the frequency-domain features. Similarly, other researchers have successfully identified and removed specific subcarriers that were heavily influenced by a neighbor's activities [3,35]. Such opportunities present options for researchers seeking to mitigate the effects of noise on CSI-based ADL recognition.

6.2.3. Using Wi-Fi-Friendly Construction Materials

As smart homes become more popular, Wi-Fi-friendliness will play an important role in the design and selection of construction materials [36]. The performance of the CSI-based ADL-recognition systems relies on the magnitude of signal variance at the receiver, so the Wi-Fi-friendliness of building layout and materials will reasonably drive building designs seeking to harness CSI-based tools. Already, some options for improving Wi-fi-signal quality exist. For example, the interference from surrounding Wi-Fi networks could be alleviated by using the anti-Wi-Fi paint—which contains aluminum-iron oxide for absorbing high-frequency wireless signals [37]—on exterior walls. In addition, a porous wall that propagates Wi-Fi signals well [38] can be used for interior walls in order to reduce the attenuation of the signal that occurs as the signal passes through interior walls [38,39]. While such design choices may render the signal-resiliency goals underpinning CSI-based ADL recognition, there remains some uncertainty as to whether using such Wi-fi-friendly construction materials will actually improve performance, since CSI-based ADL recognition relies on the change of signals as the signals fade and reflect on surrounding walls. Further research will be necessary to examine how the use of such materials impact the performance of the CSI-based ADL recognition.

6.3. Limitations

As this study intended to examine the effect of actual housing environments of existing homes on CSI-based ADL-recognition systems, various environmental factors were less controlled, and the results from such the experimental settings represent the combined effects of all possible environmental factors. Thus, the effect of each individual environmental factor is still unknown, although the experiment results confirm the significant effect of housing environment on CSI-based ADL-recognition systems. In addition, although two representative ADL-recognition algorithms were used in this study, the effect of housing environment could be largely dependent on algorithms.

7. Conclusions

The two robust CSI-based activity-recognition algorithms yielded lower accuracy in Unit B than Unit A, a result that speaks to the mediating influence of environment on CSI-based ADL-recognition systems. This result highlights that the effects of the housing environments should be considered when designing and implementing CSI-based ADL-recognition systems in various home environments. However, further research is necessary to analyze the isolated effects of various housing environmental factors and understand the impact of possible mitigation strategies on such environmental effects.

Author Contributions: Writing—original draft preparation, H.L. (Hoonyong Lee); supervision, C.R.A; resources, N.C.; investigation, T.K.; project administration, H.L. (Hyunsoo Lee)

Funding: This research was funded by a grant (18CTAP-C128499-02) from the Technology Advancement Research Program funded by the Korean Ministry of Land, Infrastructure and Transport.

Acknowledgments: The authors would like to thank the Korean Ministry of Land, Infrastructure and Transport for their financial support.

Conflicts of Interest: The authors declare no conflict of interest. The funders had no role in the design of the study; in the collection, analyses, or interpretation of data; in the writing of the manuscript, or in the decision to publish the results.

References

1. Tan, B.; Chen, Q.; Chetty, K.; Woodbridge, K.; Li, W.; Piechocki, R. Exploiting WiFi Channel State Information for Residential Healthcare Informatics. *IEEE Commun. Mag.* **2018**, *56*, 130–137. [CrossRef]
2. Yousefi, S.; Narui, H.; Dayal, S.; Ermon, S.; Valaee, S. A survey on behavior recognition using wifi channel state information. *IEEE Commun. Mag.* **2017**, *55*, 98–104. [CrossRef]
3. Wang, H.; Zhang, D.; Wang, Y.; Ma, J.; Wang, Y.; Li, S. RT-Fall: A Real-Time and Contactless Fall Detection System with Commodity WiFi Devices. *IEEE Trans. Mob. Comput.* **2017**, *16*, 511–526. [CrossRef]

4. Ali, K.; Liu, A.X.; Wang, W.; Shahzad, M. Keystroke recognition using wifi signals. In Proceedings of the Annual International Conference on Mobile Computing and Networking, Paris, France, 7–11 September 2015; ACM: New York, NY, USA, 2015; pp. 90–102.
5. Bahl, P.; Padmanabhan, V.N. RADAR: An in-building RF-based user location and tracking system. In Proceedings of the Annual Joint Conference of the IEEE Computer and Communications Societies, Tel Aviv, Israel, 26–30 March 2000; IEEE: Piscataway, NJ, USA, 2000; Volume 2, pp. 775–784.
6. Liu, H.; Darabi, H.; Banerjee, P.; Liu, J. Survey of wireless indoor positioning techniques and systems. *IEEE Trans. Syst. Manand Cybern. Part C (Appl. Rev.)* **2007**, *37*, 1067–1080. [CrossRef]
7. Chang, H.; Tian, J.; Lai, T.-T.; Chu, H.-H.; Huang, P. Spinning beacons for precise indoor localization. In Proceedings of the ACM Conference on Embedded Network Sensor Systems, Raleigh, NC, USA, 5–7 November 2008; ACM: New York, NY, USA, 2008; pp. 127–140.
8. Azizyan, M.; Constandache, I.; Roy Choudhury, R. SurroundSense: Mobile phone localization via ambience fingerprinting. In Proceedings of the Annual International Conference on Mobile Computing and Networking, Beijing, China, 20–25 September 2009; ACM: New York, NY, USA, 2009; pp. 261–272.
9. Xiong, J.; Jamieson, K. ArrayTrack: A fine-grained indoor location system. In Proceedings of the 10th USENIX Symposium on Networked Systems Design and Implementation, Lombard, IL, USA, 2–5 April 2013; USENIX: Berkeley, CA, USA, 2013.
10. Yang, Z.; Zhou, Z.; Liu, Y. From RSSI to CSI: Indoor localization via channel response. *ACM Comput. Surv. (CSUR)* **2013**, *46*, 25. [CrossRef]
11. Au, A.W.S.; Feng, C.; Valaee, S.; Reyes, S.; Sorour, S.; Markowitz, S.N.; Gold, D.; Gordon, K.; Eizenman, M. Indoor tracking and navigation using received signal strength and compressive sensing on a mobile device. *IEEE Trans. Mob. Comput.* **2013**, *12*, 2050–2062. [CrossRef]
12. Tahat, A.; Kaddoum, G.; Yousefi, S.; Valaee, S.; Gagnon, F. A look at the recent wireless positioning techniques with a focus on algorithms for moving receivers. *IEEE Access* **2016**, *4*, 6652–6680. [CrossRef]
13. Orphomma, S.; Swangmuang, N. Exploiting the wireless RF fading for human activity recognition. In Proceedings of the Electrical Engineering/Electronics, Computer, Telecommunications and Information Technology, Krabi, Thailand, 15–17 May 2013; IEEE: Piscataway, NJ, USA, 2013; pp. 1–5.
14. Adib, F.; Kabelac, Z.; Katabi, D.; Miller, R.C. 3D Tracking via Body Radio Reflections. In Proceedings of the NSDI 2014, Seattle, WA, USA, 2–4 April 2014; USENIX: Berkeley, CA, USA, 2014; Volume 14, pp. 317–329.
15. Arshad, S.; Feng, C.; Liu, Y.; Hu, Y.; Yu, R.; Zhou, S.; Li, H. Wi-chase: A WiFi based human activity recognition system for sensorless environments. In Proceedings of the 2017 IEEE 18th International Symposium on A World of Wireless, Mobile and Multimedia Networks (WoWMoM), Macau, China, 12–15 June 2017; IEEE: Piscataway, NJ, USA, 2017; pp. 1–6.
16. Halperin, D.; Hu, W.; Sheth, A.; Wetherall, D. Tool release: Gathering 802.11 n traces with channel state information. *ACM Sigcomm Comput. Commun. Rev.* **2011**, *41*, 53. [CrossRef]
17. Liu, X.; Cao, J.; Tang, S.; Wen, J. Wi-Sleep: Contactless sleep monitoring via WiFi signals. In Proceedings of the 2014 IEEE Real-Time Systems Symposium (RTSS), Rome, Italy, 2–5 December 2014; IEEE: Piscataway, NJ, USA, 2014; pp. 346–355.
18. Wang, Y.; Liu, J.; Chen, Y.; Gruteser, M.; Yang, J.; Liu, H. E-eyes: Device-free location-oriented activity identification using fine-grained wifi signatures. In Proceedings of the 20th Annual International Conference on Mobile Computing and Networking, Maui, HI, USA, 7–11 September 2014; ACM: New York, NY, USA, 2014; pp. 617–628.
19. Wang, X.; Gao, L.; Mao, S.; Pandey, S. DeepFi: Deep learning for indoor fingerprinting using channel state information. In Proceedings of the Wireless Communications and Networking Conference, New Orleans, LA, USA, 9–12 March 2015; IEEE: Piscataway, NJ, USA, 2015; pp. 1666–1671.
20. Wang, X.; Gao, L.; Mao, S.; Pandey, S. CSI-based fingerprinting for indoor localization: A deep learning approach. *IEEE Trans. Veh. Technol.* **2017**, *66*, 763–776. [CrossRef]
21. Wang, W.; Liu, A.X.; Shahzad, M. Gait recognition using wifi signals. In Proceedings of the 2016 ACM International Joint Conference on Pervasive and Ubiquitous Computing, Heidelberg, Germany, 12–15 September 2016; ACM: New York, NY, USA, 2016; pp. 363–373.
22. Wang, Y.; Wu, K.; Ni, L.M. Wifall: Device-free fall detection by wireless networks. *IEEE Trans. Mob. Comput.* **2017**, *16*, 581–594. [CrossRef]
23. Adib, F.; Katabi, D. *See through Walls with WiFi!* ACM: Hong Kong, China, 2013; Volume 43.

24. Tesserault, G.; Malhouroux, N.; Pajusco, P. Determination of material characteristics for optimizing WLAN radio. In Proceedings of the European Conference on Wireless Technologies, Munich, Germany, 8–10 October 2007; IEEE: Piscataway, NJ, USA, 2007; pp. 225–228.
25. UmaMaheswararao, M.; Kadaru, B.B. Seeing Through Walls Using Wi-Vi. *Int. Res. J. Eng. Technol.* **2017**, *4*, 2088–2091.
26. Pu, Q.; Gupta, S.; Gollakota, S.; Patel, S. Whole-home gesture recognition using wireless signals. In Proceedings of the 19th Annual International Conference on Mobile Computing & Networking, Miami, FL, USA, 30 September–4 October 2013; ACM: New York, NY, USA, 2013; pp. 27–38.
27. Wu, K.; Xiao, J.; Yi, Y.; Chen, D.; Luo, X.; Ni, L.M. CSI-based indoor localization. *IEEE Trans. Parallel Distrib. Syst.* **2013**, *24*, 1300–1309. [CrossRef]
28. Youssef, M.; Agrawala, A. The Horus location determination system. *Wirel. Netw.* **2008**, *14*, 357–374. [CrossRef]
29. Borhani, A.; Pätzold, M. A Non-Stationary Channel Model for the Development of Non-Wearable Radio Fall Detection Systems. *IEEE Trans. Wirel. Commun.* **2018**, *17*, 7718–7730. [CrossRef]
30. Wang, X.; Yang, C.; Mao, S. PhaseBeat: Exploiting CSI phase data for vital sign monitoring with commodity WiFi devices. In Proceedings of the IEEE 37th International Conference on Distributed Computing Systems, Atlanta, GA, USA, 5–8 June 2017; IEEE: Piscataway, NJ, USA, 2017; pp. 1230–1239.
31. Nandakumar, R.; Kellogg, B.; Gollakota, S. Wi-fi gesture recognition on existing devices. *arXiv* **2014**, arXiv:1411.5394.
32. Keerativoranan, N.; Haniz, A.; Saito, K.; Takada, J. Mitigation of CSI Temporal Phase Rotation with B2B Calibration Method for Fine-Grained Motion Detection Analysis on Commodity Wi-Fi Devices. *Sensors* **2018**, *18*, 3795. [CrossRef] [PubMed]
33. Pele, O.; Werman, M. Fast and robust Earth Mover's Distances. In Proceedings of the 2009 IEEE 12th International Conference on Computer Vision, Kyoto, Japan, 29 September–2 October 2009; pp. 460–467.
34. Lee, H.; Ahn, C.R.; Choi, N. Frequency-domain analysis for wi-fi based human activity recognition systems in smart homes. In Proceedings of the 18th International Conference on Construction Applications of Virtual Reality, Auckland, New Zealand, 22–23 November 2018.
35. Zheng, Y.; Yang, Z.; Yin, J.; Wu, C.; Qian, K.; Xiao, F.; Liu, Y. Combating Cross-Technology Interference for Robust Wireless Sensing with COTS WiFi. In Proceedings of the 2018 27th International Conference on Computer Communication and Networks (ICCCN), Hangzhou, China, 30 July–2 August 2018; IEEE: Piscataway, NJ, USA, 2018; pp. 1–9.
36. Rudd, R.; Craig, K.; Ganley, M.; Hartless, R. *Building Materials and Propagation*; Final Report; Ofcom: London, UK, 2014; Volume 2604.
37. Namai, A.; Sakurai, S.; Nakajima, M.; Suemoto, T.; Matsumoto, K.; Goto, M.; Sasaki, S.; Ohkoshi, S. Synthesis of an electromagnetic wave absorber for high-speed wireless communication. *J. Am. Chem. Soc.* **2008**, *131*, 1170–1173. [CrossRef] [PubMed]
38. Suherman, S. WiFi-Friendly Building to Enable WiFi Signal Indoor. *Bull. Electr. Eng. Inform.* **2018**, *7*, 264–271.
39. Jowitt, T. New paint promises low-cost Wi-Fi shielding. *Netw. World Can.* **2009**, *25*, N_A.

Article

A Multi-Agent Gamification System for Managing Smart Homes

Alicja Winnicka, Karolina Kęsik, Dawid Połap *, Marcin Woźniak and Zbigniew Marszałek

Institute of Mathematics, Silesian University of Technology, Kaszubska 23, 44-100 Gliwice, Poland; Alicja.Lidia.Winnicka@gmail.com (A.W.); Karola.Ksk@gmail.com (K.K.); Marcin.Wozniak@polsl.pl (M.W.); Zbigniew.Marszalek@polsl.pl (Z.M.)

* Correspondence: Dawid.Polap@polsl.pl

Received: 13 February 2019; Accepted: 7 March 2019; Published: 12 March 2019

Abstract: Rapid development and conducted experiments in the field of the introduction the fifth generation of the mobile network standard allow for the flourishing of the Internet of Things. This is one of the most important reasons to design and test systems that can be implemented to increase the quality of our lives. In this paper, we propose a system model for managing tasks in smart homes using multi-agent solutions. The proposed solution organizes work and distributes tasks to individual family members. An additional advantage is the introduction of gamification, not only between household members, but also between families. The solution was tested to simulate the entire solution as well as the individual components that make up the system. The proposal is described with regard to the possibility of implementing smart homes in future projects.

Keywords: heuristic; gamification; Internet of things; artificial intelligence; multi–agents solution

---

1. Introduction

Increasingly, intelligent technologies are present in our lives. Examples are telephones, televisions and other household appliances. However, along with the upcoming 5G (5th Generation) mobile network standard, large hopes are related to the Internet of Things [1]. The Internet of Things is nothing but the Internet where the recipients and senders of information are connected devices. Examples of such devices are the above-mentioned smartphones. Each device acquires data from the environment using built-in sensors, such as cameras, microphones, and motion and temperature sensors. Information often has to be processed to extract data, which are used by other objects in the network. This solution allows improving our life [2].

Smart homes are just one such example, where systems installed in the home are able to monitor the entire area in the absence of owners, or analyze and modify existing conditions. The conditions are understood as temperature analysis, water heating, or food control in a refrigerator. It all increases not only the comfort of life, but also minimizes energy consumption or the amount of food thrown away [3]. The Internet of Things can significantly change our lives, but to make this possible, there is a need to create systems that will monitor and supervise all devices. Moreover, the exchange of information by network users may be insufficient. If a current or future failure is detected, the user of the entire system should be able to receive such information. Moreover, such information should be transmitted not only within the installed technology, but anywhere. Hence, the need to design different interfaces and present data to the user using a smartphone or laptop so that he can intervene in different situations.

The vision of a home that helps users in performing tasks may be interesting, not only by controlling tasks, but also by motivating them to perform. Motivating can be solved by introducing the technique of gamification, that is, broadcasting additional content in order to gain some experience or points. This type of solutions may prove to be practical if gamification is subject to a certain group of users. In this paper, we propose a multi-agent system that is responsible for controlling the performance of simple tasks at home through installed sensors. In addition, the system introduces the aforementioned competition between the household members. In addition, children are taught duties at home and observe them.

From a technical point of view, the proposed solution is a multi-agent system for the purpose of exchanging information obtained by agents. This is important because the flat or house in which the system is installed can be very large, and tasks can be performed in different places. The advantage of the proposed idea is the use of popular equipment such as smartphones to obtain information because of their popularity.

The article is divided into several sections to increase readability. In Section 1, we introduce to the topic. In Section 2, we present the structure of the proposed system. Section 3 is devoted to the allocation of tasks for users, and then the agent's operation based on artificial intelligence methods (in particular artificial neural networks) is described. The next sections present the conducted experiments and their discussion.

2. Related Works

Intelligent technologies are strongly developed in the world of science. Above all, aspects are related to increasing the quality of our lives. The most developed branches of such systems are energy and security fields. In both cases, the proposed architectures are different in terms of not only the operation but also used technology. In the case of intelligent energy management systems, scientists focus on optimizing different parameters to reduce consumption, which in turn reduces production costs [3]. The second most popular aspect of the Internet of Things are the security features. A simple alarm informs us about a possible break-in, but does not analyze other situations. This is particularly evident in homes with a large area, the coverage of which can be expensive. For this purpose, different communication architectures between devices are created, or even information is exchanged so different cases can be analyzed. Intelligent software having access to cameras can process the image and register unplanned guests and notify owners. In terms of security, first, methods of artificial intelligence are used, which can quickly classify data. In particular, it is worth paying attention to the fact that the recording from the camera is a set of frames [4]. In terms of security, digital attacks are also an important issue, as smart homes are protected and monitored in a digital way [5]. Researchers analyze and build complex systems that can operate on data not only on given devices (edge technologies), but using external devices such as cloud or servers [6]. Scientific research is carried out not only in terms of safety, but also environmental protection and health of residents. An example is air pollution system, where contaminants can get inside by opening a window or door [7].

These systems present different approaches to implementation and construction. However, similar ideas are used in the topic analyzed in this article, i.e., a system based on a certain criterion of gamification. In [8], the authors described the idea of using gamification for rule management. First, the authors focused on the integration of user interfaces with appropriate motivating techniques. This is important due to the age of users, where it will be a big determinant not only to design the interface, but also to choose a strategy to motivate users to act. In contrast, the authors of [9] showed the technique of constructing rules in such systems to involve all participants of the house to cooperate. Again, in [10], the risks associated with the installation of such systems are described. Particularly, the aspects of motivation are important to not discourage action, which may result from assigning too difficult tasks to specific household members.

In this paper, a different approach is presented by using sensors in popular devices and their common communication by placing a single point with a knowledge database and using artificial intelligence methods. In addition, we present the method of assigning tasks to individual household members, taking into account various factors such as age, type of performed work or commitment.

3. Proposed Gamification System

The main idea is to create a home management system for the smallest tasks such as picking up garbage, to repairing items that have broken or are close to such a state. Imagine a situation that a client buys a smart home for his family. The system should supervise the situations in the home. In the case of some inaccuracies, it should inform the household members as well as make sure that everything functions optimally. Unfortunately, some tasks can be avoided by family members or be put off to be done later. A trivial example is watering flowers.

For the system to function, all tasks should be in the database from which they are broadcast, depending on the time of execution. The basic database of tasks should be available to the client, which he can be modified as a system administrator. Each task is composed of several elements, of which the basic are the name, frequency of repetition, occurrence (in which days) and the prize. The administrator can modify the frequency and repeatability, but only to a certain extent. Some tasks such as watering flowers should not occur too often, therefore any deviations from the norm should be accepted as higher, that is, above the level of one house. This is similar for the prize: within the house itself, prizes can be allocated in any way, but not if in competition, as houses on the whole street could be competing.

It is easy to see that the competition takes place on two levels: between members of a single house, as well as among different houses. This type of competition would motivate family members in all homes, especially if, after a certain time, the results are evaluated and the highest sum of points highlighted in a certain way.

The reward for completing the task is the number of points, which are calculated on the basis of several factors such as difficulty level ϵ, the minimum and maximum number of points to win θ, execution time t and the priority ζ. Some of these conditions depend on the status of specific family members, for example, some task would be more difficult to do for a child than for an adult, thus its level of difficulty would increase when the child is a user. This would result in a higher reward that can be obtained for this task. Such a way of evaluating and rewarding tasks would help encourage children to develop through performing more and more difficult and demanding home duties and other tasks included in the database.

The level of difficulty is not the only parameter on which the reward depends, but the only one whose value is determined by the user's status. The other parameters are independent of the person performing the task. They can be determined by persons supervising the system (for example, parents) or directly by a program that would determine a specific value on the basis of other coefficients.

One of independent parameters of the user's status is the maximum and minimum number of points per task. Both values are set in advance and they are in the range $\theta \in \langle 1, 10 \rangle$. The maximum amount allows preventing too fast progress, thus increasing its safety. For example, points scored too quickly by a child could result in choosing too difficult tasks for him that he could not cope with physically or mentally (e.g., he could be harmed while trying to do it). To prevent this, we set a minimum number of points. It is a protection of the program that sets up the prize under certain conditions. The simplest and most general situation is to assign a task to a user who has already performed this task many times. In this case, the level of difficulty is very low, which results in an adequately low reward. Too few points could discourage the user from completing the task (and, in the case of home duties, it would not be good because it still needs to be done). The minimum number determines the smallest number of points that a user can get for completing this task—no matter how many times he has already done it and how simple it is.

Priority is another important parameter to determine the award, which is specified in the range of $\zeta \in \langle 0, \max \rangle$. The total value is independent of the difficulty and constant for each task. It depends only on the day or time when the task has to be done—for example, certain tasks are more important on Saturday, and not on Tuesday, thus on Saturday they have the value max, and on Tuesday $0.5 \cdot \max$. For those with higher priority, there is a correspondingly higher prize, which should encourage the user to do it as soon as possible at the most appropriate time.

The time is another parameter participating in the description of the task. It is predetermined for each task and prevents the user from increasing the tendency to procrastinate. It does not directly affect the reward, but rather is its opposite. It can result in punishing the user with negative points (which affect not only his own score, but also the score of the whole house). For not completing the task within a given time, the user receives a warning (depending on the preferences) or loses points, and, as a last resort, a notification is sent to the person supervising the system. If the task is completed at a certain time, the obtained parameters are compared to the properties of the task. Depending on their fulfillment, reward points are assigned. This parameter is dependent on parameters such as the time interval—some tasks should be performed in exact hours on a specific day. An example is making dinner, which should be done in the afternoon when the family members return from work.

Obviously, some tasks should appear to be done automatically when appropriate factors occur. Laundry is an example. Depending on the amount of dirty clothes, e.g., if it exceed a certain weight, the task "laundry" should be created.

This way of assigning tasks allows earning reward points by individual members of the house, which are added to the account of the whole house. The gamification system allows various simple tasks and prevents them from being omitted. Moreover, it introduces into the life of the members additional elements such as competition or motivation.

4. A Multi-Agent Idea for Assigning Tasks

We assume that the described gamification has the right to exist at the system, which will try to search for the task itself. Obviously, those tasks should make sense. That is why we propose a system based on the operation of multi-agents. Formally, an agent is a system that is placed in a given environment. In addition, it may be autonomous, communicate with other agents, and achieve specific goals.

A multi-agent system is a system based on a set of agents who cooperate with each other to increase efficiency or achieve goals [11–15]. In our case, the system must supervise assigning tasks and exchange data between agents at home. Moreover, the additional task is to send the data from a given house to make possible competition between whole houses in a particular area.

Imagine the house in which most of the tasks are analyzed automatically. To make this possible, agents will be located in many places. It indicates that they will be responsible for the specific area. Each agent saves data about the task execution by any user, thus there is no one agent assigned to a specific user. However, the system being installed will not contain as many agents as users, but a much larger number of agents. The reason for this is the ability to control other equipment or rooms. This number depends on the functionality of the system.

To illustrate such a situation, let us assume that the system works in a house inhabited by one person. Assume that one of the basic tasks is watering flowers. The system cannot acknowledge the task to be finished if the user waters flowers only in one room, because there are sensors. A similar situation occurs when the flowers are in several rooms and the system calculates points for watering each flower. These types of things should be handled by the operation of multi-agents, where each of them will collect data and process it according to current needs.

The proposed solution is that we will treat an agent as a single instance of the program, which sends and gets the information from the database. The agent has the task to retrieve data using built-in sensors, and then to process them. The processed information is compared to the previously obtained data (the interval between downloading may depend on a specific time interval or on the occurrence of movement—for this purpose, additional sensors are required). If there are any differences, the data are sent to the database. The database is built in the simplest relational model where many triggers are created. If the information in a given cell reaches the limit set in the task, the trigger returns information to the device that recently made the query. Received data may mean that the user has completed the task, and thus, an agent sends information to the user. If the task is completed, the database sends information about the remaining ones. The agent generates a task based on the received information from the database. The illustration of these actions is presented in Figure 1.

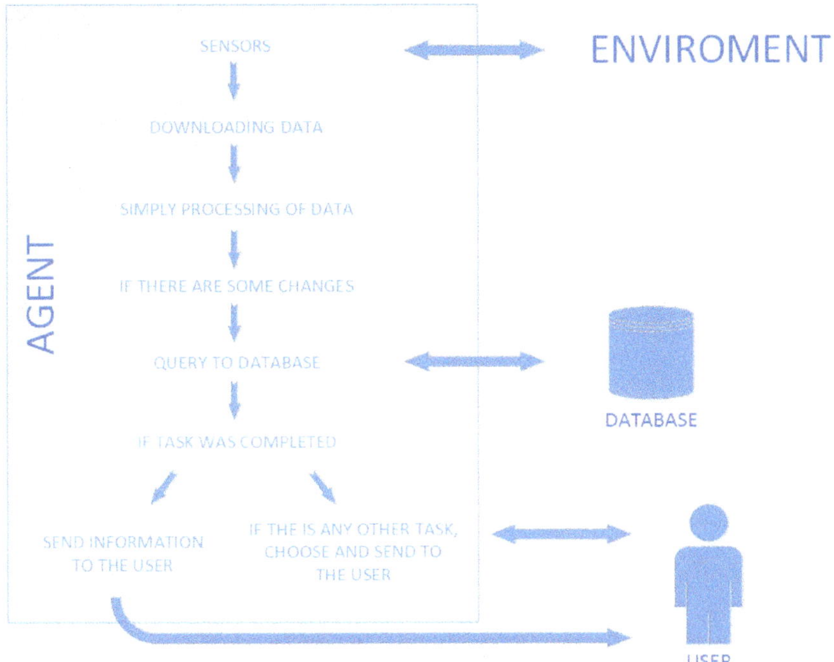

Figure 1. Agent architecture in the proposed system.

This allows creating many agents whose communication will be indirect through an external, common information database (see Figure 2). Each agent can download a list of current tasks and modify it. In addition, each agent should control the current status of the apartment with the help of built-in sensors. If the agent has access to motion sensors, in the case of any movement, it should be analyzed and the results should be saved in the database. If the data are updating, each agent is informed about it (through triggers). Each update also checks if the task is completed. In the case, when there is a change of data and the task is performed, such information is saved in the database and information about it is returned to the user. Obviously, the prizes are given.

4.1. Task Handling

In the described idea, we have several agents who register and process data and send them to the database. Communication takes place through the central point, which is the database. The great advantage of such an action is to relieve agents of additional information as well prevent their duplication throughout in the system. The description of the operation and construction of agents is presented in the following sections, where we focus on the idea of gamification and the mechanics of tasks operation.

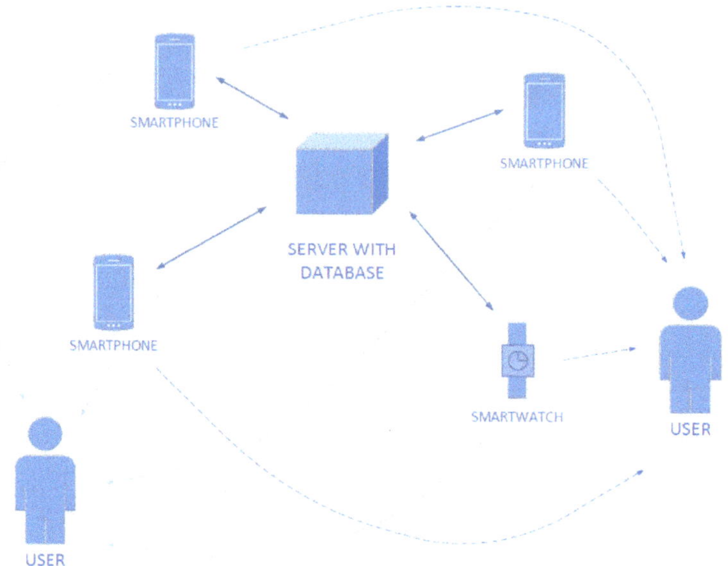

Figure 2. Visualization of communication between components in the proposed system.

4.1.1. Generating New Tasks

Generating tasks consists in entering new tasks manually by the user, or by introducing a new agent into the system. The first case is a simple operation, where the user enters the task with all data into the system, and it can be accepted or rejected by the other houses. If we create a task only for one house, no reward points will be assigned to it. The situation can be changed if there is no competition with other houses—it is possible when there are no other system users (interpreted as houses, not household members), or when there is not any shared tasks.

The second case is to introduce a new agent into the system. It is also very simple—while adding new agents, there is requirement to give information about the database, where the data should be sent. Depending on the installed functionality, the completed tasks will be built into the agent. Connection to the database is defined as sending its own tasks to database—if they already exist, the data on the agent is updated. Otherwise, the new tasks is inserted into the database.

4.1.2. Assigning Tasks

The information about users is used to assign tasks. The same task can be assigned to several people. When multiple people do one task, the reward is shared among all of them.

Every day, a list is created with all tasks that have to be done this day. Whether the task is carried out on a working day or on a day off, it will have different difficulty level ω. The basic distribution of this value may be as follows

$$\omega = \begin{cases} \text{manual labour} & \omega = 0.8 \\ \text{intellectual work} & \omega = 1 \\ \text{school} & \omega = 1.2 \end{cases} \tag{1}$$

The above values have been chosen in an empirical way to differentiate the effort of individual people. Furthermore, the age coefficient ξ of every user is taken into account. If ξ is greater than the required age limit, it is set to a constant value, for example $\xi = 10$. Additionally, the equation takes into account the health status μ of the user in the range $\langle 0, 1 \rangle$, which, respectively, means sick and healthy. Another parameter is the current point status ϕ of the user, which is initially equal to 1 (the value is not zero to prevent dividing by zero). These data are used to calculate the current state of the user by the following equation

$$\Xi = \frac{\sqrt{\xi} \cdot \xi \cdot \mu}{\omega \cdot \phi}. \tag{2}$$

It is easy to notice that the state of every user depends on the amount of earned points, which is the reason for relieving the most distinguished members of the household.

In the next step, the stack R with potential rewards is created and it is used in calculation the value of function $Q(\cdot)$ for each undone task k in database in the following way

$$x_k = \gamma_1 \cdot \frac{R[k]}{10} + \gamma_2 \cdot \frac{\max\limits_{i \neq k} R[i]}{\Xi}, \tag{3}$$

where γ_1 and γ_2 are the percentage contribution of components, where these values fulfill the equation $\gamma_1 + \gamma_2 = 1$. The obtained value is used in calculating the value of the function described as

$$Q(x_k) = \int_{x_k}^{\infty} \frac{1}{2\pi} \exp\left(\frac{-x_k^2}{2}\right) dx_k \approx \left[\frac{1}{(1-a) \cdot x_k + a \cdot \sqrt{x_k^2 + b}}\right] \frac{1}{\sqrt{2\pi}} \exp\left(\frac{-x_k^2}{2}\right), \tag{4}$$

where a and b are coefficients, mostly $\frac{1}{\pi}$ and 2π. Using these values, the best task is chosen according to

$$\max_k \eta_k \cdot Q_n(x_k) \tag{5}$$

where $\eta_k \in \langle 0, 5 \rangle$ is the priority assigned to kth task and a function $Q_n(x_k)$ [16] reflects the chart (see Figure 3) to prioritize tasks with small prizes but a high priority. It can be described as

$$Q_n(x_k) = \begin{cases} |1 - Q(x_k)| & \text{if } |1 - Q(x_k)| \geq \frac{1}{n} \sum\limits_{i=1}^{n} |1 - Q(x_i)| \\ |Q(x_k)| & \text{if } |1 - Q(x_k)| < \frac{1}{n} \sum\limits_{i=1}^{n} |1 - Q(x_i)| \end{cases}. \tag{6}$$

The use of the function $Q(\cdot)$ is for normalization of values, and then to allow assigning tasks at any priority.

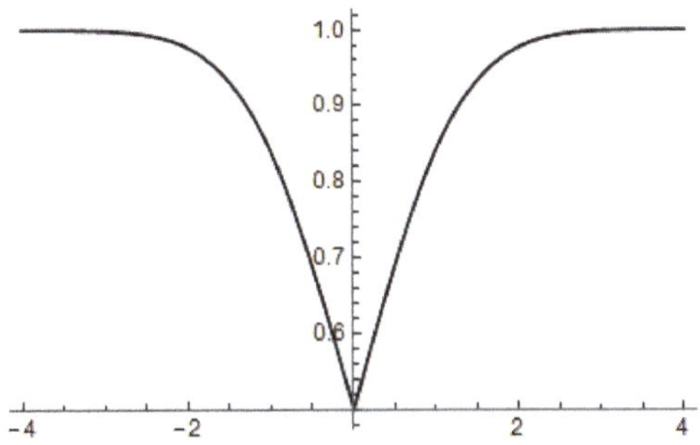

Figure 3. Chart of function $Q(\cdot)$ for assigning tasks, where $\frac{1}{n} \sum_{i=1}^{n} |1 - Q(x_i)| = 0.5$.

5. The Operation of Agent

The object, which will be placed in the house and will have sensors for analyzing the environment, will be called an agent. The data will be recorded depending on the equipment of the agent. Among the most popular and most commonly used agents, we can distinguish smartphones, cameras, motion sensors, gas sensors, etc. The choice of how the data are processed depends on the type of sensors. The most frequently used data are images (camera recording) and numerical data (weight and temperature). To speed up analysis of various data, we propose the use of machine learning, in particular artificial neural networks. In the proposed system, each agent retrieves information, processes it and communicates with the database. This solution means that the implemented software can be the same. In addition, the agent's operation and the used method will only depend on the sensors that are build-in.

The idea is based on the assumption that each agent could process the obtained data by using a neural network and the results will be placed in the database. If the task criteria are fulfilled, the task status in the database is changed to executed. Then, the reward is assigned to the appropriate person and that person is sent a notification. However, in this model, we do not consider any punishments for not carrying out the task. That is the reason for our proposition to check all records in database every few hours.

5.1. Machine Learning Approach

Unfortunately, there is no single universal method for the processing various types of data. Therefore, we propose a hierarchical structure consisting of two types of neural networks, where one of them will process graphics, and the other numerical values. The diagram of this idea is presented in Figure 4. However, there is no way to create one classifier that will process everything. Thus, we propose a following solution: one classifier for one particular problem. In fact, such an idea is possible because each agent will take care of specific tasks. For many tasks, the calculations are performed and records in the database are modified depending on the results.

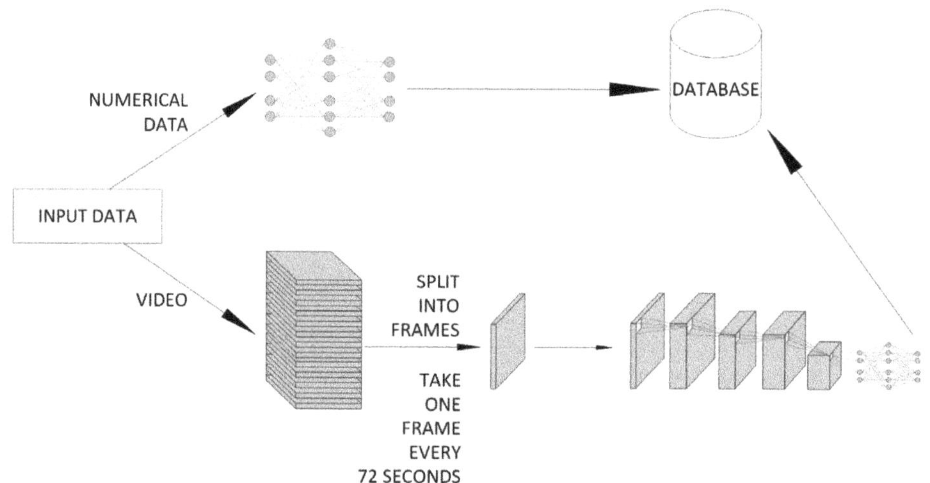

Figure 4. Graphical representation of the division of samples into classifiers.

5.1.1. Neural Network for Numerical Data

Artificial neural networks are inspired by the operations of the neurons in human brain. The idea is to create a grid of smaller objects connected to each other for data processing [17]. Formally, the network is a set of columns (called layers) that are created by smaller elements, which are neurons.

There are three types of layers: input, hidden and output. The first one is responsible for accepting input values, hidden layers process values, and the output one returns the result. Neurons between the layers are connected with each other by synapses, which are modeled as a connection with an assigned weight, i.e., a random value in the range $\langle 0, 1\rangle$. Therefore, the neuron is the smallest element of the network that receives data, processes it and returns the result of its calculations. The values of neurons from the previous layer and the set of weights assigned to the connections between them may be called as input data for next layer. Let us denote the output from the ith neuron as z_i, and the weight as w_i, then the value of the neuron will be calculated as the sum of the products of these values normalized by the sigmoidal function $f(\cdot)$ as follows

$$f\left(\sum_{i=0}^{n-1} w_i \cdot x_i\right) = \tanh\left(\sum_{i=0}^{n-1} w_i \cdot x_i\right), \tag{7}$$

where n is the number of neurons in the previous layer.

The most commonly used training algorithm is backpropagation. This process is about modifying the weights on connections in the whole network with respect to the loss function, defined as half of the square of the difference between expected value from the network and obtained one

$$E = \frac{1}{2}\sum_{i=0}^{k-1}\left(y_i^{expected} - y_i^{actual}\right)^2. \tag{8}$$

Depending on the given modified weights in each layers, the above equation can be interpreted differently. The modification of weights between jth neuron in the output layer and kth neuron in the hidden one is performed as follows:

$$\Delta w_{kj} = -\alpha \cdot \frac{\partial E}{\partial w_{kj}}, \tag{9}$$

where α is a learning rate. To derive this equation, we use the rule chain, which is defined as

$$\frac{\partial E}{\partial w_{kj}} = \frac{\partial E}{\partial y_j} \cdot \frac{\partial y_j}{\partial x_j} \cdot \frac{\partial x_j}{\partial w_{kj}}. \tag{10}$$

Thus, using the definition of a neuron and simple transformations, we obtain the following formula

$$\frac{\partial E}{\partial w_{kj}} = -(t_j - y_j) \cdot y_j \cdot (1 - y_j) \cdot y_k, \tag{11}$$

where y_j is the value of jth neuron.

If the neuron j is in the hidden layer, the calculation of weight will be as follows

$$\frac{\partial E}{\partial w_{kj}} = -\sum_{i \in I_j}(\delta_i w_{ji}) \cdot y_j \cdot (1 - y_j) \cdot y_k, \tag{12}$$

where δ_i is error value of ith layer.

5.1.2. Neural Network for Graphical Data

The image is perceived by the computer in a different way than the human eye. A two-dimensional image for a computer is a tensor, an object containing three components needed to describe it, including height, width and depth. As depth, we mean the individual layers of the image. For example, an image considered in terms of the RGB model will have three components, where each corresponds to one of the colors. The objects described in this way are used in the classification of graphics by convolutional networks. They are mathematical structures inspired by the functioning of the visual cortex [18].

The mathematical model includes the construction of three layers. The first of these is the convolutional layer, which works as an image filter. The layer has one task, i.e. to process all layers of the image. Processing is based on the displacements of the filter described as a matrix of the size $k \times k$. In fact, a single image covered with a filter matrices can be interpreted as a one large grid of rectangles. Each of these rectangles is assigned a weight w, which is a numeric value used in the training process described later. Furthermore, this part of image can be interpreted as a neuron, which value is calculated as

$$f\left(\sum_{i=0}^{k}\sum_{j=0}^{k} w_{i,j} \cdot a_{x+i,y+j}\right), \tag{13}$$

where $a_{x+i,y+j}$ is the value returned by the neuron at position (x, y).

The second type of layers is the layer called pooling, which is designed to reduce the size of images by feature extraction. A given matrix of the size $m \times m$ is moved over the image, where, depending on the chosen function, one of the values in this matrix is transferred to the new image. The mentioned function can be a maximum condition that selects the pixel with the highest value on the matrix. The third type of layers is called fully-connected, which is the classic neural network described in the previous subsection. Constructing such a network often requires the use of the dropout technique, i.e., removing neurons which value is lower than the threshold. In practice that means simplifying the neural network for removing non-essential elements.

The convolutional neural network returns the value assigned to a given class. If the network is designed to classify 10 objects, there will be 10 neurons on the output layer and the classes will be labeled with a vector consisting of single one and nine zeros. To normalize the output from the network, a function called softmax is used.

Training such a construction is possible by using various algorithms that modify the weights (which were given in random way) between neurons. It is an optimization task, where weights are modified to minimize the loss function (often the difference between the expected value and the received value). One of this algorithm is RMSProp [19]. An algorithm is based on calculating two statistical parameters, which are variation v and mean m in each iteration t as

$$m_t = \beta_1 \cdot m_{t-1} + (1 - \beta_1) \cdot \nabla f(w_{t-1}), \tag{14}$$

$$v_t = \beta_2 \cdot v_{t-1} + (1 - \beta_2) \cdot (\nabla f(w_{t-1}))^2, \tag{15}$$

where β_1 and β_2 are coefficients. These two values are used in the process of updating weights w_t as

$$w_t = w_{t-1} - \frac{\lambda}{\sqrt{v_t + \epsilon}} \odot m_t, \tag{16}$$

where λ is a training rate and ϵ is a small constant value (usually equal to 10^{-6}) that prevents division by 0.

6. Experiments

To evaluate the proposed method, two groups of tests were carried out. The first consisted in creating agents and testing various configurations of the proposed techniques due to accuracy, as well as the sense of practical use. The individual components were implemented in JAVA, C# languages, Mathematica 11 and the classic mysql database was used. After preparing the exemplary functionality, a simple system was implemented with available devices such as smartphones and motion sensors. The tests were carried out in three different apartments.

6.1. Evaluation of the Proposed Method

In the first step, the basic information that can be taken by the agents were selected and three tasks were added:

- light meter—control of light in the room; and
- video—watering plants, taking out the trash.

Neural networks were used to classify these three tasks, due to the large amount of computing power required. The classic neural network was used to detect data on the light meter. The built-in sensor was aimed at the bulb in the room. The obtained data were recorded over a period of time, for example 3 s. The data were saved to a numerical vector. In addition, two items were added: time of day (morning, 0; noon, 0.4; evening, 0.7; and night, 1), which was stamped on the current time, and the presence of people in the room. The presence was checked using a motion sensor that records infrared radiation, which was returned as a numeric value 0 (meaning no movement) or 1 (meaning movement) every 3 s. These three values were saved in the vector that was used in the training process. The time of day is an important element when sensors are in the bedrooms. We assumed that the task of saving light was sent if there was no movement in the room and the light was on. We established 50 such samples, 35 of which were used in the training process in the proportions 80:20 (training samples to verification). The classifier structure was composed of four layers, where two of them were hidden and contained four neurons. Then, it was trained to get an error of 0.01. After that, effectiveness was checked for the entire database (an example is presented in Figure 5), and the results are shown in Figure 6. The resulting efficiency for such a small database was exactly 78%, which is not a high score, although it should be noted that there were few training data. The exact results of statistical measurements are shown in Table 1. It is worth noting that the false omission rate was over 27%, which indicates the proportion of false negatives that are falsely rejected.

The analysis of the behavior of people, e.g., taking out trash or watering flowers, was much more complicated. For this purpose, there was recorded a database of 150 video files (the size of 148×148) with people taking out trash. Each video lasts on average 10 s. Each of the videos was divided into 24 frames, which were used in the process of training the convolutional network. If the person was on the image with a garbage bag, the photo was labeled as trash. Similarly, training data for flower watering were prepared—the person had to hold a bottle of water. In both cases, a network was used with a similar structure, as is shown in Table 2. The networks were trained to achieve an effectiveness of more than 90% in the 80:20 ratio (training samples to verification) with only 130 video files. The obtained training charts are shown in Figure 7. Then, the database was extended by additional 20 unused video files in the training process. Then, real effectiveness was checked along with statistical measurements in Table 1. For both networks, the achieved efficiency was over 80% and the precision was above 86%. It is worth noting that the probability of rejecting a false sample was very large.

Figure 5. An exemplary frame used in the training process to detect the execution of a task.

Table 1. Statistical parameters for all tested classifier.

	NN for Controling the Light	CNN for Watering Flowers	CNN for Taking out the Trash
Accuracy	0.78	0.818676	0.858483
Sensitivity	0.793103	0.779742	0.746325
Specificity	0.761905	0.863244	0.93415
Precision	0.821429	0.867138	0.884344
Negative predictive value	0.727273	0.773954	0.845161
Miss rate	0.206897	0.220258	0.253675
Fall-out	0.238095	0.136756	0.06585
False discovery rate	0.178571	0.132862	0.115656
False omission rate	0.272727	0.226046	0.154839
F1 score	0.807018	0.821121	0.809494

Table 2. Structure of convolutional neural network used for image classification.

Layer	Output Shape
Convolutional	(None,148,148,32)
Activation	(None,148,148,32)
MaxPooling	(None,74,74,32)
Convolutional	(None,72,72,32)
Activation	(None,72,72,32)
MaxPooling	(None,36,36,32)
Convolutional	(None,34,34,64)
Activation	(None,34,34,64)
MaxPooling	(None,17,17,64)
Flatten	(None,18496)
Dense	(None,64)
Activation	(None,64)
Dropout	(None,64)
Dense	(None,1)
Activation	(None,1)

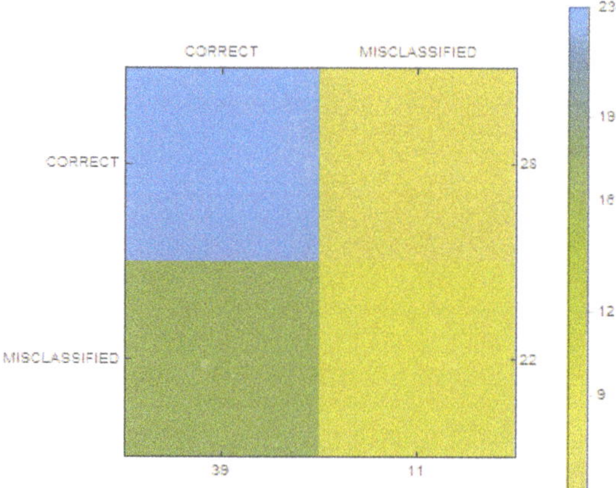

Figure 6. Confusion matrices for classic neural network.

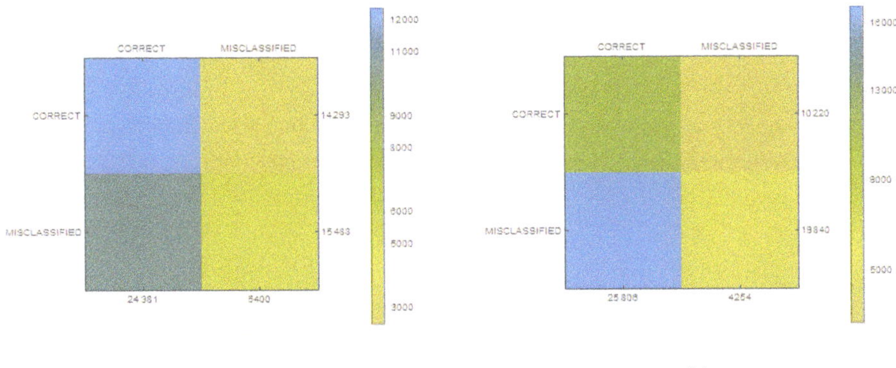

(**a**) (**b**)

Figure 7. Confusion matrices for convolutional neural network: (**a**) watering flowers; and (**b**) taking out the trash.

6.2. *Case Study*

To check the operation and impact of the system, three families including children (aged 15–18 years) were recruited for a 48 h test and to express their opinions.

6.2.1. Main goals

To check the real performance of the proposed system using available equipment, the following questions were asked

- Is the system of assigning tasks working correctly? Are the tasks assigned to children within their abilities?
- Is the proposed gamification solution useful? Does it result in better performance of basic tasks at home?
- Does the system work correctly? How does the placement of agents affect its operation?

6.2.2. Preparation of the Environment

The classifiers prepared in this way were used to check the operation of the multi-agent system in practice. Three different flats were used in the research, the structures of which are shown in Figures 8–10. In each of the flats, a server was set up where the database was maintained. Additionally, in the kitchens and salons, there aws a smartphone hung on the wall to record video data relative to watering flowers, as well as taking out the trash. In addition, in the room for children and bedrooms, smartphones were suspended and set to lamps as well as motion sensors. Each device was treated as an agent, thus the access to the network and the application in the background were set. They processed the obtained data and sent them to the database. Examples of tasks were sent in the form of a text message to smartphones of the household.

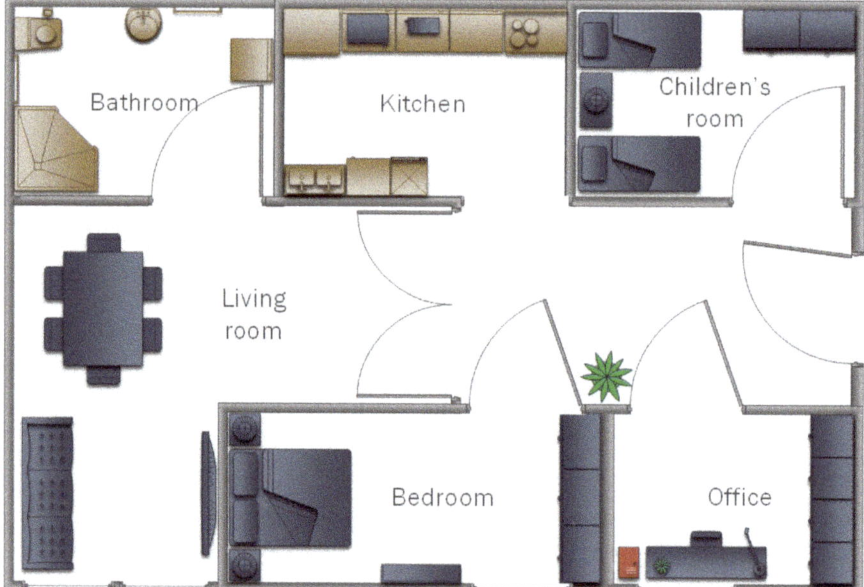

Figure 8. A model of the home that was used in testing proposed model in Experiment I.

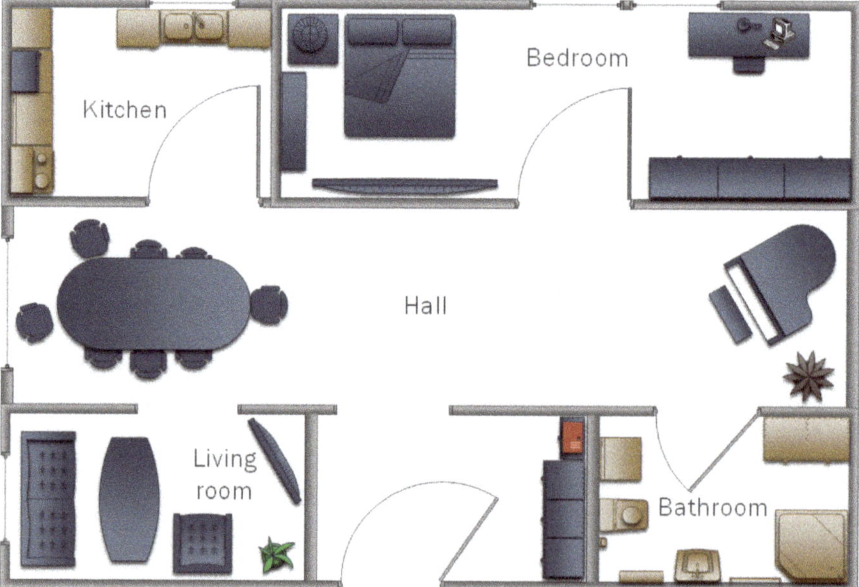

Figure 9. A model of the home that was used in testing proposed model in Experiment II.

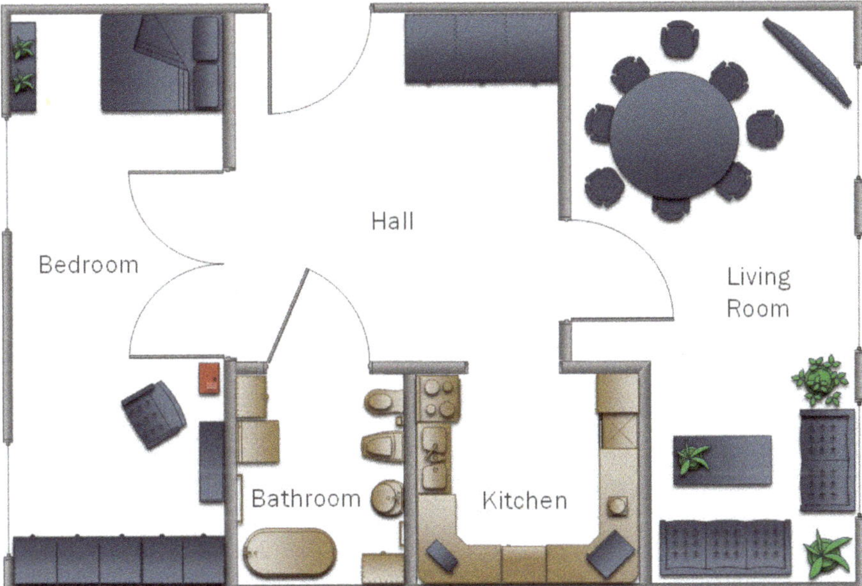

Figure 10. A model of the home that was used in testing proposed model in Experiment III.

6.2.3. System Evaluation

Three families that agreed to take part in software testing were asked to complete the questionnaire, which is presented along with the average results in Table 3.

Table 3. Question and answers questionnaire on a scale of $\langle 0, 10 \rangle$.

Question	House I	House II	House III	Average
System operation in terms of waste disposal	4	3.4	6	4.47
Operation of the system in terms of watering flowers	6	6	7	6.33
System operation in terms of light control	8.6	8	7.4	8
Automatic assignment of tasks	8	9	8	8.33
Points calculation	7	6	6	6.33
Motivation thanks to competition	9	8	8	8.33

Based on the obtained results, we can answer the questions described in Section 6.2.1. The system of assigning tasks was assessed very well, which was confirmed by the results of questionnaire as well as opinions. It was pointed out that the tasks were repeated at regular intervals, or if one was done. There was not a situation that there were two tasks. The parents pointed out that the children were primarily given the task of watering flowers, and the parents took out the rubbish. It was well received, because it was noted that smaller children should not receive such a task. The second highly valued element was the gamification, which resulted in a lot of fun during the competition for earning points and as a result of such action the lack of problems with garbage was pointed out. Unfortunately, it was also noticed that children wishing to get extra points watered flowers even without the task. This note indicates that users are eager to earn points, thus a richer task database is advisable.

The functioning system was the worst rated element. Based on the results and statements of the testers, the system does not always correctly classify the performance of tasks by using the image. The reason may be a bad sensor setting or location. In the case of agents working during light control, users were satisfied. However, they pointed out that at night the system does not work optimally: the light turned on late at night was not controlled by agents.

7. Conclusions

Intelligent technologies enter our lives faster and faster by integrating multi-agent systems into various types of equipment. A much larger growth is planned along with the arrival of the 5G network, which will increase development opportunities through a higher speed and a much larger number of connected objects to the network in a small area. In this paper, we show that the multi-agent system can be used to oversee the performance of simple household duties. An especially important aspect was the gamification scheme, which mobilized the family members to overcome them in time. This solution allowed automatic assignment of tasks depending on many parameters. This was important due to the age of the household—for the youngest, tasks will be simpler. Simpler does not mean less demanding, because the tasks should enable the child to develop. In addition, we showed that the operation of many agents is a valuable solution in the application of a large area, where different agents could analyze the same task (thanks to the use of an external database). The proposed solution was tested by simple elements, but nothing stands in the way of expanding the entire model by adding more functionality.

Author Contributions: Methodology, A.W. and K.K.; software, A.W.; validation, K.K.; formal analysis, D.P.; investigation, A.W., K.K. and M.W.; resources, A.W.; data curation, K.K.; writing–original draft preparation, A.W. and K.K.; writing–review and editing, A.W., K.K. and D.P.; visualization, A.W. and K.K.; supervision, M.W., Z.M. and D.P.; funding acquisition, M.W., D.P. and Z.M.

Funding: Authors acknowledge contribution to this project to the Diamond Grant No. 0080/DIA/2016/45 funded by the Polish Ministry of Science and Higher Education and the Rector proquality grant No. 09/010/RGJ19/0039, No. 09/010/RGJ19/0040 and No. 09/010/RGJ19/0042 at the Silesian University of Technology, Poland.

Conflicts of Interest: The authors declare no conflict of interest.

References

1. Palattella, M.R.; Dohler, M.; Grieco, A.; Rizzo, G.; Torsner, J.; Engel, T.; Ladid, L. Internet of Things in the 5G Era: Enablers, Architecture, and Business Models. *IEEE J. Sel. Areas Commun.* **2016**, *34*, 510–527. [CrossRef]
2. Połap, D.; Winnicka, A.; Serwata, K.; Kęsik, K.; Woźniak, M. An Intelligent System for Monitoring Skin Diseases. *Sensors* **2018**, *18*, 2552. [CrossRef] [PubMed]
3. Pedrasa, M.A.A.; Spooner, T.D.; MacGill, I.F. Coordinated scheduling of residential distributed energy resources to optimize smart home energy services. *IEEE Trans. Smart Grid* **2010**, *1*, 134–143. [CrossRef]
4. Zhang, Y.; Xiang, Y.; Huang, X.; Chen, X.; Alelaiwi, A. A matrix-based cross-layer key establishment protocol for smart homes. *Inf. Sci.* **2018**, *429*, 390–405. [CrossRef]
5. Ali, B.; Awad, A.I. Cyber and Physical Security Vulnerability Assessment for IoT-Based Smart Homes. *Sensors* **2018**, *18*, 817. [CrossRef] [PubMed]
6. Tao, M.; Zuo, J.; Liu, Z.; Castiglione, A.; Palmieri, F. Multi-layer cloud architectural model and ontology-based security service framework for IoT-based smart homes. *Future Gener. Comput. Syst.* **2018**, *78*, 1040–1051. [CrossRef]
7. Schieweck, A.; Uhde, E.; Salthammer, T.; Salthammer, L.; Morawska, L.; Mazaheri, M.; Kumar, P. Smart homes and the control of indoor air quality. *Renew. Sustain. Energy Rev.* **2018**, *94*, 705–718. [CrossRef]
8. Benzi, F.; Cabitza, F.; Fogli, D.; Lanzilotti, R.; Piccinno, A. Gamification techniques for rule management in ambient intelligence. In Proceedings of the European Conference on Ambient Intelligence, Athens, Greece, 11–13 November 2015; pp. 353–356.
9. Fogli, D.; Lanzilotti, R.; Piccinno, A.; Tosi, P. AmI@ Home: A game-based collaborative system for smart home configuration. In Proceedings of the International Working Conference on Advanced Visual Interfaces, Bari, Italy, 7–10 June 2016; pp. 308–309.
10. Jacobsson, A.; Boldt, M.; Carlsson, B. A risk analysis of a smart home automation system. *Future Gener. Comput. Syst.* 2018, *56*, 719–733. [CrossRef]
11. Rodríguez, S.; Gil, O.; De La Prieta, F.; Zato, C.; Corchado, J.M.; Vega, P.; Francisco, M. People detection and stereoscopic analysis using MAS. In Proceedings of the 2010 IEEE 14th International Conference on Intelligent Engineering Systems, Las Palmas, Spain, 5–7 May 2010; pp. 159–164.

12. Román, J.A.; Rodríguez, S.; de la Prieta, F. Improving the distribution of services in MAS. In Proceedings of the International Conference on Practical Applications of Agents and Multi-Agent Systems, Sevilla, Spain, 1–3 June 2016; pp. 37–46.
13. de la Prieta, F.; Navarro, M.; García, J.A.; González, R.; Rodríguez, S. Multi-agent system for controlling a cloud computing environment. In Proceedings of the Portuguese Conference on Artificial Intelligence, Azores, Portugal, 9–12 September 2013; pp. 13–20.
14. Ferber, J.; Weiss, G. *Multi-Agent Systems: An Introduction to Distributed Artificial Intelligence*; Addison-Wesley: Boston, MA, USA, 1999; Volume 1.
15. Nagata, T.; Watanabe, H.; Ohno, M.; Sasaki, H. A multi-agent approach to power system restoration. In Proceedings of the PowerCon 2000. 2000 International Conference on Power System Technology, Perth, WA, Australia, 4–7 December 2000; Volume 3, pp. 1551–1556.
16. Karagiannidis, G.K.; Lioumpas, A.S. An improved approximation for the Gaussian Q-function. *IEEE Commun. Lett.* **2007**, *11*, 644–646. [CrossRef]
17. Arulmurugan R.; Anandakumar H. Early Detection of Lung Cancer Using Wavelet Feature Descriptor and Feed Forward Back Propagation Neural Networks Classifier. In *Computational Vision and Bio Inspired Computing*; Springer: Cham, Switzerland, 2018; pp. 103–110.
18. Zhang, C.; Sun, G.; Fang, Z.; Zhou, P.; Pan, P.; Cong, J. Towards uniformed representation and acceleration for deep convolutional neural networks. *IEEE Trans. Comput.-Aided Des. Integr. Circuits Syst.* **2018**. [CrossRef]
19. Kumar Roy, S.; Mhammedi, Z.; Harandi, M. Geometry aware constrained optimization techniques for deep learning. In Proceedings of the IEEE Conference on Computer Vision and Pattern Recognition, Salt Lake City, UT, USA, 18–22 June 2018; pp. 4460–4469.

Article

Nonintrusive Appliance Load Monitoring: An Overview, Laboratory Test Results and Research Directions

Augustyn Wójcik [1,*], Robert Łukaszewski [1,*], Ryszard Kowalik [2,*] and Wiesław Winiecki [1,*]

1 Institute of Radioelectronics and Multimedia Technologies, Warsaw University of Technology, Nowowiejska 15/19, 00-665 Warsaw, Poland
2 Institute of Electrical Power Engineering, Warsaw University of Technology, Koszykowa 75, 00-662 Warsaw, Poland
* Correspondence: a.wojcik@ire.pw.edu.pl (A.W.); r.lukaszewski@ire.pw.edu.pl (R.Ł.); ryszard.kowalik@ee.pw.edu.pl (R.K.); w.winiecki@ire.pw.edu.pl (W.W.)

Received: 12 July 2019; Accepted: 17 August 2019; Published: 20 August 2019

Abstract: Nonintrusive appliance load monitoring (NIALM) allows disaggregation of total electricity consumption into particular appliances in domestic or industrial environments. NIALM systems operation is based on processing of electrical signals acquired at one point of a monitored area. The main objective of this paper was to present the state-of-the-art in NIALM technologies for the smart home. This paper focuses on sensors and measurement methods. Different intelligent algorithms for processing signals have been presented. Identification accuracy for an actual set of appliances has been compared. This article depicts the architecture of a unique NIALM laboratory, presented in detail. Results of developed NIALM methods exploiting different measurement data are discussed and compared to known methods. New directions of NIALM research are proposed.

Keywords: NIALM; smart home; electrical appliances; home events; load disaggregation; sensing technologies; intelligent algorithms; human behavior

---

1. Introduction to Appliance Load Monitoring Systems

A major problem of power systems in recent years has been the constant increase in demand for electricity. In view of this problem, saving energy and reducing its consumption are promoted. Many people have joined in the initiatives aimed at increasing the energy efficiency of the places where they live. However, the users of electricity have no opportunity for conscious control of energy consumption, because their possibilities are mostly limited to analysis of accounts after settlement periods lasting usually a few months. Moreover, in both the domestic and industrial environment, dozens of appliances with different power consumptions are switched on and off several times a day. Accounts containing total energy consumption do not give any information about electricity consumption of particular appliances. Therefore, it is hard to form habits of saving energy consciously and effectively, which is one of the most important aims of smart homes.

The first purpose of developing an appliance load monitoring (ALM) system is to provide an electricity consumer with information about the energy consumption of individual appliances. This leads to limiting electricity consumption and less atmospheric pollution. Moreover, consumers with an ALM system would be aware of the appliances consuming the most energy. From a social point of view, it is important to educate people about the habit of saving energy. Obtaining information about the most energy-consuming appliances provides potential opportunities for electricity management. Some of the most energy-consuming appliances could be switched on only when there is an excess of energy in the power system. For industry, ALM systems may suggest optimal configurations of industrial machines to limit reactive power. Future possibilities of ALM systems may include diagnostics of electrical appliances [1], including monitoring of device wear (e.g., mechanical problems in rotating motors), and detection of supply network states, including dangerous inferences, voltage spikes, weakening of insulation, etc.

The simplest way to monitor power consumption of all appliances is to equip them with individual electricity meters. There are many problems resulting from the fact that each monitored appliance needs an individual meter. Firstly, with the increase in the number of monitored appliances, the cost of the measuring system increases significantly [2]. Secondly, measuring data need to be collected in one central unit, which is another system element. Thirdly, data from meters need to be sent to a central unit, so electricity meters have to be equipped with a suitable interface, e.g., a radio interface. Moreover, current flowing through the monitored appliances has to flow through the meter too, which reduces the reliability of the power supply. Lastly, measuring devices consume energy. The more monitoring system elements, the more energy is consumed. Because of the mentioned arguments, such a system is called IALM—intrusive appliance load monitoring (see Figure 1a).

Another concept for solving the presented problem is a nonintrusive appliance load monitoring (NIALM) system, which also determines the energy consumption of particular appliances turning on and off in local domestic or industrial power grids. First concept of NIALM system was introduced by Hart [3], and is also known as energy disaggregation [4]. It is called nonintrusive because measurements are made solely near the energy meter, in contrast to intrusive systems where every socket or load should be equipped with a suitable sensor [5]. When new appliances are plugged into the area monitored by nonintrusive system, the hardware does not need to be expanded. Measured values are typically current and voltage of the total load [6]. A measuring system of this type is presented in Figure 1b.

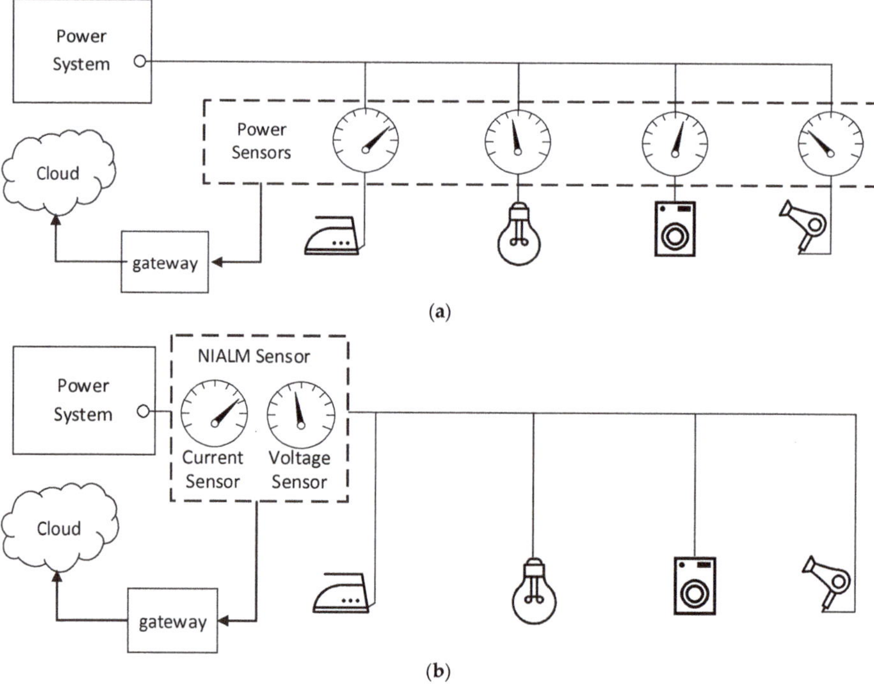

Figure 1. Appliance load monitoring systems: (**a**) intrusive appliance load monitoring (IALM); (**b**) nonintrusive appliance load monitoring (NIALMS).

Electricity is transferred in the form of voltage. Under the action of voltage, the current flows through a specific receiver. The goal of electrical signal processing algorithms is to determine individual values of currents on the basis of characteristic parameters of current, voltage, or any combination of both of these signals. Information about the operating states of individual appliances and estimations of their energy consumption are obtained by a sophisticated analysis of collected waveforms. The main challenge of NIALM is to develop methods of signal analysis which provide knowledge about the behavior of electrical appliances. This is crucial for proper system operation. The hardware of a complete NIALM system does not need to be extended with new devices in the monitored area.

The general architecture of a nonintrusive appliance load monitoring system is presented in Figure 2.

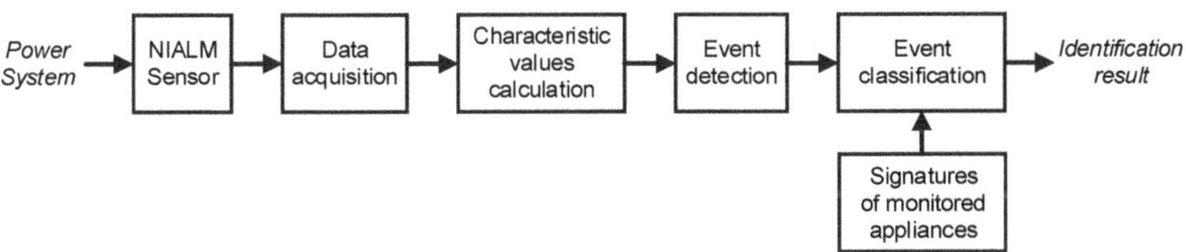

Figure 2. General architecture of a nonintrusive appliance load monitoring system.

The first stage is a NIALM sensor, the main purpose of which is to transform the quantity of interest (e.g., current or voltage) into voltage with a range appropriate for the acquisition equipment. In the data acquisition stage, measuring signals are converted from analog to digital samples and saved in the memory. The sampling frequency of the recording hardware determines the type of characteristic values extracted from measured signals. Examples of characteristic values are: average power, harmonics of current, wavelet transform coefficients, root mean square (RMS) power. Characteristic value is calculated on the basis of recorded samples. The majority of methods employ more than one characteristic value. The set of characteristic values should be unique for every monitored appliance. The third stage is event detection based on analysis of the characteristic value vector. In the simplest scheme, an event is detected when the difference in two subsequent characteristic values is above the determined detection threshold. A characteristic value used often for event detection is the envelope of the current signal. Because multiple appliances work simultaneously during online operation, most methods identify only the device that has recently changed its state. The sequence of correct identifications allows which appliances were operating in the household in the particular moment of time to be determined. The characteristic values calculated after the event detection are calculated for the group of currently operating appliances, and cannot be used instantly for identification. Instead, the system also stores the set of characteristic values prior to the analyzed event. In this way, it is possible to calculate the features for the particular appliance as the difference between two vectors, calculated after the previous $f(t{-}1)$ and the last event $f(t)$:

$$f_i(t) = f(t) - f(t-1), \tag{1}$$

The event classification stage compares the calculated vector of characteristic values $f_i(t)$ with the signatures of the monitored appliances. A signature is the vector of characteristic values calculated under controlled appliance operating conditions, i.e., when each appliance was turned on individually. As the result of event classification stage, the set of appliances actually operating is determined and individual energy consumption is calculated.

It should be noted that the signal analysis methods can be varied. They take into account characteristic values calculated either when the operating state of an appliance changes, or when operating state is determined. Analyses at both mentioned types of operating state can be performed using methods belonging to one of three groups:

- Group I (fundamental 50 Hz harmonic group, LF, RMS)—analysis methods exploiting RMS of signals or the amplitude of the first current and voltage harmonic collected with a sampling frequency from fractions of Hz to several Hz.
- Group II (fundamental 50 Hz and its harmonics group, MF, harmonics)—analysis methods exploiting instantaneous values of signals (samples) collected with a sampling frequency from 1 kHz to dozens of kHz.
- Group III (high-frequency group, HF)—methods exploiting instantaneous values of signals collected with a sampling frequency from dozens of kHz to several dozen MHz.

Within each of the above groups, current and/or voltage signals are analyzed in several stages, similar to those presented in Figure 2. Analysis methods can be grouped by the duration of the analysis window. Moreover, there are methods exploiting steady-state operating states (SS) and transient states (TS) when appliance operating states change [7].

The aforementioned methods are described in the following sections with the use of the measurement setup shown in Figure 3. Power supply voltage and aggregate current were measured and recorded. Measured voltage was applied for a set of three appliances switched on and off apart from each other.

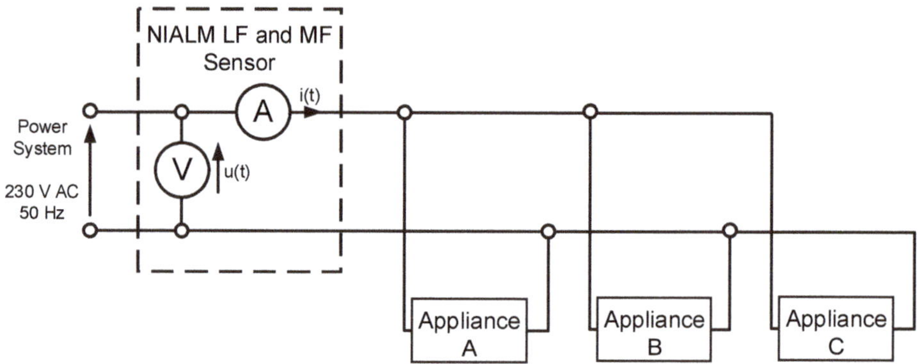

Figure 3. Block diagram of the measurement setup.

2. Group I (LF) Measurement Methods

Signals analyzed with the use of methods of group I (LF) can be evaluated using the following Equations:

$$I_{RMS} = \int_{t_0}^{t_0+T} i^2(t)\, dt, \tag{2}$$

$$P_{RMS} = \int_{t_0}^{t_0+T} i(t)u(t)\, dt, \tag{3}$$

where *i(t)* is the current signal; *u(t)* is the voltage signal; *T* is the time of analysis, which lasts one period of the fundamental harmonic or a multiple of one period; I_{RMS} is the RMS value of the current; and P_{RMS} is the RMS value of the power. It should be mentioned that RMS values derived from power meter are a mean value of many RMS evaluated in a 1 second or 0.5 second time interval. Signals analyzed using methods of group I are presented in Figure 4. It presents the result of measurements performed in the set of three electrical appliances detailed in Table 1, supplied from one common voltage of 230 V AC (50 Hz).

Table 1. Tested appliances.

#	Name	Producer	Type	Max Power
A	LED bulb	Osram	AB30526	17 W
B	Hair dryer	Apollo	SUA-2000-SIL	2000 W
C	Vacuum cleaner	Zelmer	ZVC425HT	1000 W

The LED bulb (A) was switched at the moment 1. Next, at the moment 2, hair dryer (B) is switched on. Moment 3 indicates the vacuum cleaner (C) switching on. Moments 4, 5, and 6 indicate the time points at which the appliances were switched off, respectively: LED bulb, hair dryer, and vacuum cleaner.

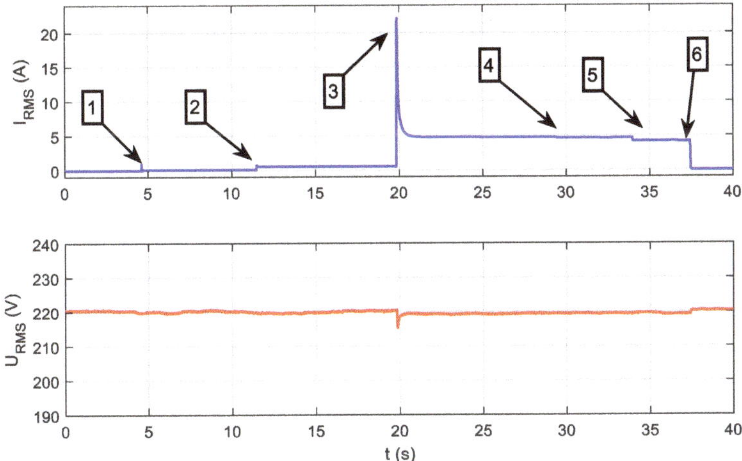

Figure 4. The sequence of switching on and off tested appliances.

By using the presented signals, analyses using the following methods were performed. In relation to the power grid voltage frequency of 50 or 60 Hz [8–10], analysis in a low-frequency range allows extraction of the parameters of the waveforms averaged over several periods. Generally, LF methods are based on analysis of power changes between steady states [11]. The most frequently used parameters of power are the average power [8,12–16], reactive power [17,18], and power factor [12]. The detected changes are often plotted as points in two-dimensional space. The obtained points can be assigned to groups with approximately the same changes of power. Groups with the same value but the opposite sign of change correspond to turning on and off the same appliance. In numerous publications, hidden Markov models (HMM) and their variants [19–23] have been applied for analyzing power changes. Baranski and Voss [11] presented an identification method based on fuzzy clustering and genetic algorithm, while a temporal mining approach was proposed in Reference [24]. The idea of representing a dataset of power values using a graph defined by a set of nodes and a weighted adjacency matrix has been introduced in References [13,25], among others. Appliances with a finite number of states are often considered finite state machines (FSM) [3,26]. The main constraint of LF methods is their inability to distinguish appliances with similar power. For this purpose, more sophisticated acquisition hardware is exploited.

3. Group II (MF) Measurement Methods

If signals u(t) and i(t) (see Figure 3) are measured with the use of an appropriate measuring device with a sampling frequency in the range 1000 S/s to 20 kS/s, these signals will be observed as is presented in Figure 5. The waveforms presented in Figure 5 were acquired in moment 3 (see Figure 4), and the only difference is that the scale of time is more accurate. Thanks to this approach, there were visible changes in signals completely invisible for LF methods, which were particularly evident in the example of the LED bulb signals (moment 1), presented in Figure 6. LF methods are not suitable for analyzing signals that are so distorted with respect to a 50 Hz sinusoidal signal.

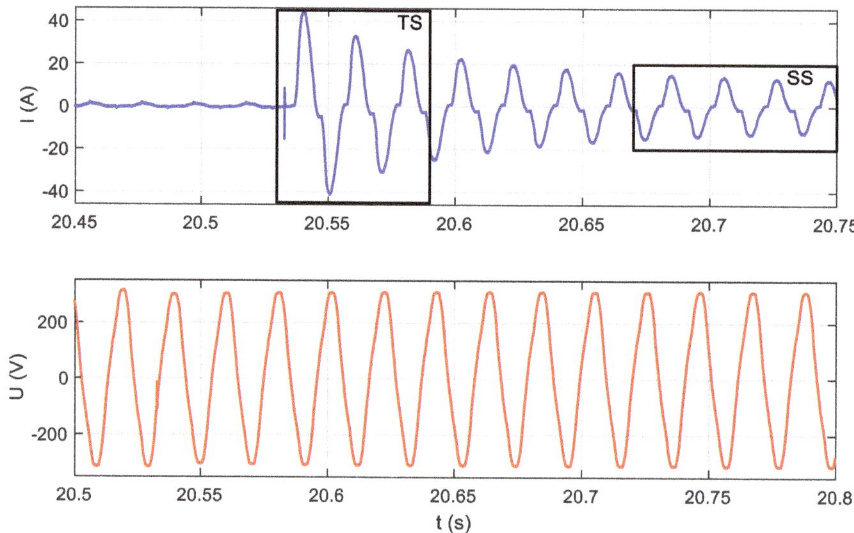

Figure 5. Moment 3, the vacuum cleaner switching on, in the scale of milliseconds.

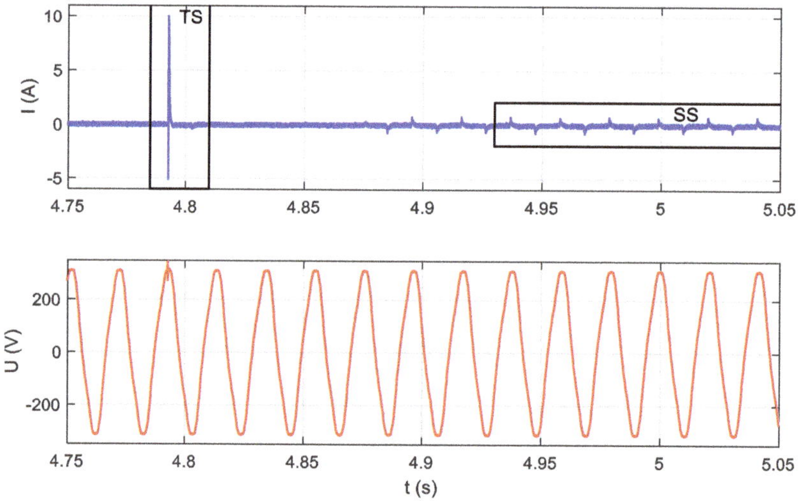

Figure 6. Moment 1, switching on the LED bulb (**A**).

The majority of household appliances are nonlinear, e.g., computer or LED lamp. Hence, the current waveforms of these appliances contain significant harmonic distortion. The fact of harmonic distortion was employed in the design of NIALM methods. Parameters characterizing appliances are commonly determined on the basis of current harmonics obtained from the complex Fourier spectrum. Usually, the first 11–13 harmonics are used in the analysis, so in order to capture the microscopic parameters, it is necessary to have measurements with a sampling frequency of at least 1.5 kHz. The sample signature of an electrical appliance contains information about the amplitude [27] and the phase [28] of the subsequent current harmonics. In some works, power parameters have also been included [29]. The distribution of amplitudes and phases of individual harmonics depends on the set of appliances currently operating. The basis for identification algorithms is a comparison of signature parameters calculated from measurement data and labels of individual appliances collected during system training. Neural networks are often used as a classifier. In the input layer of the neural network, information about the harmonics of the electric current is entered, while in the output layer, the appropriate appliance category is obtained.

In Reference [30], the real and imaginary parts of the odd harmonics of the current were features characterizing devices. A number of experiments were carried out comparing the operation of three types of neural networks: radial basis function (RBF), multi-layer perceptron (MLP), and support vector machine (SVM). The presented experimental results show that average classification accuracy of developed method was about 85%.

In Reference [31], a slightly different approach was applied. The spectrum distribution in sub-bands distributed linearly or logarithmically in the range from 0 Hz–5 kHz was calculated. Jaccard distance was used to assess the similarity of classified examples with the training data. A high identification accuracy was achieved. However, the algorithm was only tested with devices switched on individually.

A spectral envelope for estimating the power of variable speed drive (VSD) with a bulb operating in background was used in Reference [32]. The spectral envelope can also be used in the analysis of transient states. Shaw et al. [1] proposed a method of transient spectrum analysis which may be applied both in AC power grids and DC power supply in automotive applications.

A different approach was proposed by Lang, Fung, and Lee. They proposed the taxonomy of electrical appliances based on voltage–current (V-I) trajectories [33]. In this method, graphs of current are drawn as a function of voltage. The features distinguishing devices are the shape parameters of these graphs. The authors described some features of trajectory shape which corresponded to parameters of electrical signals (e.g., the number of self-intersections as related to harmonic content). An investigation of a trajectory-based load monitoring system was presented in Reference [34]. An identification accuracy of about 75% was achieved.

4. Group III (HF) Measurement Methods

If we look more precisely at the phenomena occurring when appliances change their state, we will see that current and voltage measured according to Figure 3 contain high-frequency (HF) components not harmonic with the 50 Hz fundamental frequency of the power supply. Information about appliances' state changes may be included in these HF components. HF components are the response of appliance impedance to the instantaneous voltage of a power grid present at the moment of switching on an appliance. Therefore, a wide frequency band should be used for analysis. The waveforms presented in Figure 4 are expanded in Figure 7.

One group of HF methods are those characterizing transient states during switching on/off of electrical appliances. Due to the high sampling frequency of recorded data, it is necessary to apply an appropriate acquisition system that allows for initial reduction of the data, e.g., by detecting the moment of state change and recording only a significant part of the signal for further analysis [35]. Waveforms registered in transient states (TSs) very often differ significantly in shape, even if they have been measured for the same device. This makes it necessary to collect many examples of TSs, so that it might be possible to characterize them with universal and robust features. Therefore, before developing an identification system, it is necessary to prepare an appropriate database containing many TSs registered for different devices [36].

In Reference [37], a NIALM system sampling current and voltage with a frequency of 1 MHz was presented. Authors observed that TSs after switching on appliances lasted for different times, depending on the type of device. A vector of harmonics comprised from alternating real and imaginary parts of the complex Fourier transform and partial least-square regression applied to raw waveform data were recognized as the most robust features. Classification was directly related to simple statistical hypothesis testing. To improve the classification accuracy, a fusion of classifiers using both types of features was used. In the laboratory, the method allowed perfect accuracy to be obtained, but in a real household environment, a decrease of classification accuracy was observed, especially for low-power devices.

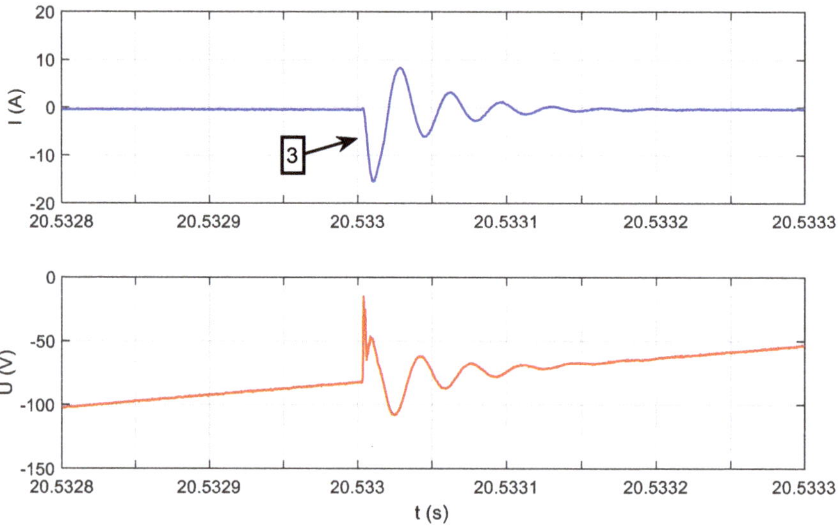

Figure 7. Moment 3, the vacuum cleaner switching on, in the scale of microseconds, transient state (TS).

The voltage signal sampled at 5.21 MHz was used to analyze the energy consumption in Reference [38]. As the method is based solely on the analysis of transient states, the basic component (60 Hz) and its harmonics were filtered with the Notch IIR filter. Two sets of features were prepared. The first one, based on continuous wavelet transform (CWT), is a vector containing energy for selected scales and signal duration. The second feature set constitutes parameters calculated from short-term Fourier transform (STFT): averages of the individual complex Fourier spectrum components and the number of time windows. In the experiments performed, three selected appliances were tested. The STFT based method achieved an accuracy of 70%, while the use of CWT ensured 80% identification accuracy.

In most electronic appliances, switched mode power supplies (SMPS) are used. Such a solution is more effective than a classical transformer in terms of size, weight, and cost. Moreover, in the case of a transformer, the power is constant, whereas electronic appliances with SMPS get current pulses containing a large number of harmonics. Therefore, important signals contained in the high-frequency band also appear in steady states (SS). SMPS generate a high-frequency electrical noise (EMI—electromagnetic interference) during operation. The measuring system should be adjusted to extract EMI components from steady-state signals. The measurement setup from Figure 3 was modified by adding a high-pass filter in the voltage measurement path cutting of the fundamental 50 Hz frequency of the power grid voltage. It is presented in Figure 8. An example of signals measured in this measurement setup is presented in Figure 9. High-pass filter inclusion causes a phase delay of the voltage signal.

The EMI noise is specific for different electronic appliances and can be used for the identification [39]. The continuous noise is a narrowband with characteristic center frequency for different appliances. The frequency for which the level of noise is the highest is called the switching frequency and can be modeled with a normal distribution. For appliance identification, in this method, the use of switching frequency and normal distribution parameters has been proposed. These parameters are different for each appliance and are saved in an appliance library. The appliances are identified using k-Nearest Neighbor and parameters saved in the appliance library. The NIALM system that uses electromagnetic interference for appliance identification was proposed by Gupta and co-workers [39] and expanded by operating state type identification in Reference [40].

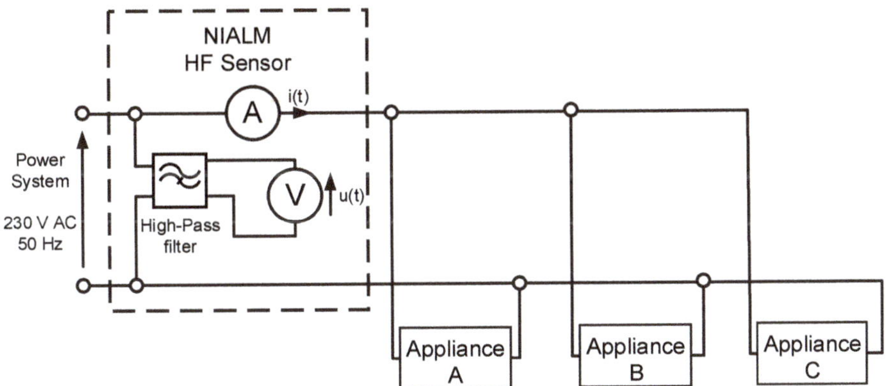

Figure 8. The block diagram of electromagnetic interference (EMI) measurement setup.

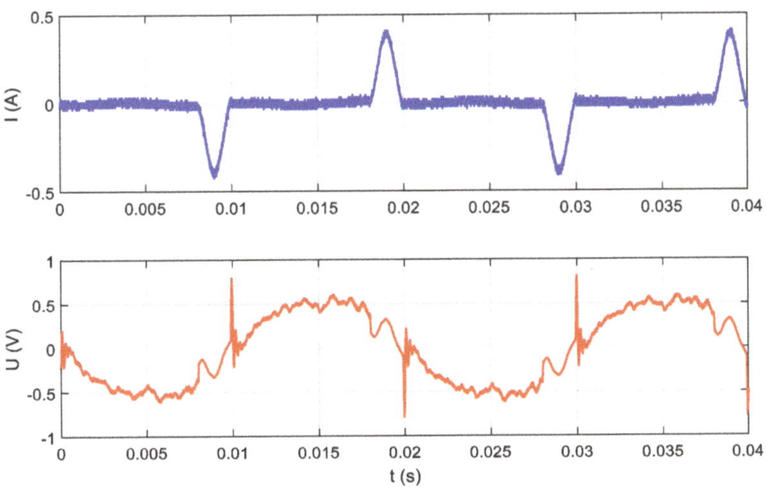

Figure 9. Current and high-pass (HP)-filtered voltage measured during LED bulb operation.

5. Extra-High-Frequency (EHF) Measurements

Signals measured in the measurement setup presented in Figure 3 contain components with frequencies higher than those discussed in Section 4. Other frequency components up to GHz are included in current and voltage signals. Some measurable amplitudes of extra-high-frequency (EHF) components originate from very-high-frequency (VHF) radio transmitters. It can be suspected that information characteristic for electrical appliances is included in this frequency range. So far, no one has conducted such research. Current and voltage signals were measured with a sampling frequency of 2 GHz using an Agilent InfiniiVision oscilloscope. Figure 10 presents the steady-state operation of an LED bulb (A), while Figure 11 presents a transient state during LED bulb (A) switching on. The scale of current magnitude in Figures 10 and 11 is different.

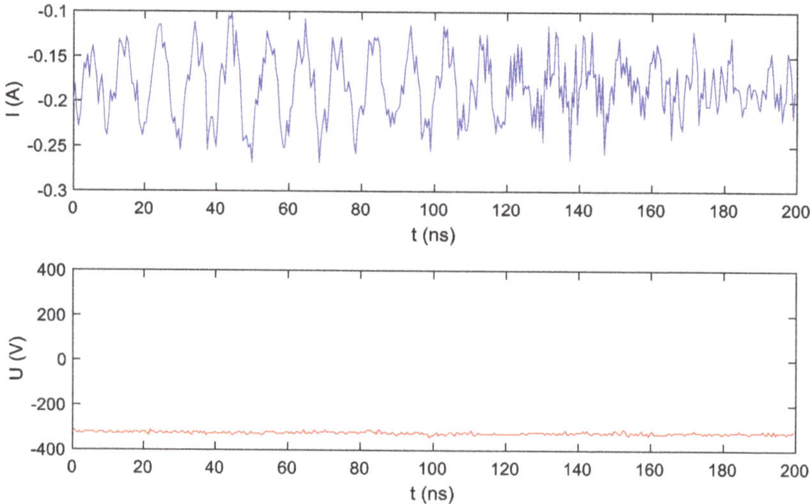

Figure 10. Extra-high-frequency (EHF) voltage and current recorded with frequency sampling of 2 Gs/s during LED bulb (A) operation in steady state (SS).

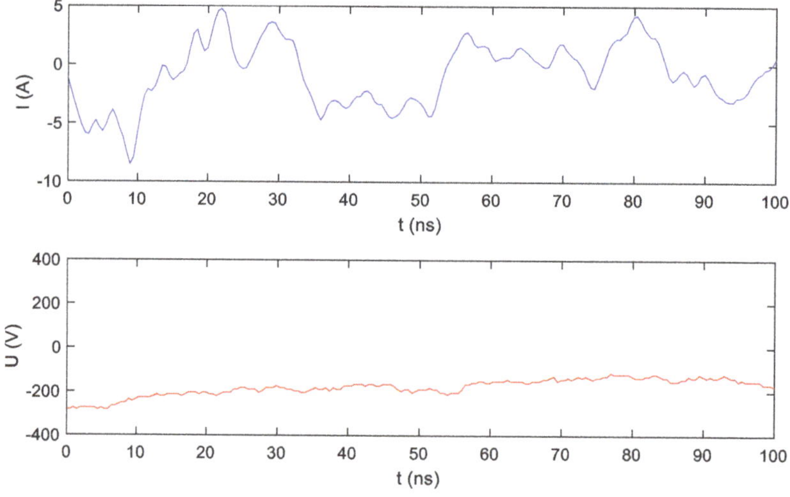

Figure 11. EHF voltage and current recorded with frequency sampling of 2 Gs/s during LED bulb (A) switching on (TS).

It is also worth mentioning that there are alternative methods of estimating electricity consumption without measuring electrical signals. The concept of device activity detection using radio frequency interference emission was presented in Reference [41]. This method takes advantage of the fact that electrical appliances during operation generate electromagnetic fields around their enclosures. A suitable broadband antenna sensor and data acquisition system allows this disturbance to be recorded and characterized [42]. Similarly, in Reference [43], the acoustic noise produced during operation was used to identify the type of device.

6. Comparison of Nonintrusive Appliance Load Monitoring Methods

The accuracy of load disaggregation methods is strongly related to the number of monitored appliances. The smaller the number of devices used for the experiment, the greater the probability of a correct energy disaggregation. Similarly, a larger set of tested appliances requires more complex processing algorithms. Therefore, in order to obtain a complete overview of the state of the art in NIALM, the classification accuracy of individual appliances should be analyzed with regard to the type and number of all tested appliances in each experiment. Only this approach allows the suitability of methods in practical applications to be assessed.

Various quality measures have been used in published studies. There are various objective evaluation metrics under the umbrella of classification accuracy. The authors do not always precisely define how the accuracy of identification was determined. The quality of the load disaggregation in NIALM can be assessed on the basis of:

- accuracy of classification of events in the monitored area, where an event is a change of operating state of a particular appliance,
- accuracy of reconstructing operating states of particular appliances in every observed moment of time,
- accuracy of electricity consumption disaggregation expressed in physical quantity, e.g., in kWh, in each considered observation interval.

The most common method of accuracy evaluation in the classification issue is the confusion matrix. Confusion matrices afford the number of examples classified correctly and the number of examples assigned to an inappropriate class. In the field of NIALM, class means usually the same as appliance. Evaluation measures calculated from confusion matrices are: precision, recall, and F-measure. A detailed introduction to these measures, used widely in data mining and machine learning, is provided in Reference [44]. Application of a confusion matrix in NIALM issue was proposed in Reference [6], while in Reference [45], a study of the classifier parameters adaptation to NIALM-specific problems was provided. Depending on the processing algorithms used, evaluation metrics may be applied to the overall system performance or to the individual processing steps. Examples of evaluation metrics distinguishing event detection and classification are described in Reference [46].

It should be noted that the introduced evaluation metrics apply to the classification of detected events. The imperfection of them is that all consider only the moments of event occurrence, i.e., moments of operating state change. Under certain NIALM operating conditions, operating states may change at large intervals, e.g., at night. In this case, more important for evaluation would be the accuracy of reconstruction of the power of particular appliances in each observation interval, because one of the main goals of NIALM research is to determine which devices consume the most energy. Evaluation of NIALM by comparing the disaggregated energy consumption expressed in kWh to the total energy consumption was proposed in Reference [23]. A comprehensive approach to quality measures used in NIALM was presented in Reference [47].

Another issue is the preparation of testing data. NIALM methods are often tested initially when appliances are turned on and off individually, and no appliance is operating in the background [34]. However, this is not a sufficient testing scheme. Testing data should be aggregate signals of power, current, or voltage and current recorded during operation of tested appliances switched on/off in a certain scenario. This scenario of switching on and off appliances is often referred to as a sequence.

Scenarios should include a variety of appliances state change combinations. This ensures that the considered method is evaluated in most possible operating conditions. Unfortunately, testing scenarios have often been collected in houses or flats, where appliances were switched on/off randomly. Moreover, appliances were often switched on/off when one or even zero appliances were operating in the background. As an example, in Reference [48], the washing machine identification was performed with high accuracy, but the figure presenting the testing sequence shows that only one appliance was operating in the background.

The mentioned issues suggest that reliable evaluation of NIALM method is demanding. Moreover, comparing different methods is more demanding. However, we have tried to determine groups of appliances which may be classified with the use of particular NIALM methods.

A literature review has been carried out in order to identify groups of appliances that can be classified in one area with the use of a given method. Due to the fundamental differences in the architecture of systems using data with different sampling frequencies, the analysis was divided into three groups: LF, MF, and HF. On the basis of the literature review, the methods used for load disaggregation in relation to the types of devices have been presented. The results are collected in Tables 2 and 3 (LF and MF accordingly), containing, on one hand, a set of appliances monitored successfully in one area, and on the other hand, the methods used for load disaggregation listed as references to source articles. If more than one testing set was used by the authors, it was marked with consecutive letters in brackets "(a)" etc. No publication that would meet the criteria adopted during the analysis was found within the HF group.

Table 2. Groups of appliances possible to classify properly using selected low-frequency (LF) methods.

Source Appliance	[49] **(a)**	[49] **(b)**	[21]	[50] **(a)**	[50] **(b)**	[50] **(c)**	[15]	[25] **(a)**	[25] **(b)**	[25] **(c)**	[25] **(d)**	[25] **(e)**	[51]	[19]	**Our Study** [52]
Air conditioner	+	N.A.	N.A.	N.A.	N.A.	N.A.	N.A.	N.A.	N.A.	+	N.A.	N.A.	N.A.	N.A.	N.A.
Bath outlet	N.A.	N.A.	N.A.	N.A.	N.A.	N.A.	N.A.	N.A.	N.A.	N.A.	N.A.	N.A.	N.A.	+	N.A.
Bulb	N.A.	N.A.	N.A.	N.A.	N.A.	N.A.	N.A.	N.A.	N.A.	N.A.	N.A.	N.A.	N.A.	N.A.	+
Dishwasher	N.A.	+	+	N.A.	+	+	+	N.A.	+	N.A.	+	N.A.	+	N.A.	N.A.
Dryer	N.A.	N.A.	N.A.	+	+	+	N.A.	N.A.	N.A.	N.A.	N.A.	N.A.	N.A.	N.A.	N.A.
Electronics	N.A.	N.A.	N.A.	N.A.	N.A.	N.A.	N.A.	N.A.	N.A.	+	N.A.	N.A.	N.A.	N.A.	N.A.
Fridge	N.A.	N.A.	+	N.A.	+	+	N.A.	+	+	+	N.A.	+	+	N.A.	N.A.
Fan	N.A.	N.A.	N.A.	+	N.A.	N.A.	N.A.	N.A.	N.A.	N.A.	N.A.	N.A.	N.A.	N.A.	N.A.
Furnace	+	N.A.	N.A.	N.A.	N.A.	+	+	N.A.	N.A.	N.A.	N.A.	N.A.	N.A.	+	N.A.
Freezer	N.A.	N.A.	N.A.	N.A.	N.A.	N.A.	N.A.	N.A.	N.A.	N.A.	+	+	N.A.	N.A.	N.A.
Heater	N.A.	N.A.	N.A.	+	N.A.	N.A.	N.A.	N.A.	N.A.	N.A.	N.A.	N.A.	N.A.	N.A.	N.A.
Iron	N.A.	N.A.	N.A.	N.A.	N.A.	N.A.	+	N.A.	N.A.	N.A.	N.A.	N.A.	N.A.	N.A.	N.A.
Kettle	N.A.	N.A.	+	N.A.	N.A.	N.A.	N.A.	N.A.	N.A.	N.A.	+	+	N.A.	N.A.	+
Kitchen outlet	N.A.	N.A.	N.A.	N.A.	N.A.	N.A.	N.A.	N.A.	+	+	N.A.	N.A.	+	N.A.	N.A.
Lighting	N.A.	+	N.A.	N.A.	N.A.	N.A.	N.A.	N.A.	N.A.	N.A.	N.A.	N.A.	N.A.	N.A.	N.A.
Microwave	N.A.	N.A.	+	N.A.	+	+	N.A.	+	+	+	+	+	+	+	+
Oven	N.A.	N.A.	N.A.	N.A.	+	N.A.	+	N.A.	N.A.	N.A.	N.A.	N.A.	+	N.A.	N.A.
Stove	N.A.	N.A.	N.A.	N.A.	N.A.	N.A.	+	N.A.	+	+	N.A.	N.A.	N.A.	N.A.	N.A.
Washing Machine (washer-dryer)	N.A.	+	N.A.	N.A.	N.A.	N.A.	+	+	N.A.	N.A.	+	+	+	+	N.A.
Number of +	2	3	4	3	5	5	6	3	5	6	5	5	6	4	3

Because the authors used different evaluation metrics, no percentage accuracy results are presented. It was only specified for each appliance whether it was possible to classify the appliance using a given method in the specific background according to authors' conclusions. Therefore, for more information, please see the source papers. A review of the literature covered about 100 sources. Only a few of them, listed in the tables, contained evaluations of classification of particular appliances. The table legend is as follows: "+" means that appliance was in the testing set and was classified properly. "N.A." means that the particular appliance was not tested.

The presented results show there is a lack of universal methods able to recognize dozens of any type of appliance, which is crucial in practical applications. Therefore, we still need further study in the field of NIALM.

7. Our Approach to Low-Frequency NIALM

In our first approach to NIALM systems, active power was measured with a frequency sampling of 1 Hz using a Christ CLM 1000PP ELEKTRONIK power sensor. Power samples were processed using hidden Markov models (HMM) adopted to NIALM, among others in References [19,20]. In our study, additive and differential factorial HMMs and an approximate inference algorithm (AFAMAP—additive factorial approximate MAP) were used. Cplexqp function from IBM ILOG CPLEX was used for optimization. The architecture of the proposed NIALM is presented in Figure 12. We performed experiments using many appliances. The results led to the following conclusions. Firstly, accurate identification is possible when the nominal power of the tested appliance is high enough (>40 W). Secondly, the difference in power draw of appliances must be noticeable. Examples of appliances for which this method operates properly are: electric kettle, light bulb, and microwave. We have recognized the following problems in the LF NIALM approach:

- low-power appliances, e.g., energy-saving light bulbs, were not recognized properly,
- the system was confused when two appliances with similar nominal power were tested,
- when variable power load was operating in the background, identification was inaccurate.
- appliances with variable power, e.g., washing machine, induced false event detection as the result of rough current and power draw.

For the reasons presented, we decided to design a universal laboratory that would be able to perform experiments with methods exploiting both current and voltage signals sampled with frequency modifiable in wide range.

Table 3. Groups of appliances possible to classify properly using selected medium-frequency (MF) methods.

Source Appliance	[53]	[48]	[54]	[55]	Our Study [56]
CFL (compact fluorescent lamp)	N.A.	N.A.	+	N.A.	+
Charger	+	N.A.	N.A.	N.A.	N.A.
Electronic	N.A.	+	N.A.	N.A.	N.A.
Fan	N.A.	N.A.	+	+	N.A.
Fluorescent light	+	N.A.	N.A.	N.A.	N.A.
Furnace	N.A.	+	N.A.	N.A.	N.A.
Hairdryer	N.A.	N.A.	N.A.	N.A.	+
Heater	N.A.	N.A.	+	N.A.	N.A.
Incandescent light	+	N.A.	N.A.	+	+
Iron	N.A.	N.A.	+	N.A.	+
Kettle	N.A.	N.A.	N.A.	+	+
Kitchen outlet	N.A.	+	N.A.	N.A.	N.A.
Laptop	N.A.	N.A.	+	N.A.	N.A.
LED	N.A.	N.A.	N.A.	N.A.	+
Microwave	N.A.	+	N.A.	N.A.	N.A.
PC	+	N.A.	N.A.	+	N.A.
Vacuum cleaner	N.A.	N.A.	N.A.	N.A.	+
Washing Machine	N.A.	+	N.A.	N.A.	N.A.
Number of +	4	5	5	4	7

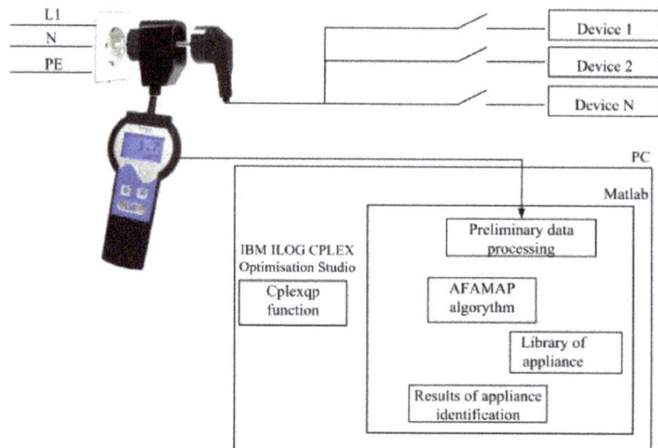

Figure 12. Low-frequency NIALM architecture [52].

8. Architecture of the NIALM Laboratory

As a part of our research, the NIALM laboratory was created. The creation was a process that took more than a year, and required a couple of different approaches, starting from the use of digital fault recorders (DFR) normally utilized in high-voltage networks for recording of currents and voltages appearing during various kinds of disturbances. At the beginning of our research, dual speed DFR (BEN6000) was used. It allows recording of instantaneous values of current and voltage signals with a sample rate up to 12 kHz. RMS values of currents, voltages, and their derivatives (active power, reactive power, power factor etc.) were recorded with a speed of 50 samples per second. Numerous tests performed led us to conclusion that there was a need for other types of recorders that would be able to record samples with frequencies of tenths of kHz and few MHz simultaneously. Therefore, we developed two recorders (NIALMREC), one (MF) with a sample rate up to a few hundred kHz, and the second (HF) operating with frequencies up to 10 MHz. Both devices, based on PC, DAQ cards, and the LabVIEW program, work according to our wishes. To have the possibility of performing tests in many locations, we decided to set up a complete measuring system consisting of two recorders in one movable 19 inch rack, equipped with many electrical sockets for plugging in household appliances.

Moreover, we also decided to have the same power supply environment and the same set of devices for testing purposes. Because of mentioned reasons, we set up our laboratory in a dedicated room located in Warsaw University of Technology, where the new supply circuits (electrical installation) were made according to our requirements. For the purpose of this stationary laboratory, the second measuring system, composed of two recorders (data acquisition hardware), was made and installed in a 19 inch rack. As the result, the laboratory consisted of both demonstrator and measurement system functions to verify the designed system model and automatic identification algorithms contained therein.

8.1. Electrical Installation

The architecture of the electrical installation is presented in Figure 13. Power supply in the form of three phase voltages (3 × 230 V/400 V), N and PE was connected to socket Z1. The main differential protection relay DI1 was then installed. One phase voltage (L1) was connected through switch S1 to the main line, supplying 24 single phase sockets (S1 to S24) via 24 dedicated switches (SW1 to SW24). Each Sx socket also had connections to N and PE. The current transducer model SCT-013-020 was used as a current sensor (T7 to T30). In serial to each Sx socket, one current sensor Tx allowing individual measurement of current was connected. In order to measure the summary currents in each phase, three current sensors (T4 to T6)—current transducer model SCT-013-020—were connected in serial to the main supply lines L1, L2, and L3. The MF measuring system could also measure three phase voltages using voltage sensors V31 to V33. The MF system also used a very flexible solution of measuring the channel switching matrix (PP) utilizing an Ethernet network STP (shielded twisted pair) patch panel, to which all voltages from the current sensors (signals: Is1, Is2, Is3, I1 to I24) and voltage sensors (signals: U1, U2, U3) were connected in the form of F/UTP cables. In each of those cables, only one twisted pair of wires (among four) was used to transfer the signal. The signals were then connected to the desired channels (U0 to U15) of the middle-frequency data acquisition system (DAQ MF) using an STP patch cord.

The HF measuring system consisted of dedicated current (T1, T2, T3) and voltage (HFV) sensors. Sensors' output voltage signals were connected by coaxial cables to the inputs of the high-frequency data acquisition system (DAQ HF).

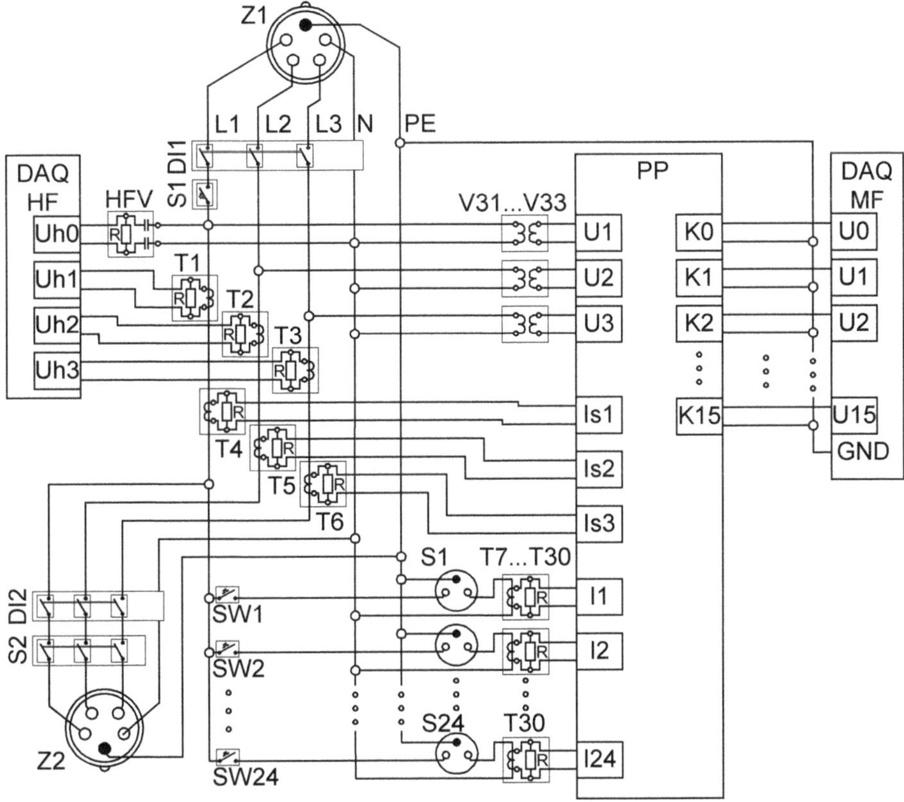

Figure 13. Architecture of the NIALM laboratory.

8.2. Data Acquisition Hardware

One of the assumptions made while developing the laboratory was to provide the possibility of testing methods based on medium-frequency (MF) and high-frequency (HF) measurements. For this reason, the data acquisition was performed in two paths using two PCs class Intel i5. The computers were equipped with fast SSDs for online data processing and large capacity HDDs for data storage. One computer was equipped with a 16 bit Advantech PCIE-1816H card, sampling in 16 channels up to 65 kS/s for MF measurements. In the second one, a 12 bit Advantech PCIE-1744AE card was installed. This Advantech DAQ card allowed sampling with frequencies up to 10 MS/s on two channels simultaneously, or up to 20 MS/s on a single channel for HF measurements. The NIALM high-frequency measurement setup equipped with a NIALM high-frequency sensor (NHFS) is presented in Figure 14a. Following the measurement setup presented in Reference [39], a high-pass filter (Figure 14b) was installed in the HF voltage measurement path to eliminate the fundamental component of high-amplitude voltage when steady-state (SS) characteristics of appliances are extracted. It should be noted that the PC case via the power outlet was connected to a protective conductor (PE). Consequently, the acquisition card connector armor providing the measurement signal was also connected to the protective conductor (Figure 14a). The result is that one of the branches of the filter was always shorted. The cut-off filter frequency in this system was 25 kHz. The filter consisted of three CR stages (Figure 14b). The values of elements were, accordingly: C1, C2, C3, C4, C5, C6 = 100 nF, R1 = 90 Ω, R2 = 600 Ω, R3 = 1000 Ω. In order to ensure the safety of the measurement system, high-voltage capacitors were used. For measurements of transient states (TS), the filter was removed.

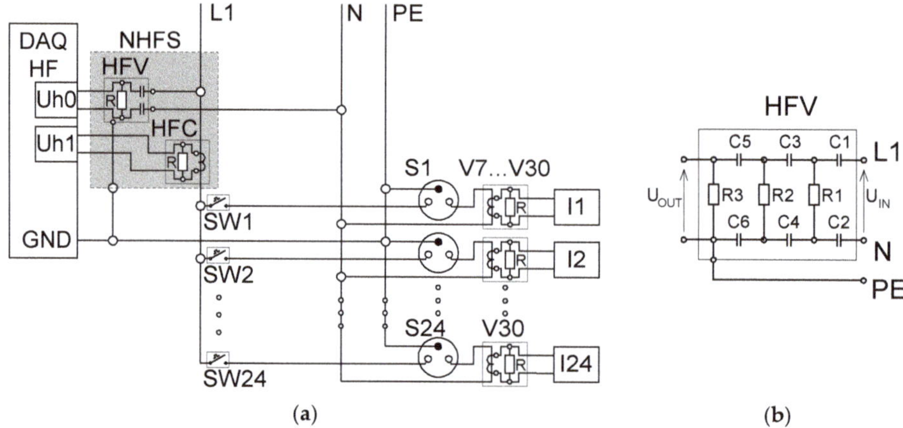

Figure 14. (**a**) Schematic diagram of the high-frequency voltage and current measurement setup; (**b**) schematic diagram of NIALM high-frequency voltage sensor (HFV).

8.3. Software Supporting Measurement Layer

Software supporting the acquisition cards was implemented in the LabVIEW environment. The basic function of the software was to save to the file buffers all samples obtained during acquisition from the current and voltage sensors. In the case of the MF card, it was possible to simultaneously verify the identification algorithms based on the energy consumption calculated from the current measurements for individual devices. The software included the following functions:

- Simultaneous acquisition of measurement data from at least two input channels configured to measure voltage and current,
- Presenting the waveform of the signal from several channels (virtual oscilloscope),
- Presenting Fourier spectra of two measured signals,
- The ability to adjust FFT analysis parameters,
- Writing to files all observed samples of signals as well as calculated spectra.

The main window of the prepared software is presented in Figure 15.

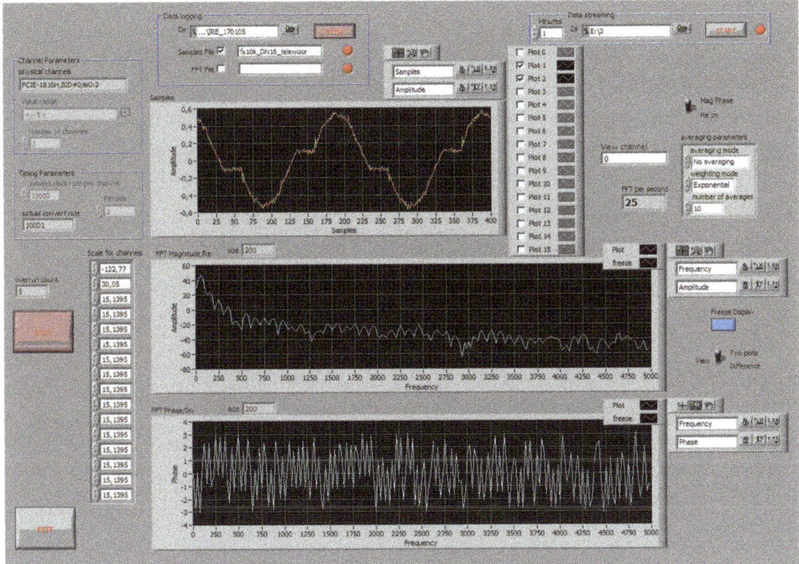

Figure 15. Main window of data acquisition software.

9. Medium-Frequency Characteristic Values

Using the prepared measurement setup, we were able to verify and develop the implemented algorithms for data processing and identification in an environment similar to a real place in which a NIALM system might be installed. In our NIALM system [Patent 1], presented in Figure 16, voltage and current were recorded with frequency sampling of 2 kHz.

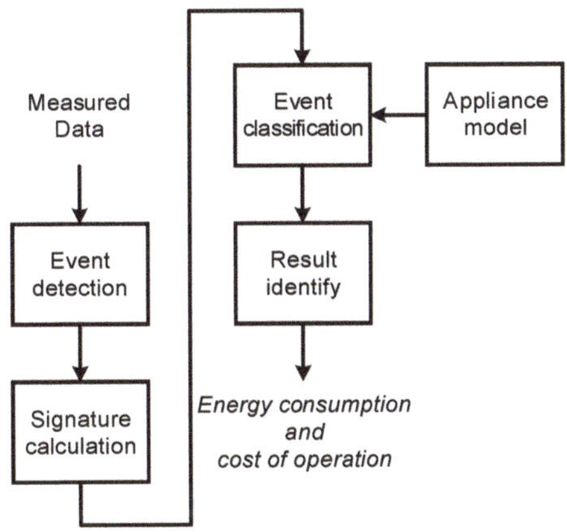

Figure 16. The proposed NIALM architecture.

Appliance models were developed during the training stage, when devices are turned on individually. Two waveforms representing 500 ms of current and voltage signals were processed to obtain a vector of parameters constituting the pattern (label) of an appliance. For better characterization of appliances, patterns were calculated one hundred times for each appliance. Therefore, the appliance model consists of many patterns obtained in similar operating conditions. During the experiments, we observed that device models based only on current and power parameters did not describe unambiguously electrical appliances, especially if the same power supply circuit supplied a number of different appliances. If we exploit only the AC current to determine characteristics of a device, we assume that the grid is an ideal voltage source, what is not truthful. Therefore, it was necessary to apply a method for determining parameters of electrical equipment which were not the features of the current flowing through it, because the current value depends on the supply voltage. Finally, we decided to use the following pattern, consisting of 133 parameters calculated on the basis of measured voltage and current signals (the brackets contain their numbers):

- Amplitudes of the first 16 harmonic components of the current (50 Hz, 100 Hz, etc.) (1–16),
- Phase shifts of the first 16 harmonic components of the current (17–32),
- Root mean square (RMS) of the current (33),
- Mean value of the current (34),
- Maximum current value (35),
- DC component in the current (36),
- Mean power (37),
- Values of the active power in the first 16 harmonics (38–53),
- Values of the reactive power in the first 16 harmonics (54–69),
- The first 16 harmonics of the real current components (70–85),
- The first 16 harmonics of the imaginary current components (86–101),
- The first 16 harmonics of conductances (102–117),
- The first 16 harmonics of susceptances (118–133).

Selected features of some appliances are presented in Figures 17 and 18. Some appliances, like the hairdryer or vacuum cleaner, had unique harmonic content, allowing characteristic parameters to be distinguished even visually, while appliances with similar power and impedance like the electric kettle (2200 W) and iron (2200 W) differed slightly in the magnitude of the first current harmonic only (50 Hz). Another interesting issue was the content of conductance harmonics. Presented in Figure 18b, the graph for the incandescent lamp had approximately constant magnitudes of conductance harmonics because resistance content in its impedance was dominant. What is worth emphasizing is that the LED bulb had a visible increase in the conductance amplitude for higher harmonics, and, moreover, a negative magnitude of the 450 Hz harmonic. This negative conductance indicates that at this frequency, the LED bulb generates energy [Patent 2]. This indicator can be used also in recognizing appliances generating higher harmonics, and hence worsening quality parameters of the power supply voltage and thus electric energy quality.

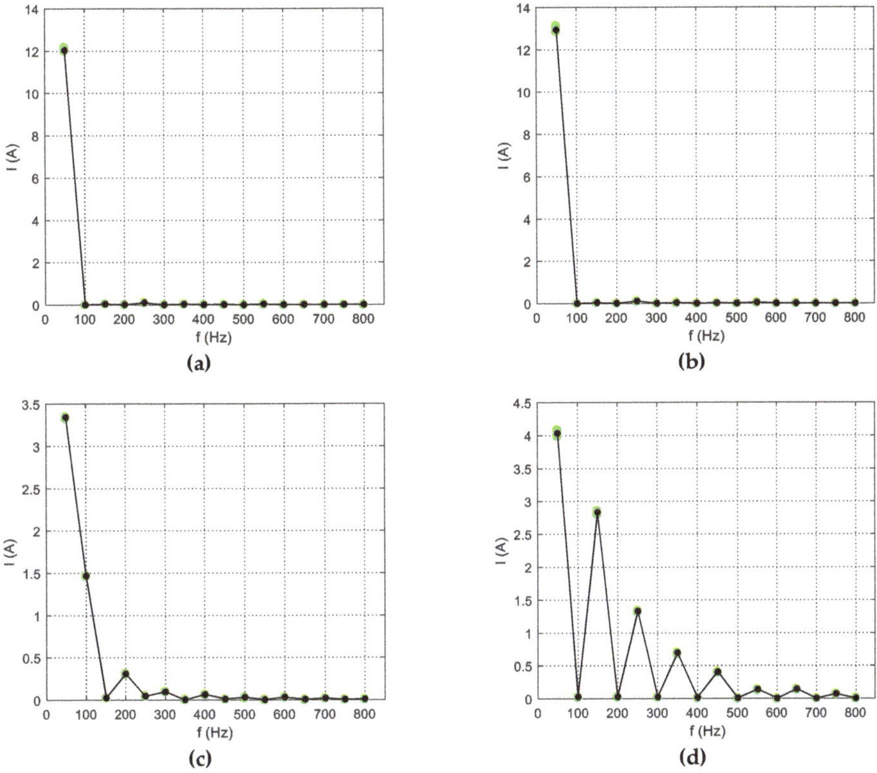

Figure 17. Current harmonic components (1–16) of (**a**) electric kettle; (**b**) iron; (**c**) hairdryer; (**d**) vacuum cleaner.

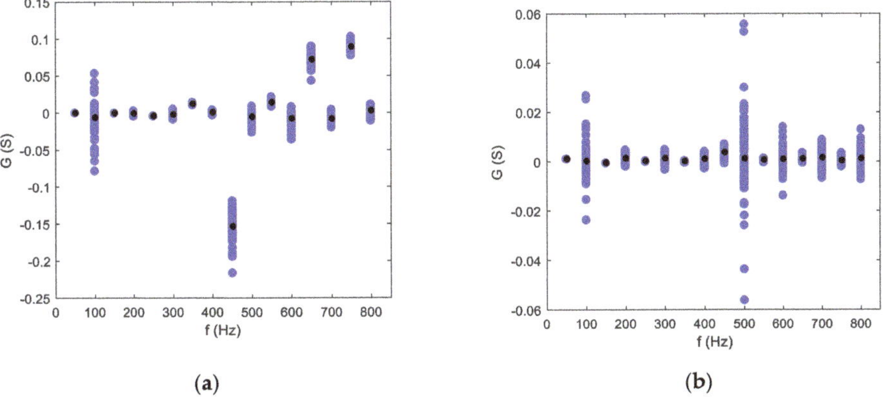

Figure 18. Conductance harmonics (102–117) of (**a**) LED bulb; (**b**) incandescent bulb.

Besides acquisition data for particular devices, we also recorded sequences of appliance operation. Overall current and voltage signals during the operation of multiple appliances switched on and off in different combinations imitated events in the real environment of system installation. Information about device states at specific times were attached to waveforms. This allowed verification of the correct operation of the identification algorithms. Sequences lasted from 5 to 100 min. Figure 19 presents an example of such a sequence.

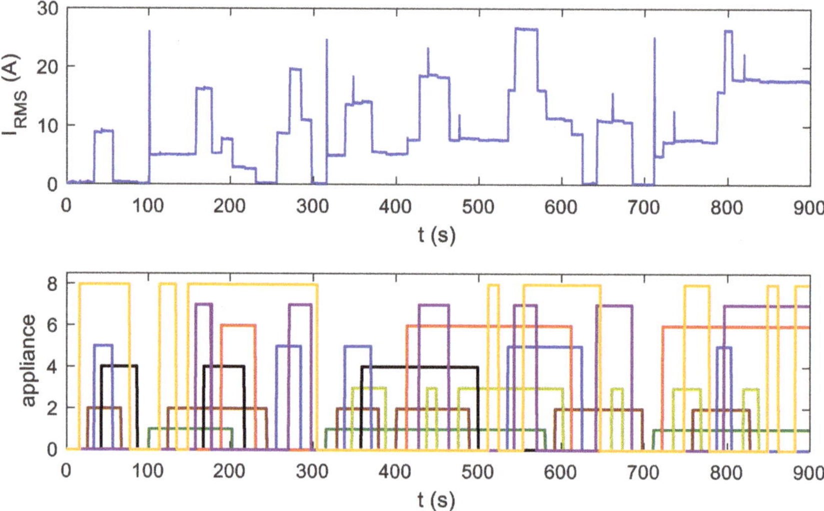

Figure 19. Testing sequences of appliance operation. Different colors in the bottom refer to different appliances.

10. Identification Results—MF Methods

The sets of recorded and processed data described above enabled us to perform a number of experiments aimed at verifying the quality of prepared patterns. We tested several artificial intelligence algorithms, adjusting their parameters in such a way as to obtain the best results for appliance identification. Moreover, we prepared a complete online NIALM system able to perform disaggregation automatically. Below, we briefly present the results of some experiments. In all experiments, the same six two state devices were considered: power-saving lightbulb, dryer, vacuum cleaner, mixer, juicer, and kettle. Table 4 lists the identification results for these experiments.

In experiments A and B, because of the feature additive criterion described in Reference [46], among others, non-additive features designed previously were removed from the set of features used for algorithm testing. Experiments C and D were performed to adjust the optimal parameters of Random Forest and Rule Induction algorithms [57]. Results listed in [4 were reached when Random Forest was trained with 13 trees, while optimal training configuration for the Rule Induction algorithm was obtained for $\alpha = 10$ and $\beta = 0.05$. The number of rules was equal to 21. Because there was no knowledge about parameter significance for the device identification, in experiment E, feature selection methods limiting the number of parameters to important ones only were applied. Therefore, two algorithms were exploited: one for reducing the set of features and the second for verifying the reduction efficiency. The classification accuracy was expected to be the same for both cases, but with a shorter training time for the second set. The applied Decision Tree classifier achieved an accuracy of 92.57% for the optimal set of features. In experiment F, a fusion of three classification algorithms was implemented. Fusion of three classifiers (Decision Tree, Rules Induction, and Random Forest) minimized the threat of the decision uncertainty when outcomes of the subsequent modules were equally distributed among different appliance identifiers. The overall ensemble achieved an overall accuracy of 92.96%.

Table 4. Overall identification accuracy for several algorithms.

Experiment	Identification Algorithm	Number of Events	Applied Features	Identification Accuracy
A	kNN	128	34, 36–133	85.69%
B	Random Forest	128	34, 36–133	85.69%
C	Rule Induction	84	1–69	65.47%
D	Random Forest	84	1–69	87.69%
E	Decision Tree	175	34, 36–39, 46, 49, 55, 56, 60, 70, 71, 74, 77, 79, 80, 87, 88, 92, 94	92.57%
F	Ensemble (DT, RI, RF)	128	1–69	92.96%

11. High-Frequency Experiments

The genesis of our experiments was the research presented in Reference [39]. The aim of the experiments was to record high-frequency components of a voltage signal, which are probably not the harmonics of the 50 Hz power supply network frequency, but electromagnetic inference generated by electrical appliances. We repeated the experiment carried out by American team using the measurement method proposed in their paper [39]. Although the fundamental harmonic of the voltage signal was filtered by analog high-pass filter, we were not able to perceive high-frequency voltage components characteristic for every tested appliance. We observed distinctive amplitudes at some frequencies only during operation of a few devices (e.g., an old type of LCD monitor). Components of all other devices were invisible because of high level of the high-frequency signals in the background. Therefore, we modified the measurement setup by introducing inductance into the electricity transmission line in order to separate interferences generated by other appliances powered by the same low-voltage grid. Houses in USA are powered from a single transformer, while, in a European low-voltage grid, one transformer supplies appliances in several houses, flats etc. Modification of measurement setup decreased the level of emissions in the background, enabling us to observe differences in the spectra obtained for different devices. Signals were processed using Wavelet Transform. An example of the obtained results is presented in Figure 20.

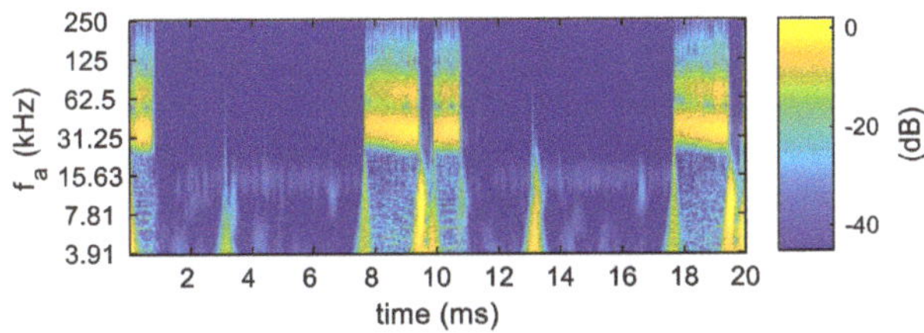

Figure 20. Scalogram of voltage signal during fluorescent lamp (CFL) and pedestal fan operation.

The conclusion drawn from the performed experiments was that high-frequency electromagnetic inferences generated by appliances may contain components characteristic for some appliances. Unfortunately, in a European power grid, it was impossible to observe them because of the high level of emissions in the background. Therefore, we conclude that the actual challenge is to detect and identify operation of low-power appliances like energy-saving lamps, LEDs, or switched-mode power suppliers.

Due to the limited volume of the article, we were not able to include the results of further experiments based on high-frequency signals recorded during electrical appliance operation. We developed methods of signal processing allowing different types of appliances to be distinguished using both SS high-frequency components [Patent 3] and TS high-frequency components [Patent 4] of voltage or current signals. Acquisition and processing of signals during appliance switching on allows precise characterization of some appliances, because the amplitudes of transient signals are significant.

However, steady states may be also a source of high-frequency features characterizing appliances, provided that the time of occurrence of the high-frequency oscillation is not lost. In both cases, it is necessary to perform signal analysis at a particular moment of time. That enforces the application of methods other than the previously applied time–frequency analysis. The introduced topic, along with a description of the measurement method and examples of our analysis results, will be presented in our next article.

12. Conclusions

The NIALM system may be an important instrument allowing improvement of effective use of energy by particular consumers. The NIALM system solutions proposed so far do not allow all types of devices to be identified in all operating conditions. In particular, there is a lack of experimental results in which the accuracy of particular appliances was evaluated. In our research, a multilateral approach [56] analyzing data obtained using various measurement setups was applied. We prepared a comprehensive NIALM laboratory adapted to perform various experiments. Tests based on a set of over 100 features and many methods of artificial intelligence were conducted. We achieved identification results similar to those presented by other authors. The performed experiments showed that the use of frequency analysis of EMI interference in appliance identification devices cannot be implemented in the European power grid.

We also analyzed high-frequency signals that appear in both steady and in transient states. The methods of time–frequency analysis that we will present in the next paper will show new features of devices that can be used to characterize electrical appliances. We have developed a method that uses high-frequency waveforms caused by introducing an artificial energy receiver for a short time period, in order to cause disturbance and measure the system's response to this disturbance. On the basis of this response, we can specify which appliance was currently switched on.

13. Patents

Patent 1: Kowalik R., Winiecki, W., Januszewski M., Nogal Ł., Łukaszewski R., Bilski P., Bobiński P., "Method and the device for identification of electrical energy receivers", PL patent no. PAT.229626, 12.03.2018.

Patent 2: Winiecki W., Kowalik R., Łukaszewski R., Bilski P., Januszewski, M., Nogal Ł., Wójcik A., "Method and the system for identification of a source of harmonic distortions of power network, preferably caused by energy receivers", PL patent no. PAT.229627, 12.03.2018

Patent 3: Kowalik R., Winiecki W., Dowalla K., Łukaszewski R., Wójcik A., Bilski P., Nogal Ł., "Device for the identification of the electrical appliances in the power grid and the method of the appliances identification in the power grid", PL patent no. PAT.232306, 6.02.2019.

Patent 4: Winiecki W., Kowalik R., Wójcik A., Łukaszewski R., Dowalla K., Bilski P., "Device for detecting changes of the operation mode and identification of appliances in the power grid and the method for detecting changes in the operation mode of appliances in the power grid", PL patent no. PAT.232305, 4.02.2019

Author Contributions: Conceptualization, W.W. and R.Ł.; Data curation, A.W.; Formal analysis, R.K.; Funding acquisition, R.Ł. and W.W.; Investigation, A.W. and R.K.; Methodology, R.K.; Project administration, W.W.; Resources, A.W. and R.Ł; Visualization, A.W.; Writing -original draft, A.W. and R.Ł.; Writing - review & editing, W.W. and R.K.

Funding: This research was funded by statutory grant funded by Warsaw University of Technology, Poland.

Conflicts of Interest: The authors declare no conflict of interest.

References

1. Shaw, S.R.; Leeb, S.B.; Norford, L.K.; Cox, R.W. Nonintrusive Load Monitoring and Diagnostics in Power Systems. *IEEE Trans. Instrum. Meas.* **2008**, *57*, 1445–1454. [CrossRef]

2. Liszewski, K.; Łukaszewski, R.; Kowalik, R.; Łukasz, N.; Winiecki, W. Different appliance identification methods in Non-Intrusive Appliance Load Monitoring. In *Advanced Data Acquisition and Intelligent Data Processing*; Haasz, V., Madani, K., Eds.; River Publishers: Aalborg, Denmark, 2014; pp. 31–58.
3. Hart, G.W. Nonintrusive Appliance Load Monitoring. *Proc. IEEE* **1992**, *80*, 1870–1891. [CrossRef]
4. Tang, G.; Wu, K.; Lei, J. A Distributed and Scalable Approach to Semi-Intrusive Load Monitoring. *IEEE Trans. Parallel Distrib. Syst.* **2016**, *27*, 1553–1565. [CrossRef]
5. He, D.; Lin, W.; Liu, N.; Harley, R.G.; Habetler, T.G. Incorporating non-intrusive load monitoring into building level demand response. *IEEE Trans. Smart Grid* **2013**, *4*, 1870–1877.
6. Zeifman, M.; Roth, K. Nonintrusive appliance load monitoring: Review and outlook. *IEEE Trans. Consum. Electron.* **2011**, *57*, 76–84. [CrossRef]
7. Najafi, B.; Moaveninejad, S.; Rinaldi, F. Data Analytics for Energy Disaggregation: Methods and Applications. In *Big Data Application in Power Systems*; Arghandeh, Reza, Zhou, Yuxun, Eds.; Elsevier Science: Amsterdam, The Netherlands, 2018; pp. 377–408.
8. Liu, B.; Luan, W.; Yu, Y. Dynamic time warping based non-intrusive load transient identification. *Appl. Energy* **2017**, *195*, 634–645. [CrossRef]
9. Cardenas, A.; Agbossou, K.; Guzmán, C. Development of real-time admittance analysis system for residential load monitoring. In Proceedings of the 2016 IEEE 25th International Symposium on Industrial Electronics (ISIE), Santa Clara, CA, USA, 8–10 June 2016; pp. 696–701.
10. Nguyen, M.; Alshareef, S.; Gilani, A.; Morsi, W.G. A Novel Feature Extraction and Classification Algorithm Based on Power Components Using Single-point Monitoring for NILM. In Proceedings of the IEEE 28th Canadian Conference on Electrical and Computer Engineering (CCECE), Halifax, NS, Canada, 3–6 May 2015; pp. 37–40.
11. Baranski, M.; Voss, J. Detecting patterns of appliances from total load data using a dynamic programming approach. In Proceedings of the Fourth IEEE International Conference on Data Mining (ICDM'04), Brighton, UK, 1–4 November 2004; pp. 327–330.
12. Wang, Z.; Zheng, G. Residential appliances identification and monitoring by a nonintrusive method. *IEEE Trans. Smart Grid* **2012**, *3*, 80–92. [CrossRef]
13. Zhao, B.; Stankovic, L.; Stankovic, V. On a Training-Less Solution for Non-Intrusive Appliance Load Monitoring Using Graph Signal Processing. *IEEE Access* **2016**, *4*, 1784–1799. [CrossRef]
14. Li, D.; Dick, S. Whole-house Non-Intrusive Appliance Load Monitoring via multi-label classification. In Proceedings of the 2016 International Joint Conference on Neural Networks (IJCNN), Vancouver, BC, Canada, 24–29 July 2016; pp. 2749–2755.
15. Henao, N.; Agbossou, K.; Kelouwani, S.; Dube, Y.; Fournier, M. Approach in Nonintrusive Type I Load Monitoring Using Subtractive Clustering. *IEEE Trans. Smart Grid* **2017**, *8*, 812–821. [CrossRef]
16. Kong, W.; Dong, Z.; Xu, Y.; Hill, D. An Enhanced Bootstrap Filtering Method for Non- Intrusive Load Monitoring. In Proceedings of the 2016 IEEE Power and Energy Society General Meeting (PESGM), Boston, MA, USA, 2016, 17–21 July; pp. 1–5.
17. Ducange, P.; Marcelloni, F.; Antonelli, M. A novel approach based on finite-state machines with fuzzy transitions for nonintrusive home appliance monitoring. *IEEE Trans. Ind. Informatics* **2014**, *10*, 1185–1197. [CrossRef]
18. Sultanem, F. Using appliance signatures for monitoring residential loads at meter panel level. *IEEE Trans. Power Deliv.* **1991**, *6*, 1380–1385. [CrossRef]
19. Kolter, Z.; Jaakkola, T.; Kolter, J.Z. Approximate Inference in Additive Factorial HMMs with Application to Energy Disaggregation. In Proceedings of the Fifteenth International Conference on Artificial Intelligence and Statistics (AISTATS), PMLR 22:1472-1482, La Palma, Canary Islands, 21–23 April 2012; pp. 1472–1482.
20. Makonin, S.; Popowich, F.; Bajic, I.V.; Gill, B.; Bartram, L. Exploiting HMM Sparsity to Perform Online Real-Time Nonintrusive Load Monitoring. *IEEE Trans. Smart Grid* **2016**, *7*, 2575–2585. [CrossRef]
21. Bonfigli, R.; Felicetti, A.; Principi, E.; Fagiani, M.; Squartini, S.; Piazza, F. Denoising autoencoders for Non-Intrusive Load Monitoring: Improvements and comparative evaluation. *Energy Build.* **2018**, *158*, 1461–1474. [CrossRef]
22. Kong, W.; Dong, Z.Y.; Hill, D.J.; Luo, F.; Xu, Y. Improving Nonintrusive Load Monitoring Efficiency via a Hybrid Programing Method. *IEEE Trans. Ind. Informatics* **2016**, *12*, 2148–2157. [CrossRef]

23. Aiad, M.; Lee, P.H. Unsupervised approach for load disaggregation with devices interactions. *Energy Build.* **2016**, *116*, 96–103. [CrossRef]
24. Shao, H.; Tech, V.; Marwah, M. A Temporal Motif Mining Approach to Unsupervised Energy Disaggregation. In Proceedings of the 1st International Workshop on Non-Intrusive Load Monitoring, Pittsburgh, PA, USA, 7 May 2012; pp. 1–2.
25. He, K.; Stankovic, L.; Liao, J.; Stankovic, V. Non-Intrusive Load Disaggregation using Graph Signal Processing. *IEEE Trans. Smart Grid* **2016**, *9*, 1739–1747. [CrossRef]
26. Wójcik, A.; Winiecki, W. The method of identification operating states of multi-state electrical devices with complex modes of operation. *Przegląd Elektrotech.* **2016**, *92*, 87–90. [CrossRef]
27. Reinhardt, A.; Burkhardt, D.; Zaheer, M.; Steinmetz, R. Electric appliance classification based on distributed high resolution current sensing. In Proceedings of the 37th Annual IEEE Conference on Local Computer Networks-Workshops, Clearwater, FL, USA, 22–25 October 2012; pp. 999–1005.
28. Yoshimoto, K.; Nakano, Y.; Amano, Y.; Kermanshahi, B. Non-Intrusive Appliances Load Monitoring System Using Neural Networks. *Inf. Electron. Technol.* **2000**, *2*, 183–194.
29. Agyeman, K.A.; Han, S.; Han, S. Real-time recognition non-intrusive electrical appliance monitoring algorithm for a residential building energy management system. *Energies* **2015**, *8*, 9029–9048. [CrossRef]
30. Srinivasan, D.; Ng, W.S.; Liew, A.C. Neural-network-based signature recognition for harmonic source identification. *IEEE Trans. Power Deliv.* **2006**, *21*, 398–405. [CrossRef]
31. Bouhouras, A.S.; Milioudis, A.N.; Labridis, D.P. Development of distinct load signatures for higher efficiency of NILM algorithms. *Electr. Power Syst. Res.* **2014**, *117*, 163–171. [CrossRef]
32. Wichakool, W.; Avestruz, A.T.; Cox, R.W.; Leeb, S.B. Modeling and estimating current harmonics of variable electronic loads. *IEEE Trans. Power Electron.* **2009**, *24*, 2803–2811. [CrossRef]
33. Lam, H.Y.; Fung, G.S.K.; Lee, W.K. A novel method to construct taxonomy electrical appliances based on load signatures. *IEEE Trans. Consum. Electron.* **2007**, *53*, 653–660. [CrossRef]
34. De Baets, L.; Ruyssinck, J.; Develder, C.; Dhaene, T.; Deschrijver, D. Appliance classification using VI trajectories and convolutional neural networks. *Energy Build.* **2018**, *158*, 32–36. [CrossRef]
35. Duarte, C.; Delmar, P.; Barner, K.; Goossen, K. A signal acquisition system for non-intrusive load monitoring of residential electrical loads based on switching transient voltages. In Proceedings of the 2015 Clemson University Power Systems Conference (PSC), Clemson, SC, USA, 10–13 March 2015.
36. Meziane, M.N.; Ravier, P.; Lamarque, G.; Abed-meraim, K.; Le Bunetel, J.; Raingeaud, Y.; Blois, D. Modeling and Estimation of Transient Current Signals. In Proceedings of the 23rd European Signal Processing Conference, Nice, France, 31 August–4 September 2015; pp. 1960–1964.
37. Zeifman, M.; Akers, C.; Roth, K. Nonintrusive monitoring of miscellaneous and electronic loads. In Proceedings of the 2015 IEEE International Conference on Consumer Electronics (ICCE), Las Vegas, NV, USA, 9–12 January 2015; pp. 305–308.
38. Duarte, C.; Delmar, P.; Goossen, K.W.; Barner, K.; Gomez-Luna, E. Non-intrusive load monitoring based on switching voltage transients and wavelet transforms. In Proceedings of the 2012 Future of Instrumentation International Workshop, Gatlinburg, TN, USA, 8–9 October 2012; pp. 101–104.
39. Gupta, S.; Reynolds, M.S.; Patel, S.N. ElectriSense: single-point sensing using EMI for electrical event detection and classification in the home. In Proceedings of the 12th ACM international conference on Ubiquitous computing, Copenhagen, Denmark, 26–29 September 2010; pp. 139–148.
40. Chen, K.Y.; Gupta, S.; Larson, E.C.; Patel, S. DOSE: Detecting user-driven operating states of electronic devices from a single sensing point. In Proceedings of the 2015 IEEE International Conference on Pervasive Computing and Communications (PerCom), St. Louis, MO, USA, 22–27 March 2015; pp. 46–54.
41. Gulati, M.; Singh, V.K.; Agarwal, S.K.; Bohara, V.A. Appliance activity recognition using radio frequency interference emissions. *IEEE Sens. J.* **2016**, *16*, 6197–6204. [CrossRef]
42. Kulkarni, A.S.; Harnett, C.K.; Welch, K.C. EMF signature for appliance classification. *IEEE Sens. J.* **2015**, *15*, 3573–3581. [CrossRef]
43. Pathak, N.; Khan, M.A.A.H.; Roy, N. Acoustic based appliance state identifications for fine-grained energy analytics. In Proceedings of the 2015 IEEE International Conference on Pervasive Computing and Communications (PerCom), St. Louis, MO, USA, 22–27 March 2015; pp. 63–70.
44. Fawcett, T. An introduction to ROC analysis. *Pattern Recognit. Lett.* **2006**, *27*, 861–874. [CrossRef]

45. Kim, H.; Marwah, M.; Arlitt, M.; Lyon, G.; Han, J. Unsupervised Disaggregation of Low Frequency Power Measurements. In Proceedings of the 2011 SIAM international conference on data mining, Mesa, AZ, USA, 28–30 April 2011; pp. 747–758.
46. Liang, J.; Ng, S.K.K.; Kendall, G.; Cheng, J.W.M. Load Signature Study—Part I: Basic Concept, Structure and Methodology. *IEEE Trans. Power Deliv.* **2010**, *25*, 6551. [CrossRef]
47. Makonin, S.; Popowich, F. Nonintrusive load monitoring (NILM) performance evaluation. *Energy Effic.* **2014**, *8*, 809–814. [CrossRef]
48. Nguyen, T.K.; Dekneuvel, E.; Jacquemod, G.; Nicolle, B.; Zammit, O.; Nguyen, V.C. Development of a real-time non-intrusive appliance load monitoring system: An application level model. *Int. J. Electr. Power Energy Syst.* **2017**, *90*, 168–180. [CrossRef]
49. Singhal, V.; Maggu, J.; Majumdar, A. Simultaneous Detection of Multiple Appliances from Smart-meter Measurements via Multi-Label Consistent Deep Dictionary Learning and Deep Transform Learning. *IEEE Trans. Smart Grid* **2019**, *10*, 2969–2978. [CrossRef]
50. Bhotto, M.Z.A.; Makonin, S.; Bajić, I.V. Load Disaggregation Based on Aided Linear Integer Programming. *IEEE Trans. Circuits Syst. II Express Briefs* **2017**, *64*, 792–796. [CrossRef]
51. Egarter, D.; Bhuvana, V.P.; Elmenreich, W. PALDi: Online load disaggregation via particle filtering. *IEEE Trans. Instrum. Meas.* **2015**, *64*, 467–477. [CrossRef]
52. Lukaszewski, R.; Liszewski, K.; Winiecki, W. Methods of electrical appliances identification in systems monitoring electrical energy consumption. In Proceedings of the IEEE 7th International Conference on Intelligent Data Acquisition and Advanced Computing Systems (IDAACS), Berlin, Germany, 12–14 September 2013.
53. Gillis, J.M.; Member, S.; Morsi, W.G. Non-Intrusive Load Monitoring Using Semi-Supervised Machine Learning and Wavelet Design. *IEEE Trans. Smart Grid* **2017**, *8*, 2648–2655. [CrossRef]
54. Dhananjaya, W.A.K.; Rathnayake, R.M.M.R.; Samarathunga, S.C.J.; Senanayake, C.L.; Wickramarachchi, N. Appliance-level demand identification through signature analysis. In Proceedings of the 2015 Moratuwa Engineering Research Conference (MERCon), Moratuwa, Sri Lanka, 7–8 April 2015; pp. 70–75.
55. Jiang, L.; Luo, S.; Li, J. Automatic power load event detection and appliance classification based on power harmonic features in nonintrusive appliance load monitoring. In Proceedings of the IEEE 8th Conference on Industrial Electronics and Applications (ICIEA), Melborne, Australia, 19–21 June 2013; pp. 1083–1088.
56. Bilski, P.; Winiecki, W. Generalized Algorithm for Non-intrusive Identification of Electrical Appliances in the Household. In Proceedings of the 9th IEEE International Conference on Intelligent Data Acquisition and Advanced Computing Systems: Technology and Applications (IDAACS), Bucharest, Romania, 21–23 September 2017; pp. 730–735.
57. Bilski, P.; Winiecki, W. The rule-based method for the non-intrusive electrical appliances identification. In Proceedings of the 8th International Conference on Intelligent Data Acquisition and Advanced Computing Systems: Technology and Applications (IDAACS, Warsaw, Poland, 24–26 September 2015.

Article

eHomeSeniors Dataset: An Infrared Thermal Sensor Dataset for Automatic Fall Detection Research

Fabián Riquelme [1,2,*], Cristina Espinoza [3], Tomás Rodenas [1], Jean-Gabriel Minonzio [1,2] and Carla Taramasco [1,2,*]

1 Escuela de Ingeniería Civil Informática, Universidad de Valparaíso, Valparaíso 2340000, Chile; tomas.rodenas@uv.cl (T.R.); jean-gabriel.minonzio@uv.cl (J.-G.M.)
2 Centro de Investigación y Desarrollo en Ingeniería en Salud, Universidad de Valparaíso, Valparaíso 2340000, Chile
3 Independent Researcher, Valparaíso 2340000, Chile; cristina.espinozalai@gmail.com
* Correspondence: fabian.riquelme@uv.cl (F.R.); carla.taramasco@uv.cl (C.T.)

Received: 29 July 2019; Accepted: 25 September 2019; Published: 21 October 2019

Abstract: Automatic fall detection is a very active research area, which has grown explosively since the 2010s, especially focused on elderly care. Rapid detection of falls favors early awareness from the injured person, reducing a series of negative consequences in the health of the elderly. Currently, there are several fall detection systems (FDSs), mostly based on predictive and machine-learning approaches. These algorithms are based on different data sources, such as wearable devices, ambient-based sensors, or vision/camera-based approaches. While wearable devices like inertial measurement units (IMUs) and smartphones entail a dependence on their use, most image-based devices like Kinect sensors generate video recordings, which may affect the privacy of the user. Regardless of the device used, most of these FDSs have been tested only in controlled laboratory environments, and there are still no mass commercial FDS. The latter is partly due to the impossibility of counting, for ethical reasons, with datasets generated by falls of real older adults. All public datasets generated in laboratory are performed by young people, without considering the differences in acceleration and falling features of older adults. Given the above, this article presents the eHomeSeniors dataset, a new public dataset which is innovative in at least three aspects: first, it collects data from two different privacy-friendly infrared thermal sensors; second, it is constructed by two types of volunteers: normal young people (as usual) and performing artists, with the latter group assisted by a physiotherapist to emulate the real fall conditions of older adults; and third, the types of falls selected are the result of a thorough literature review.

Keywords: fall detection; public dataset; thermal sensor; infrared sensor; smart home

---

1. Introduction

The continually aging population worldwide [1] represents a huge challenge for the care and prevention systems of accidents within the home, especially for the elderly living alone. A permanent risk in older people are falls [2] (In specialized literature, it is indicated that, on average, about one third of adults over 65 suffer a fall a year. Actually, although we know that falls are very frequent in older adults, after looking for the origin of this sentence, we have not been able to arrive at a concrete and updated reference where this is proven). The risk of falls and their negative effects on health increase with age. A study of 110 adults older than 90 years showed that only one half who fall are capable of getting up on their own [3]. Falls can lead to various health problems in the short and long terms, such as fractures [4], carpet burns, dehydration, hypothermia, pneumonia [3], volume depletion, internal infections and bleeding, cellulitis, ulcers, chest pain, syncope, heart attacks, and even death [5].

From a psychological point of view, many elderly people after falling develop a fear of falling again, which limits their daily activities [6].

Due to the above, during the last ten years, various fall-detection systems (FDSs) have been developed, both for the detection and early assistance of falls for the elderly and for the prevention and prediction of falls in their activities of daily living (ADL) [7]. FDSs are computational algorithms usually based on either a predictive or a machine-learning approach. Therefore, they require a training dataset, which allows them to differentiate a real fall from normal activities out of risk, such as walking, standing, sitting, etc.

The main ways to collect fall datasets are wearable devices and ambient-based sensors. Table 1 illustrates the main positive and negative aspects of each type of device [8,9]. Among the different ambient-based sensors, infrared thermal sensors allow to capture data even during no-light conditions. Moreover, some studies have concluded that it is easier to analyze thermal rather than normal images [8]. For data analysis, the collection of quality data is often a costly problem. In the context of elderly falls, there are additional ethical issues, the most critical of which is that one cannot ask an old person to fall voluntarily due to the high risk of injury. As we shall see in Section 2, since 2008, some public datasets on falls have being published to use as benchmarking and training of new FDS. This has undoubtedly been a great help for research in the area. However, fall datasets still present some general deficiencies:

1. Fall datasets are still few, as we will see in Section 2.
2. Due to the ethical problems mentioned above, the datasets do not include elderly falls but falls of healthy young volunteers, who fall differently compared to older adults. The most noticeable difference is that young people fall with a greater acceleration than the elderly [9]. Other kinesthetic differences will be described in more detail in Section 3.2. Because of this, the performance of many algorithms could drastically decrease by changing their laboratory environment to that of a real environment (i.e., the elderly home).
3. Many fall datasets are based on acceleration data, which has been shown to be insufficient on its own as predictors of falls. In fact, it has been proven that FDSs based on acceleration amplitude produce a large number of false alerts unless post-fall posture identification is also considered [10].
4. Although datasets based on video recordings often use low-resolution images (e.g., depth images with 320×240 resolution from Kinect sensors), these resolutions still allow for the identification of certain characteristics of people (e.g., height, texture, and gender) and, therefore, present privacy problems.
5. Finally, there is no standardized format for presenting fall data. This makes it difficult to use different datasets for application development.

This article presents a new public dataset, which is innovative in at least three aspects. First, it collects data from two different privacy-friendly infrared thermal sensors, with a very low resolution. The low-cost sensors used for this purpose are an Omron D6T-8L-06 and a Melexis MLX90640. Both sensors can be purchased commercially at an approximate value of $52 and $49 dollars, respectively. As far as we know, these sensors have only been used for the detection of falls in older adults [11], but other investigations based on similar sensors have also been carried out [12,13]. Second, it is constructed by two types of volunteers: normal young people (as usual) and performing artists, with the latter group assisted by a physiotherapist to emulate the real fall conditions of older adults. Finally, the types of falls are selected as a result of a thorough literature review. Note that the dataset is limited to the case of one person.

Table 1. The main types of devices that collect fall datasets and their positive and negative aspects.

Devices	Examples	Type of Data	Positive	Negative
Wereable devices	smartwatch, smartphone (compass, accelerometer, magnetometer, and gyroscope), inertial measurement unit (IMU), and EEG	acceleration, orientation data, rotation data, angular velocity, magnetic signal, and brain electrical activity	privacy-friendly, rich data, and highly accurate performance	invasive and depends on both the user's memory and abilities to to use them all the time.
Ambient-based sensors	camera, Kinect sensor, infrared thermal sensor, and pressure sensor (on the floor),	low-resolution video, low-resolution image (RGB, depth, or skeleton data), and ambient light	noninvasive, user independence, and long battery life	intrusive (it depends on resolution and data quality); only suitable for closed spaces; noise from other objects, people, or pets.

The paper continues as follows. In Section 2, we review 18 public datasets on falls obtained since 2008 from ambient-based sensors. As far as we know, so far, this is the largest survey of datasets based on a vision/camera approach. This gives us an idea of the usual size of the datasets and their main characteristics. In Section 3, we describe the materials (i.e., the two different infrared thermal sensors) and methods used to build the dataset. Here, we also present the details of the new eHomeSeniors dataset. In Section 4, we describe a brief experiment that compares the data obtained by the two thermal sensors and the two types of volunteers. In Section 5, we discuss the main results, and in Section 6, we present the conclusions as well as some ideas of future work.

2. Related Work

Although public datasets on falls are still scarce, from 2008 onwards, more and more public datasets have appeared, as well as numerous surveys related to automatic fall detection. Only between 1998 and 2012, a systematic review on automatic FDS using body-worn sensors gathered 96 research papers [14]. In 2015, a survey collected five vision-based public datasets on falls [15], while in 2017, twelve wearable-based public datasets on falls were surveyed [16]. Additionally, only in 2019 have surveys about techniques for abnormal human activity recognition [17], healthcare monitoring systems for elderly people [18], and fall prediction with sensors in smart homes appeared [7].

Using Google Scholar, we collected all the citations of the 2015 survey [15] found until June 2019, as well as all citations to the corresponding datasets included in that survey. The search results were filtered with the keywords "public dataset" + "fall". From the results obtained, we selected only those publications that published new public datasets on falls obtained from ambient-based sensors. This search process was repeated for each new article found in this way, using a snowballing literature review approach. Thus, we found a total of 18 public datasets on falls based on ambient sensors, published between 2008 and 2019. In addition, it was found the YouTube Fall Dataset (YTFD), created in 2016, but until June 2019, it has not been already published online [19]. As far as we know, this is the largest collection that exists to date on this type of datasets. The results of this search are described in Table 2. Details of the falls collected, of the participants involved in the sample, and of the data collection system used for each case are included.

It is observed that all available fall data have been made by young adults in good health, falling according to their physiognomy and without emulating falls of an older adult. There are other fall datasets that are simply not intended for the fall detection of older adults. For example, in Reference [36], the authors use accelerometers to collect data of falls simulated by practitioners of the athletic discipline parkour. In general, it is also observed that most investigations construct datasets to be used as benchmarks in FDS based on traditional supervised techniques (e.g., threshold based and machine learning). Therefore, together with fall actions, several of these datasets also include data of activities of daily living, useful for training of their algorithms. These additional activities usually involve actions such as walking, sitting, standing, etc., and they do not imply additional technical or ethical difficulties. In fact, they are very simple data to generate and emulate automatically. That is why, in this article, we focus exclusively on the actions of falls.

Note that, on average, datasets include 121 falls of 4 types (they may be different from each other) made by 12 volunteers. The median is 60 falls, 4 fall types, and 10 volunteers. Among the ambient-based devices used, the most common are Kinect sensors (9 datasets), followed by cameras (6 datasets), and infrared thermal sensors (4 datasets).

Table 2. Public datasets on falls obtained from ambient-based sensors.

Year	Dataset Name	ref.	Falls #	Falls #types	Participants #	Participants #F	Participants #M	Participants Age	Data Collection Systems
2019	UP-Fall	[20]	255	5	17	8	9	18–24	6 infrared sensors, 2 cameras (18 fps), 5 IMUs with accelerometer, gyroscope, ambient light, 1 EEG
2018	CMDFALL	[21]	400	8	50	20	30	21–40	7 overlapped Kinect sensors and 2 WAX3 wireless accelerometers
	FALL-UP	[22]	255	5	17	?	?	?	6 infrared sensors; 2 cameras; 1 EEG; 5 wearable inertial sensors on left ankle, right wrist, neck, waist, and right pocket with accelerometer, angular velocity, and luminosity
	UP-Fall	[23]	60	5	4	2	2	22–58	4 ambient infrared presence/absence sensors, 1 RaspberryPI3, 4 IMUs with accelerometer, ambient light, angular velocity, 1 EEG
2017	Dataset-D	[24]	95	2	4	?	?	30–40	4 Kinect sensors (RGB, depth, skeleton data; 20 fps, 640 × 480)
	MICAFALL-1	[24]	40	2	20	?	?	25–35	*idem*
	Thermal Simulated Fall	[8]	35	?	?	?	?	?	9 FLIR ONE thermal cameras (640 × 480) mounted to Android phone
2016	KUL Simulated Fall	[25]	55	?	10	?	?	?	5 web-cameras (12 fps, 640 × 480)
2015	–	[26]	21	4	?	?	?	?	IP camera (Dlink DCS-920) through Wi-Fi connection (MJPEG, 320 × 240)
	EDF	[15]	320	?	10	?	?	?	2 Kinect sensors (depth maps, 320 × 240, 30 fps)
2014	OCCU	[27]	60	1	5	?	?	?	*idem*
	SDU Fall	[28]	30	1	10	2–8	2–8	young	1 Kinect sensor
	TST	[29]	132	4	11	?	?	22–39	1 Kinect sensor (depth maps); 2 IMUs on waist and right wrist with accelerometer
	UR Fall	[30]	30	2	5	0	5	>26	2 Kinect sensors (depth maps); 1 IMU on waist (near the pelvis) with accelerometer
2013	Le2i fall	[31]	143	3	11	?	?	?	1 video camera in 4 different locations (25 fps, 320 × 240)
2012	Le2i fall	[32]	192	3	11	?	?	?	*idem*
	vlm1	[33]	26	?	6	?	?	?	2 Kinect sensors (RGB, depth; 10 fps, 320 × 240)
2008	Multi camera fall	[34,35]	22	8	1	0	1	adult	8 video cameras
		average	121	4	12				
		median	60	4	10				

In Table 3, we include the 30 types of falls used by these datasets for those cases in which more than one type of fall is specified. The classification of the first column was created for this work in order to better organize the different types of falls. In the same table, we also include 14 additional fall types chosen for the SisFall dataset [9]. This is a well-known dataset based on wearable devices. We include it here because it uses 15 types of falls, chosen from 41 types of falls used by another study [37], crossed with a survey of 15 seniors living alone and 17 administrative people from retirement homes. This article is the only one from the table that considers more detailed falls caused by fainting (syncope or falling asleep). On average, each dataset uses 6 types of falls, with a median of 5. The most commonly used types of falls are backward (from standing), lateral (from standing), backward when trying to sit down (empty chair), and forward (from standing).

Table 3. Classification of different types of falls considered in the literature.

Fall by	ID	Description	Reference [34]	[32]	[29]	[30]	[26]	[9]	[24]	[23]	[22]	[21]	[20]	#
general	F1	Fall (from standing)	✗	✗	✗	✓	✗	✗	✓	✗	✗	✗	✗	2
	F2	Backward (from standing)	✗	✗	✗	✓	✓	✗	✗	✓	✓	✗	✓	5
	F3	Forward (from standing)	✗	✓	✗	✓	✓	✗	✗	✗	✓	✗	✗	4
	F4	Lateral (from standing)	✗	✗	✗	✓	✓	✗	✗	✓	✓	✗	✓	5
	F5	Backward (from walking)	✗	✗	✗	✗	✗	✗	✗	✗	✗	✓	✗	1
	F6	Forward (from walking)	✗	✗	✗	✗	✗	✗	✗	✗	✗	✓	✗	1
	F7	Leftward (from walking)	✗	✗	✗	✗	✗	✗	✗	✗	✗	✓	✗	1
	F8	Rightward (from walking)	✗	✗	✗	✗	✗	✗	✗	✗	✗	✓	✗	1
cause	F9	Forward while walking caused by a slip	✗	✗	✗	✗	✗	✓	✗	✗	✗	✗	✗	1
	F10	Backward while walking caused by a slip	✗	✗	✗	✗	✗	✓	✗	✗	✗	✗	✗	1
	F11	Lateral while walking caused by a slip	✗	✗	✗	✗	✗	✓	✗	✗	✗	✗	✗	1
	F12	Forward while walking caused by a trip	✗	✗	✗	✗	✗	✓	✗	✗	✗	✗	✗	1
	F13	Forward while jogging caused by a trip	✗	✗	✗	✗	✗	✓	✗	✗	✗	✗	✗	1
	F14	Cause by fainting/syncope/loss of balance	✗	✓	✗	✗	✓	✗	✗	✗	✗	✗	✗	2
	F15	Vertical fall while walking, by fainting	✗	✗	✗	✗	✗	✓	✗	✗	✗	✗	✗	1
	F16	Forward while sitting, caused by fainting	✗	✗	✗	✗	✗	✓	✗	✗	✗	✗	✗	1
	F17	Backward while sitting, caused by fainting	✗	✗	✗	✗	✗	✓	✗	✗	✗	✗	✗	1
	F18	Lateral while sitting, caused by fainting	✗	✗	✗	✗	✗	✓	✗	✗	✗	✗	✗	1
	F19	Fall while walking caused by fainting (use of hands in a table to dampen fall)	✗	✗	✗	✗	✗	✓	✗	✗	✗	✗	✗	1
	F20	Forward when trying to get up	✗	✗	✗	✗	✗	✓	✗	✗	✗	✗	✗	1
	F21	Lateral when trying to get up	✗	✗	✗	✗	✗	✓	✗	✗	✗	✗	✗	1
	F22	Forward when trying to sit down	✗	✗	✗	✗	✗	✓	✗	✗	✗	✗	✗	1
	F23	Backward when trying to sit down	✓	✓	✗	✗	✗	✓	✗	✓	✗	✗	✓	5
	F24	Lateral when trying to sit down	✗	✗	✗	✗	✗	✓	✗	✗	✗	✗	✗	1
	F25	Leftward when trying to sit down	✗	✗	✗	✗	✗	✗	✗	✗	✗	✓	✗	1
	F26	Rightward when trying to sit down	✗	✗	✗	✗	✗	✗	✗	✗	✗	✓	✗	1
location	F27	On bed (then leftward)	✗	✗	✗	✗	✗	✗	✗	✗	✗	✓	✗	1
	F28	On bed (then rightward)	✗	✗	✗	✗	✗	✗	✗	✗	✗	✓	✗	1
	F29	From chair	✗	✗	✗	✓	✗	✗	✓	✗	✗	✗	✗	2
impact	F30	Fall (impact on hands and elbows)	✗	✗	✗	✗	✗	✗	✗	✓	✗	✗	✗	1
	F31	Forward (impact on hands and elbows)	✗	✗	✗	✗	✗	✗	✗	✗	✗	✗	✓	1
	F32	Forward (impact on knee)	✗	✗	✗	✗	✗	✗	✗	✓	✗	✗	✓	2
termination	F33	Backward (end up sitting)	✗	✗	✓	✗	✗	✗	✗	✗	✗	✗	✗	1
	F34	Backward (end up lying)	✗	✗	✓	✗	✗	✗	✗	✗	✗	✗	✗	1
	F35	Forward (end up lying)	✗	✗	✓	✗	✗	✗	✗	✗	✗	✗	✗	1
	F36	Lateral (end up lying)	✗	✗	✓	✗	✗	✗	✗	✗	✗	✗	✗	1
	F37	Forward on knees (stay down)	✗	✗	✗	✗	✗	✗	✗	✗	✓	✗	✗	1
articulation	F38	Fall (legs straight)	✓	✗	✗	✗	✗	✗	✗	✗	✗	✗	✗	1
	F39	Fall Backward (legs straight)	✓	✗	✗	✗	✗	✗	✗	✗	✗	✗	✗	1
	F40	Fall Forward (legs straight)	✓	✗	✗	✗	✗	✗	✗	✗	✗	✗	✗	1
	F41	Fall Leftward (legs straight)	✓	✗	✗	✗	✗	✗	✗	✗	✗	✗	✗	1
	F42	Fall Rightward (legs straight)	✓	✗	✗	✗	✗	✗	✗	✗	✗	✗	✗	1
	F43	Fall Fall (knee flexion)	✓	✗	✗	✗	✗	✗	✗	✗	✓	✗	✗	2
	F44	Fall Rightward (knee flexion)	✓	✗	✗	✗	✗	✗	✗	✗	✗	✗	✗	1
		# fall types	8	3	4	5	4	15	2	5	5	8	5	

3. Materials and Methods

In this section, the process carried out to build the eHomeSeniors dataset is described in detail. Section 3.1 describes the two different infrared thermal sensors used to collect data. Section 3.2 describes the methodological process for data collection, including the selection of sample size, number of fall types, and volunteers. Finally, Section 3.3 describes the dataset in detail, including its information for download and operation.

3.1. Data Collection Systems

The first sensor used in the dataset is a Melexis MLX90640 Far Infrared Thermal Sensor. It is a low-cost sensor that contains 768 FIR (Far Infrared) pixels and provides a privacy-friendly, low-resolution image of 32 × 24 pixels, with a frame rate of approx. 16 fps. It has an operational temperature range between −40 °C and 85 °C and can measure object temperatures between −40 °C and 300 °C. Figure 1 shows two example frames collected by this sensor, painted according to the temperature range of each pixel. Note that the image quality clearly distinguishes a fall, and at the same time, it fully conserves the privacy of the person. This sensor was fixed to a wall at a height of 1.2 m, so that the viewing angle is distributed equally from the center of the sensor, forming a vertical angle of 55° and a horizontal angle of 37.5° as shown in Figure 2.

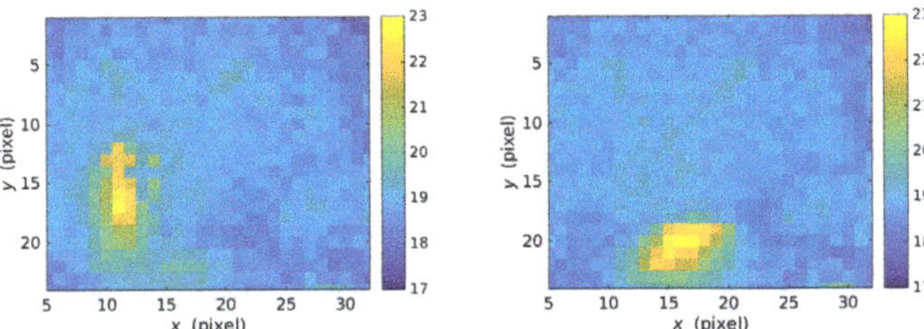

Figure 1. Heat maps of two 32 × 24 frames generated by the Melexis MLX90640 sensor for a standing body (left) and a fallen body (right).

The second sensor is simpler than the previous one, since the returned heat maps are distributed in linear arrays instead of two-dimensional arrays. It is an Omron D6T-8L-06 infrared thermal sensor. It provides a very low-resolution image of just 1 × 8 pixels. It has an operational temperature range between 0 °C and 50 °C and can measure object temperatures between −10 °C and 60 °C. In order to identify falls and to expand the opening range, we use a system of four of these sensors: two sensors at half height (1 m from the floor) and two sensors at ground level (10 cm from the floor). The four sensors are connected to an ATMEGA328P microcontroller, which reads sensor data and sends it to an ODROID-C1+ via UART interface with a baud rate of 115, 200 and a sample rate of 5 Hz. In this way, a fall is recognized as a decrease in the temperature identified by the upper sensors and an increase in the temperature of the lower sensors. The dataset collects data from this four-sensor system, with a frame rate of approx. 5 fps. This sensor system was previously used in preliminary laboratory experiments, obtaining 93% accuracy in fall detection for a neural network based on a bi-LSTM model (bidirectional long-term memory) [11]. Figure 3 illustrates the temperatures detected by this sensor system for a standing body (left) and a fallen body (right). Note that, when the body reaches the ground, the temperature is concentrated in the lower sensors. In this case, the image quality also conserves the privacy of the person. However, to interpret a fall here is necessary to analyze timestamps.

A schema of the data-collection environment with the two types of sensors is illustrated in Figure 2. Figure 4 shows the real environment where the experiments were performed.

3.2. Methodology Description

A fall is often seen as an abnormal movement of the activities of daily living (ADL) [38]. Therefore, to train FDS algorithms, datasets usually include both falls and ADLs. In this article, we have focused on the collection of falls, since the ADLs do not require a greater effort and are easily replicable. As a matter of fact, the data collected at the time before each fall can be considered a common ADL, such as standing, walking, sitting, or lying down. In addition, for any dataset based on video images, it is possible to increase the data using a combination of translation, repetition, and rotation effects. Since a fall can occur anywhere in the room, volunteers simulated falls at different distances between 1 and 5 m from the sensors. The room where the experiment was developed has 6 m × 5 m of space and had no furniture inside, except for a chair and a settee bed in the center used in some tests for simulating falls.

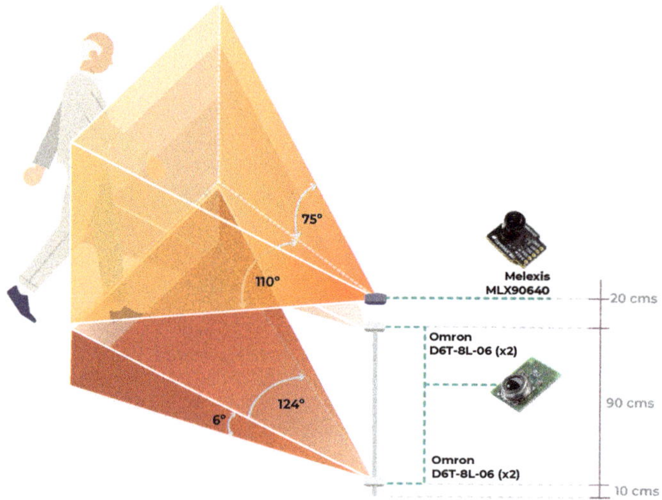

Figure 2. A schema of the data-collection environment with the two types of sensors used for the eHomeSeniors dataset.

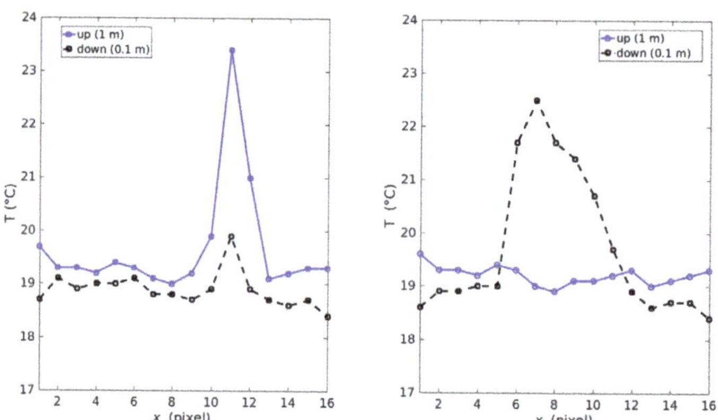

Figure 3. Temperatures detected by the Omron D6T-8L-06 sensor system for a standing body (left) and a fallen body (right): The continuous line corresponds to the 16 pixels placed at 1 meter from the floor and the dashed one corresponds to that at 0.1 m from the floor.

In order to choose the types of falls to be included in the dataset, Table 3 and expert knowledge of a physiotherapist with experience in elderly care were taken as the starting point. After a first review of the list, 25 types of falls were discarded (57% of the list): F1–F8, F20, F21, F33–F36, F38, and F43 for being too general; F13 for being very unlikely in the context of an older adult in a closed space; F24–F26 for being more unnatural at the kinesthetic level; F27 and F28 because they are much less risky than falls on the ground; and F30–F32 because the severity of the impact is irrelevant for the purpose of detecting falls (we must recognize them all, regardless of their severity). However, from these discarded fall types, three new types emerge, not considered in the table: "Backward (from walking backward)" (from F5), "Falling from bed" (from F27 and F28), and "Backward (from standing; knee flexion; slow)" (from F43). From the three, the first two have been considered in other studies, related to wearable devices [37,39], and the third one is a combination of "Backward (from standing; slow)" [40] and "Backward (knee flexion)" [37].

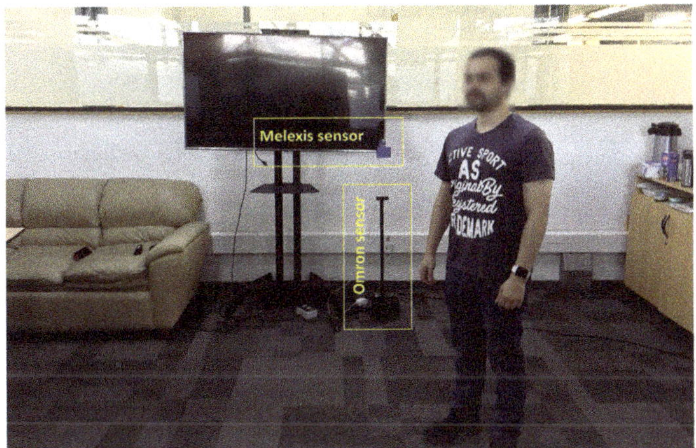

Figure 4. Laboratory where the experiments were performed.

To choose among the remaining 19 types of falls, a small pilot experiment was performed under laboratory conditions with the two infrared thermal sensors to observe the images generated by each type of fall. In this way, considering the image similarity between some types of falls, some of them were merged: F9 and F12; F22 and F23; F10 and F39; F11, F41, and F42; and F14, F15, F18, and F19. Finally, of the six remaining types of falls, F40 remained unchanged; F16, F17, F29, and F44 were adjusted; and F37 was divided into two (at normal speed and at slow speed, the latter being more typical in older adults [40]). Note that, although falls with stretched legs may seem forced for young and healthy people, they are very common in older adults with mobility problems. In conclusion, for the new dataset, the following 15 types of falls are considered:

1. Backward (from walking backward)
2. Forward while walking caused by a trip
3. Cause by fainting (slow lateral)
4. Backward when trying to sit down (empty chair)
5. From bed
6. Backward (legs straight)
7. Forward (legs straight)
8. Forward (knee flexion)
9. Backward (from standing; knee flexion, slow)
10. Forward (from standing; knee flexion; slow)
11. Lateral (from standing; legs straight)
12. Lateral (from standing; knee flexion, slow)
13. Cause by fainting or falling asleep (slow backward)
14. Cause by fainting or falling asleep (slow forward)
15. From chair, caused by fainting or falling asleep

Note that the number of fall types considered for this dataset equals the maximum number of those considered in Table 3 and exceeds diversity of all datasets based on ambient-based sensors summarized in Table 2.

Regarding the volunteers for the falls, we used two groups of three people each (see Table 4). Group 2 is made up of young and healthy people, as usual. Group 1, instead, is made up of three performing artists who work on a contemporary dance piece related to the concept of "falling". Group 2 did not receive any type of instruction. Only each type of fall they had to make were mentioned. Group 1 was assisted by the physiotherapist to instruct them about the differences in the way of falling for an older adult with respect to a young and healthy person.

Table 4. General volunteer features for the falls collection: Group 1 is formed by performing artists and group 2 is formed by normal, healthy, young people.

	#	Gender	Age	Weight	Height
	1	F	37	59	1.64
group 1	2	F	34	51	1.62
	3	M	35	62	1.80
	4	F	27	49	1.52
group 2	5	M	28	89	1.73
	6	M	29	66	1.65

Since falls occur due to a mismatch between an individual's physiological function, environmental requirements, and the individual's behaviour [41], we considered the differences of an older person in two of these three aspects to perform a likely elder's fall. In the physiological function, all senses are involved in maintaining an active attention required to prevent a fall, so sensory decline may result in impaired perception of environmental challenges [41,42]. The proprioception or awareness of where body parts are in space and the reaction time to respond to unexpected perturbations may be altered, changing the capacity to react. Also, muscle strength may be diminished or altered due to inactivity, which may lower the ability to extend the legs against gravity, making regaining an upright position in the case of a trip more difficult. Function of the various components of successful postural control can be adversely affected by physiological aging and low levels of appropriate physical activity due to disuse. Patients with osteopenia may have bone fractures or injuries as a result of low-energy trauma, typically a fall from standing height or less [43]. Furthermore, there is a disproportionately higher number of deaths in elderly compared to young people as a consequence of a same-level fall [44].

Even with standard bone density, the most common serious injury associated with the fall of an elderly person is a hip fracture, which is associated with up to 20% chance of death and 25% chance of long-term institutionalization [45]. Acute medical problems like infections, chronic conditions such as diabetes, and progressive conditions such as Parkinson's disease can also affect postural control/balance. There is an impact of medications on successful postural control, with psychoactive medications being particularly associated with falls. Another important aspect is cardiovascular and respiratory correct function, which ensures oxygen transport to the muscles and the brain to enable these functions to occur, and these can be also altered in elders due to disease or as an effect of aging. About environmental requirements, an older person with impaired physiology may fall in an unchallenging environment, which is considered in the way the fall was performed. Since individual's behaviours are specially subjective, we considered it indirect to the performance of falls in a laboratory environment and not measurable for this particular case.

Each volunteer made 5 falls of each type. From all of them, only the last two falls (volunteer 6, group 2) taken with the Omron sensors presented problems and had to be discarded. This is enough to obtain a total of $15 \times 5 \times 6 - 2 = 448$ falls in total, which makes it a larger fall dataset than all of those summarized in Table 2. Recall that the average between all the datasets summarized in Table 2 is 121 falls, with a median of 60.

3.3. eHomeSeniors Dataset Description

The dataset is publicly available (see the files in Supplementary Materials). It is made up of 180 files in `.csv` format, one for each fall type. The name of each file follows the form "`sensor_name-G`X`-`Y`-f`ZZ", where `sensor_name` is either `omron` or `melexis`; X is either 1 or 2, so that

GX represents the number of the group; Y is a number between 1 and 6, the number of the volunteer; and ZZ is a number between 1 and 15, so that fZZ represents the number of the fall type. Each file contains five falls, except `omron-G2-3-f15`, that only contains three.

The Omron sensor files are simpler. Each row contains 33 values separated by semicolons. The first value contains the date and time of collection of the data. Each of the following 16 values includes a decimal value, which represents the temperature collected by the pixels of the two upper Omron sensors, and the last 16 values contain the temperature collected by the lower sensors. Thus, each row represents a frame, which when visualized as a heat map constitutes a different moment of a fall.

The Melexis sensor files are a bit more complex. For each row, the first value contains the data collection time and the second value is information about the sensor model. The following 768 values contain the temperature of each of the pixels that make up a 32×24 pixel heat image. Finally, each row contains several additional data with the raw data from which the temperatures are obtained through formulas documented for the sensor.

Since the Melexis sensor files contain raw data in addition to temperatures, have more pixels, and also have more fps than the Omron sensor, the files are much heavier. The files of the Omron sensors total 8.18 MB, while that of the Melexis totals 802 MB. In addition to these files, the dataset also contains the same data in Matlab numeric matrix `.mat` format.

4. Experimental Results: Estimation of the Fall Duration

In order to investigate the differences between group 1 (artists mimicking elderly falls) and group 2 (young and healthy people), a preliminary heuristic approach has been tested. This approach focuses on the Melexis sensor and is divided in different steps. First of all, pixel positions associated with the volunteer were obtained by considering only those with temperatures above a threshold higher than the background average. The aim of this section is to propose a rather simple approach to verify if there is a statistically significant difference in the fall duration between the two groups.

Figure 5 illustrates the retained pixels of a volunteer in three frames during the realization of a fall. In this case, the values of the threshold and the background were equal to 21 °C and 18.9 ± 0.5 respectively. Then, the barycenter coordinates $[x(t), y(t)]$ of the retained pixels, corresponding to the median position in both direction, was calculated for each time frame. Note that the median was preferred to the average in order to reduce the influence of possible isolated caloric pixels not eliminated in the previous step. It can be observed how the barycenter, indicated with a circle on Figure 5, moved to the right and decreased along the vertical axis. The next step consisted in the smoothing of the barycenter temporal trajectory using a 10-time-step averaging moving window. The final step corresponded to the actual fall detection associated to negative time derivation of the filtered vertical position, i.e., $dy/dt < 0$. Note that falls smaller than 2 pixels were removed. Finally, the fall duration corresponds to the number of successive temporal frames, associated with negative derivation, multiplied by the sampling times, i.e., $1/16$ s.

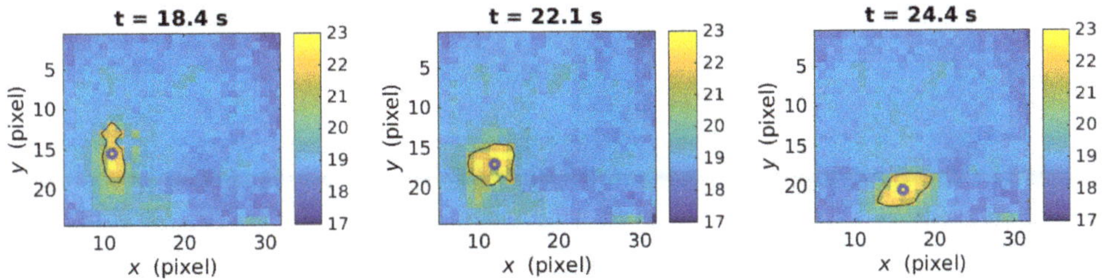

Figure 5. Three different moments during a fall. The blue circle is the barycenter.

Figure 6 shows the temporal variations of the barycenter positions of five successive type 8 (forward, knee flexion) falls made by volunteers 2 (from group 1) and 5 (from group 2). The upper frame represents the variation of the horizontal barycenter position $x(t)$ through time, while the middle frame represents the variation of the vertical barycenter position $y(t)$. The lower frame represents the spatial trajectories of each falls, the vertical position in function of the horizontal one.

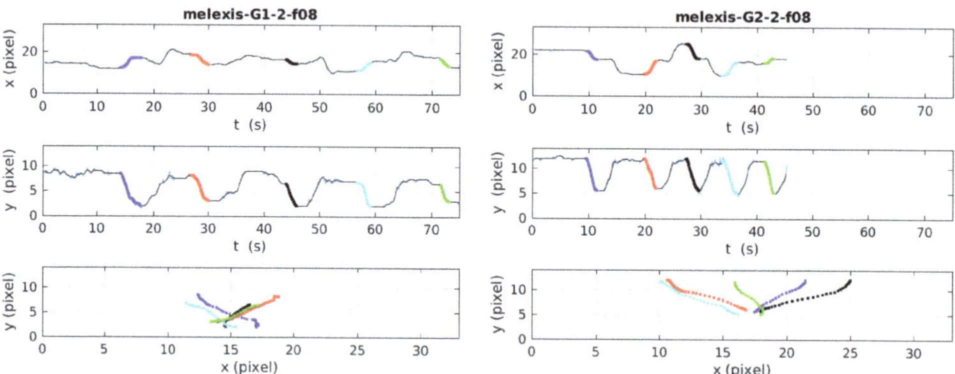

Figure 6. Trajectories of two different volunteers during the falls of type 8 (forward, knee flexion).

Finally, the histograms of Figure 7 were obtained with the fall durations for all falls of each group. It can be observed that the falls of group 1 (mean 2.62 s) are longer than those of group 2 (mean 2.20 s), in agreement with the fact that volunteers of group 1 were mimicking elderly falls. The two normal distributions were found statistically different ($p < 10^{-8}$) using a two-sample t-test.

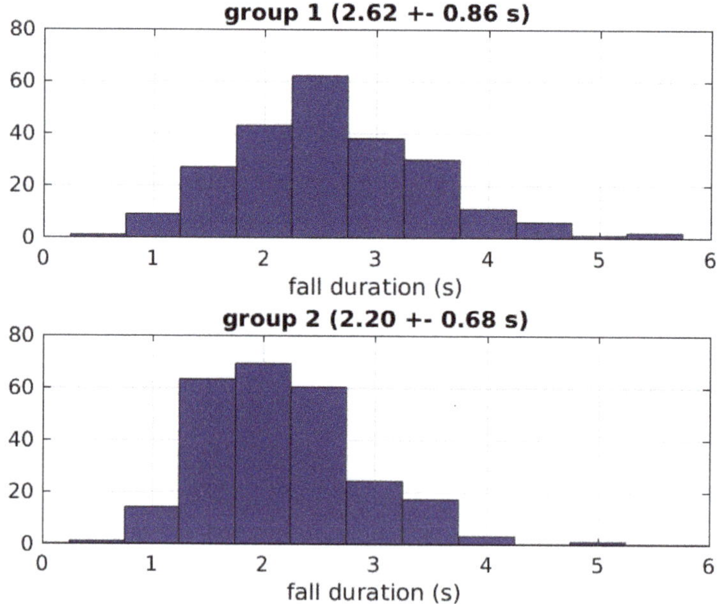

Figure 7. Seconds per fall for each group.

5. Discussion

Now we shall discuss the results obtained in Section 4. As we mentioned, the first group of volunteers was formed by three performing artists imitating the movements of older adults, while the second group of volunteers were healthy, young people, as is usual in all fall datasets. Through an analysis of the data, it was possible to show that the volunteers of group 1 fell on average more slowly than the volunteers of group 2, which corresponds to the acceleration differences between a healthy, young person and an older adult. It is important to clarify that the presented dataset involves only one person in the sensor's range of vision, since it is intended to be used by algorithms that interact with people living alone like many elderly people today.

It is worth mentioning that this type of sensor generate a lot of noise if there are other sources of heat in the radius of vision, such as stoves or pets. In the same way, the temperature of an individual cannot be recognized if his/her body is hindered by another object. Furthermore, it is necessary to mention that infrared sensors are very sensitive to several factors, such as the person's temperature, his/her clothing, his/her position with respect to the sensors, etc.

6. Conclusions

In this article, we have built a public dataset on falls obtained by two different types of thermal sensors. This dataset is novel in several ways. First, unlike many other datasets, the low resolution of these sensors prevents distinguishing physiological characteristics of individuals, which favors their privacy. Secondly, the selected falls were obtained from an exhaustive state of the art added to expert knowledge by a physiotherapist with experience in working with older adults. Third, half of the volunteers chosen for the data collection of falls are performing artists with experience in body work and who were told how to represent the falls of an older adult. As far as we know, this is the first public dataset on falls built by performing artists emulating falls of older people.

As future work, additional comparative analyses between both groups could be performed. It would also be interesting to include obstacles and other heat sources in the images to see how this affects the calculation of the barycenters and the trajectories of the falls. In addition, with the advancement of technology, sensors with improved sensitivities and very small sizes can be used in the detection of falls or ADL, such as barometers of which sensitivity has reached the magnitude of millimeters [46]. Having a dataset with this new generation of sensors can help improve systems and algorithms to help older adults.

Supplementary Materials: The dataset is available online at http://www.mdpi.com/1424-8220/19/20/4565/s1.

Author Contributions: Conceptualization, F.R. and J.-G.M.; data curation, F.R. and T.R.; formal analysis, F.R. and J.-G.M.; funding acquisition, C.T.; investigation, F.R. and C.E.; methodology, F.R. and C.E.; project administration, C.E. and F.R.; resources, T.R.; software, J.-G.M. and T.R.; validation, J.-G.M.; visualization, J.-G.M.; writing—original draft, F.R.

Funding: This research was funded by Fondef project number ID18I10212 from Consejo Nacional de Innovación, Ciencia y Tecnología (CONICYT).

Acknowledgments: We thank the volunteers who helped us collect the fall data as well as the members of the Colectivo Insinú, creation and research group in performing arts.

Conflicts of Interest: The authors declare no conflict of interest. The funders had no role in the design of the study; in the collection, analyses, or interpretation of data; in the writing of the manuscript; or in the decision to publish the results.

Abbreviations

The following abbreviations are used in this manuscript:

ADL	Activities of Daily Living
FDS	Fall Detection Systems
IMU	Inertial Measurement Unit
fps	Frames per second

References

1. Sanderson, W.C.; Scherbov, S.; Gerland, P. Probabilistic population aging. *PLoS ONE* **2017**, *12*, e0179171. [CrossRef] [PubMed]
2. World Health Organization. *WHO Global Report on Falls Prevention in Older Age*; WHO Press: Geneva, Switzerland, 2007.
3. Fleming, J.; Brayne, C. Inability to get up after falling, subsequent time on floor, and summoning help: Prospective cohort study in people over 90. *BMJ* **2008**, *337*. [CrossRef]
4. Melton, L.J., 3rd; Christen, D.; Riggs, B.L.; Achenbach, S.J.; Müller, R.; van Lenthe, G.H.; Amin, S.; Atkinson, E.J.; Khosla, S. Assessing forearm fracture risk in postmenopausal women. *Osteoporos. Int.* **2010**, *21*, 1161–1169. [CrossRef] [PubMed]
5. Tinetti, M.E.; Liu, W.L.; Claus, E.B. Predictors and Prognosis of Inability to Get Up After Falls Among Elderly Persons. *JAMA* **1993**, *269*, 65–70. [CrossRef] [PubMed]
6. Ambrose, A.F.; Paul, G.; Hausdorff, J.M. Risk factors for falls among older adults: A review of the literature. *Maturitas* **2013**, *75*, 51–61. [CrossRef]
7. Forbes, G.; Massie, S.; Craw, S. Fall prediction using behavioural modelling from sensor data in smart homes. *Artif. Intell. Rev.* **2019**. [CrossRef]
8. Vadivelu, S.; Ganesan, S.; Murthy, O.V.R.; Dhall, A. Thermal Imaging Based Elderly Fall Detection. In Proceedings of the Computer Vision—ACCV 2016 Workshops—ACCV 2016 International Workshops, Taipei, Taiwan, 20–24 November 2016; Chen, C., Lu, J., Ma, K., Eds.; Revised Selected Papers, Part III; Lecture Notes in Computer Science; Springer: Berlin/Heidelberg, Germany, 2017; Volume 10118, pp. 541–553. [CrossRef]
9. Sucerquia, A.; López, J.D.; Vargas-Bonilla, J.F. SisFall: A Fall and Movement Dataset. *Sensors* **2017**, *17*, 198. [CrossRef]
10. Wang, F.T.; Chan, H.L.; Hsu, M.H.; Lin, C.K.; Chao, P.K.; Chang, Y.J. Threshold-based fall detection using a hybrid of tri-axial accelerometer and gyroscope. *Physiol. Meas.* **2018**, *39*, 105002. [CrossRef]
11. Taramasco, C.; Rodenas, T.; Martinez, F.; Fuentes, P.; Munoz, R.; Olivares, R.; De Albuquerque, V.H.C.; Demongeot, J. A Novel Monitoring System for Fall Detection in Older People. *IEEE Access* **2018**, *6*, 43563–43574. [CrossRef]
12. Taniguchi, Y.; Nakajima, H.; Tsuchiya, N.; Tanaka, J.; Aita, F.; Hata, Y. A falling detection system with plural thermal array sensors. In Proceedings of the 2014 Joint 7th International Conference on Soft Computing and Intelligent Systems (SCIS) and 15th International Symposium on Advanced Intelligent Systems (ISIS), Kitakyushu, Japan, 3–6 December 2014; pp. 673–678. [CrossRef]
13. Adolf, J.; Macas, M.; Lhotska, L.; Dolezal, J. Deep neural network based body posture recognitions and fall detection from low resolution infrared array sensor. In Proceedings of the 2018 IEEE International Conference on Bioinformatics and Biomedicine (BIBM), Madrid, Spain, 3–6 December 2018; pp. 2394–2399. [CrossRef]
14. Schwickert, L.; Becker, C.; Lindemann, U.; Maréchal, C.; Bourke, A.; Chiari, L.; Helbostad, J.L.; Zijlstra, W.; Aminian, K.; Todd, C.; et al. Fall detection with body-worn sensors. *Z. Für Gerontol. Geriatr.* **2013**, *46*, 706–719. [CrossRef]
15. Zhang, Z.; Conly, C.; Athitsos, V. A survey on vision-based fall detection. In Proceedings of the 8th ACM International Conference on PErvasive Technologies Related to Assistive Environments, PETRA 2015, Corfu, Greece, 1–3 July 2015; Makedon, F., Ed.; ACM: New York, NY, USA, 2015; pp. 46:1–46:7. [CrossRef]
16. Casilari, E.; Santoyo-Ramón, J.A.; Cano-García, J.M. Analysis of Public Datasets for Wearable Fall Detection Systems. *Sensors* **2017**, *17*, 1513. [CrossRef] [PubMed]
17. Dhiman, C.; Kumar-Vishwakarma, D. A review of state-of-the-art techniques for abnormal human activity recognition. *Eng. Appl. Artif. Intell.* **2019**, *77*, 21–45. [CrossRef]
18. Mardini, M.T.; Iraqi, Y.; Agoulmine, N. A Survey of Healthcare Monitoring Systems for Chronically Ill Patients and Elderly. *J. Med Syst.* **2019**, *43*, 50:1–50:21. [CrossRef] [PubMed]
19. Fan, Y.; Levine, M.D.; Wen, G.; Qiu, S. A deep neural network for real-time detection of falling humans in naturally occurring scenes. *Neurocomputing* **2017**, *260*, 43–58. [CrossRef]
20. Martínez-Villaseñor, L.; Ponce, H.; Brieva, J.; Moya-Albor, E.; Núñez-Martínez, J.; Peñafort-Asturiano, C. UP-Fall Detection Dataset: A Multimodal Approach. *Sensors* **2019**, *19*, 1988. [CrossRef]

21. Tran, T.; Le, T.; Pham, D.; Hoang, V.; Khong, V.; Tran, Q.; Nguyen, T.; Pham, C. A multi-modal multi-view dataset for human fall analysis and preliminary investigation on modality. In Proceedings of the 24th International Conference on Pattern Recognition, ICPR 2018, Beijing, China, 20–24 August 2018; pp. 1947–1952. [CrossRef]
22. Peñafort-Asturiano, C.J.; Santiago, N.; Núñez-Martínez, J.P.; Ponce, H.; Martínez-Villaseñor, L. Challenges in Data Acquisition Systems: Lessons Learned from Fall Detection to Nanosensors. In Proceedings of the 2018 Nanotechnology for Instrumentation and Measurement (NANOfIM), Mexico City, Mexico, 7–8 November 2018; pp. 1–8. [CrossRef]
23. Martínez-Villaseñor, M.; Ponce, H.; Espinosa-Loera, R.A. Multimodal Database for Human Activity Recognition and Fall Detection. In Proceedings of the 12th International Conference on Ubiquitous Computing and Ambient Intelligence, UCAmI 2018, Punta Cana, Dominican Republic, 4–7 December 2018; Bravo, J., Baños, O., Eds.; MDPI: Basel, Switzerland, 2018; Volume 2, p. 1237. [CrossRef]
24. Tran, T.; Le, T.; Hoang, V.; Vu, H. Continuous detection of human fall using multimodal features from Kinect sensors in scalable environment. *Comput. Methods Progr. Biomed.* **2017**, *146*, 151–165. [CrossRef]
25. Baldewijns, G.; Debard, G.; Mertes, G.; Vanrumste, B.; Croonenborghs, T. Bridging the gap between real-life data and simulated data by providing a highly realistic fall dataset for evaluating camera-based fall detection algorithms. *Health Technol. Lett.* **2016**, *3*, 6–11. [CrossRef]
26. Chua, J.; Chang, Y.C.; Lim, W.K. A simple vision-based fall detection technique for indoor video surveillance. *Signal Image Video Process* **2015**, *9*, 623–633. [CrossRef]
27. Zhang, Z.; Conly, C.; Athitsos, V. Evaluating Depth-Based Computer Vision Methods for Fall Detection under Occlusions. In Proceedings of the Advances in Visual Computing, 10th International Symposium, ISVC 2014, Las Vegas, NV, USA, 8–10 December 2014; Bebis, G., Boyle, R., Parvin, B., Koracin, D., McMahan, R., Jerald, J., Zhang, H., Drucker, S.M., Kambhamettu, C., Choubassi, M.E., et al. Eds.; Proceedings, Part II; Lecture Notes in Computer Science; Springer: Berlin/Heidelberg, Germany, 2014; Volume 8888, pp. 196–207. [CrossRef]
28. Ma, X.; Wang, H.; Xue, B.; Zhou, M.; Ji, B.; Li, Y. Depth-Based Human Fall Detection via Shape Features and Improved Extreme Learning Machine. *IEEE J. Biomed. Health Inform.* **2014**, *18*, 1915–1922. [CrossRef]
29. Gasparrini, S.; Cippitelli, E.; Spinsante, S.; Gambi, E. A Depth-Based Fall Detection System Using a Kinect® Sensor. *Sensors* **2014**, *14*, 2756–2775. [CrossRef]
30. Kwolek, B.; Kepski, M. Human fall detection on embedded platform using depth maps and wireless accelerometer. *Comput. Methods Progr. Biomed.* **2014**, *117*, 489–501. [CrossRef] [PubMed]
31. Charfi, I.; Mitéran, J.; Dubois, J.; Atri, M.; Tourki, R. Optimized spatio-temporal descriptors for real-time fall detection: Comparison of support vector machine and Adaboost-based classification. *J. Electron. Imaging* **2013**, *22*, 041106. [CrossRef]
32. Charfi, I.; Mitéran, J.; Dubois, J.; Atri, M.; Tourki, R. Definition and Performance Evaluation of a Robust SVM Based Fall Detection Solution. In Proceedings of the Eighth International Conference on Signal Image Technology and Internet Based Systems, SITIS 2012, Sorrento, Naples, Italy, 25–29 November 2012; Yétongnon, K., Chbeir, R., Dipanda, A., Gallo, L., Eds.; IEEE: Piscataway, NJ, USA, 2012; pp. 218–224. [CrossRef]
33. Zhang, Z.; Liu, W.; Metsis, V.; Athitsos, V. A viewpoint-independent statistical method for fall detection. In Proceedings of the 21st International Conference on Pattern Recognition, ICPR 2012, Tsukuba, Japan, 11–15 November 2012; pp. 3626–3630.
34. Auvinet, E.; Reveret, L.; St-Arnaud, A.; Rousseau, J.; Meunier, J. Fall detection using multiple cameras. In Proceedings of the 2008 30th Annual International Conference of the IEEE Engineering in Medicine and Biology Society, Vancouver, BC, Canada, 20–24 August 2008; pp. 2554–2557. [CrossRef]
35. Auvinet, E.; Multon, F.; Saint-Arnaud, A.; Rousseau, J.; Meunier, J. Fall Detection With Multiple Cameras: An Occlusion-Resistant Method Based on 3-D Silhouette Vertical Distribution. *IEEE Trans. Inf. Technol. Biomed.* **2011**, *15*, 290–300. [CrossRef] [PubMed]
36. Bulotta, S.; Mahmoud, H.; Masulli, F.; Palummeri, E.; Rovetta, S. Fall Detection Using an Ensemble of Learning Machines. In *Neural Nets and Surroundings. Smart Innovation, Systems and Technologies*; Apolloni, B., Bassis, S.E.A., Morabito, F., Eds.; Springer: Berlin/Heidelberg, Germnay, 2013; Volume 19._9. [CrossRef]
37. Pannurat, N.; Thiemjarus, S.; Nantajeewarawat, E. Automatic Fall Monitoring: A Review. *Sensors* **2014**, *14*, 12900–12936. [CrossRef] [PubMed]

38. Medrano, C.; Igual, R.; Plaza, I.; Castro, M. Detecting Falls as Novelties in Acceleration Patterns Acquired with Smartphones. *PLoS ONE* **2014**, *9*, e94811. [CrossRef] [PubMed]
39. Liu, K.; Hsieh, C.; Hsu, S.J.; Chan, C. Impact of Sampling Rate on Wearable-Based Fall Detection Systems Based on Machine Learning Models. *IEEE Sens. J.* **2018**, *18*, 9882–9890. [CrossRef]
40. Mauldin, T.R.; Canby, M.E.; Metsis, V.; Ngu, A.H.H.; Rivera, C.C. SmartFall: A Smartwatch-Based Fall Detection System Using Deep Learning. *Sensors* **2018**, *18*, 3363. [CrossRef] [PubMed]
41. Sherrington, C.; Tiedemann, A. Physiotherapy in the prevention of falls in older people. *J. Physiother.* **2015**, *61*, 54–60. [CrossRef]
42. Edelberg, H.K. Falls and function. How to prevent falls and injuries in patients with impaired mobility. *Geriatrics* **2001**, *56*, 41–45.
43. Palvanen, M.; Kannus, P.; Parkkari, J.; Pitkäjärvi, T.; Pasanen, M.; Vuori, I.; Järvinen, M. The injury mechanisms of osteoporotic upper extremity fractures among older adults: A controlled study of 287 consecutive patients and their 108 controls. *Osteoporos. Int.* **2000**, *11*, 822–831. [CrossRef]
44. Sterling, D.A.; O'Connor, J.A.; Bonadies, J. Geriatric falls: Injury severity is high and disproportionate to mechanism. *J. Trauma Inj. Infect. Crit. Care* **2001**, *50*, 116–119. [CrossRef] [PubMed]
45. Parkkari, J.; Kannus, P.; Palvanen, M.; Natri, A.; Vainio, J.; Aho, H.; Järvinen, M. Majority of Hip Fractures Occur as a Result of a Fall and Impact on the Greater Trochanter of the Femur: A Prospective Controlled Hip Fracture Study with 206 Consecutive Patients. *Calcif. Tissue Int.* **1999**, *65*, 183–187. [CrossRef] [PubMed]
46. Pierleoni, P.; Belli, A.; Maurizi, L.; Palma, L.; Pernini, L.; Paniccia, M.; Valenti, S. A Wearable Fall Detector for Elderly People Based on AHRS and Barometric Sensor. *IEEE Sens. J.* **2016**, *16*, 6733–6744. [CrossRef]

Article

Design of a New Method for Detection of Occupancy in the Smart Home Using an FBG Sensor

Jan Vanus [1,*], Jan Nedoma [2], Marcel Fajkus [2] and Radek Martinek [1]

[1] Department of Cybernetics and Biomedical Engineering, Faculty of Electrical Engineering and Computer Science, VSB–Technical University of Ostrava, 708 33 Ostrava, Czech Republic; radek.martinek@vsb.cz

[2] Department of Telecommunications, Faculty of Electrical Engineering and Computer Science, VSB–Technical University of Ostrava, 708 33 Ostrava, Czech Republic; jan.nedoma@vsb.cz (J.N.); marcel.fajkus@vsb.cz (M.F.)

* Correspondence: jan.vanus@vsb.cz

Received: 29 November 2019; Accepted: 6 January 2020; Published: 10 January 2020

Abstract: This article introduces a new way of using a fibre Bragg grating (FBG) sensor for detecting the presence and number of occupants in the monitored space in a smart home (SH). CO_2 sensors are used to determine the CO_2 concentration of the monitored rooms in an SH. CO_2 sensors can also be used for occupancy recognition of the monitored spaces in SH. To determine the presence of occupants in the monitored rooms of the SH, the newly devised method of CO_2 prediction, by means of an artificial neural network (ANN) with a scaled conjugate gradient (SCG) algorithm using measurements of typical operational technical quantities (indoor temperature, relative humidity indoor and CO_2 concentration in the SH) is used. The goal of the experiments is to verify the possibility of using the FBG sensor in order to unambiguously detect the number of occupants in the selected room (R104) and, at the same time, to harness the newly proposed method of CO_2 prediction with ANN SCG for recognition of the SH occupancy status and the SH spatial location (rooms R104, R203, and R204) of an occupant. The designed experiments will verify the possibility of using a minimum number of sensors for measuring the non-electric quantities of indoor temperature and indoor relative humidity and the possibility of monitoring the presence of occupants in the SH using CO_2 prediction by means of the ANN SCG method with ANN learning for the data obtained from only one room (R203). The prediction accuracy exceeded 90% in certain experiments. The uniqueness and innovativeness of the described solution lie in the integrated multidisciplinary application of technological procedures (the BACnet technology control SH, FBG sensors) and mathematical methods (ANN prediction with SCG algorithm, the adaptive filtration with an LMS algorithm) employed for the recognition of number persons and occupancy recognition of selected monitored rooms of SH.

Keywords: smart home (SH); prediction; artificial neural network (ANN); fiber bragg grating (FBG); occupancy; number of person recognition; scaled conjugate gradient (SCG)

1. Introduction

Recognizing the occupancy, number of individuals, location-and-movement recognition and activity recognition of an individual in an indoor space is one the key functionalities of a smart home (SH), as a prerequisite to providing services to support independent living of elderly SH occupants and has a great influence on internal loads and HVAC (heating, ventilation and air conditioning) requirement, thus increasing the energy consumption optimization. Azghandi et al. focused on the particular case of an SH with multiple occupants, they developed a location-and-movement recognition method using many inexpensive passive infrared (PIR) motion sensors and a small number of more costly radio frequency identification (RFID) readers [1]. Benmansour et al. provided an overview of existing approaches and current practices for activity recognition in multi-occupant SHs [2]. Braun et al. reported the investigation of two categories of occupancy sensors with the requirements of supporting wireless communication and a focus on the low cost of the systems (capacitive proximity sensors and accelerometers that are placed below the furniture) with a classification accuracy between 79% and 96% [3]. Chan et al. proposed the methodology and design of a voice-controlled environment, with an emphasis on speech recognition and voice control, based on Amazon Alexa and Raspberry Pi in an SH [4]. Chen et al. proposed an activity recognition system guided by an unobtrusive sensor (ARGUS) with a facing direction detection accuracy, resulting from manually defined features, that reached 85.3%, 90.6%, and 85.2% [5]. Khan et al. developed a low-cost heterogeneous radar-based activity monitoring (RAM) system for recognizing fine-grained activities in an SH with detecting accuracy of 92.84% [6]. Lee et al. investigated the use of cameras and a distributed processing method for the automated control of lights in an SH, which provided occupancy reasoning and human activity analysis [7]. Mokhtari et al. proposed a new human identification sensor, which can efficiently differentiate multiple residents in a home environment to detect their height as a unique bio-feature with three sensing/communication modules: pyroelectric infrared (PIR) occupancy, ultrasound array, and Bluetooth low-energy (BLE) communication modules [8]. A new recognition algorithm for household appliances, based on a Bayes classification model, is presented by Yan et al., in which sequential appliance power consumption data from intelligent power sockets is used and for the generalization and extraction of the characteristics of occupant behavior and power consumption of typical household appliances [9]. Yang et al. proposed a novel indoor tracking technique for SHs with multiple residents by relying only on non-wearable, environmentally deployed sensors such as passive infrared motion sensors [10]. Feng et al. presented a novel real-time, device-free, and privacy-preserving WiFi-enabled Internet of Things (IoT) platform for SH-occupancy sensing, which can promote a myriad of emerging applications with an accuracy of 96.8% and 90.6% in terms of occupancy detection and recognition, respectively [11]. Traditionally, in building energy modeling (BEM) programs, occupant behavior (OB) inputs are deterministic and less indicative of real-world scenarios, contributing to discrepancies between simulated and actual energy use in buildings. Yin et al. (2016) presented a new OB modeling tool, with an occupant behavior functional mock-up unit (obFMU) that enables co-simulation with BEM programs implementing a functional mock-up interface (FMI) [12]. Occupants are involved in a variety of activities in buildings, which drive them to move among rooms, enter or leave a building. Hong et al. (2016) defined SH occupancy using four parameters and showed how they varied with time. The four occupancy parameters were as follows: (1) the number of occupants in a building, (2) occupancy status of space, (3) the number of occupants in a space, and (4) the location of an occupant [13].

In order to detect the occupancy of the SH by indirect methods (without using cameras), common operational and technical sensors are used in this article to measure the indoor temperature, the indoor relative humidity and the CO_2 indoor concentration within the BACnet technology for HVAC control. A fibre Bragg grating (FBG) sensor will be used to detect the number of occupants in the monitored space of room R104 (ground floor). The method devised for the prediction of the CO_2 waveform using artificial neural network (ANN) scaled conjugate gradient (SCG) will be verified during the experiments conducted to detect the occupancy of rooms R104, R203 and R204. The input quantities measured to ANN SCG were obtained from the indoor temperature and relative indoor humidity sensors. One of the objectives of the article is to verify the possibility of minimizing investment costs by using cheaper temperature and relative humidity sensors instead of a more expensive CO_2 sensor to detect the occupancy of monitored SH spaces. The other objectives of this article are the following:

1. Experimental verification of FBG sensor use for the recognition of the number occupants in SH room R104.
2. Experimental verification of the CO_2 concentration measurement in an SH by means of common operational sensors for the occupancy status of the SH space.

3. Experimental verification of the method with ANN SCG that was devised for CO_2 concentration prediction (more one-day measurements in the period from 25 June 2018, to 28 June 2018) to locate an occupant (in rooms R104, R203, and R204) in an SH with the highest possible accuracy.
4. Experimental verification of the possibility of ANN learning for one room only (R203) in order to predict CO_2 concentrations in other rooms (R104, R204).

The experimental measurements of objective parameters of the internal environment and thermal comfort evaluation were conducted in selected SH rooms R204, R203, and R104 in a wooden building of the passive standard located in the Faculty of Civil Engineering, VSB—TU Ostrava. (Figure 1) [14].

Figure 1. The wooden building of the passive standard located in the Faculty of Civil Engineering, VSB—TU Ostrava with selected smart home (SH) rooms R204, R203, and R104 [15].

2. Materials and Methods

2.1. Fiber Bragg Grating (FBG) Sensor Using for Recognition of Number Occupants in Smart Home (SH) Room R104

Bragg gratings (FBG) are special structures created by an ultraviolet (UV) laser inside the core of a photosensitive optical fibre. This structure consists of a periodic structure of changes in the refractive index, where the layers of the refractive index of the core n_1 alternate with the layers of the increased refractive index n_3 (1):

$$n_3 = n_1 + \delta n \tag{1}$$

where δn is the refractive index induced by UV radiation [16].

When a broad-spectrum light is introduced into the optical fibre, the Bragg grating reflects a narrow spectral portion and all the other wavelengths pass through the structure without damping (Figure 2).

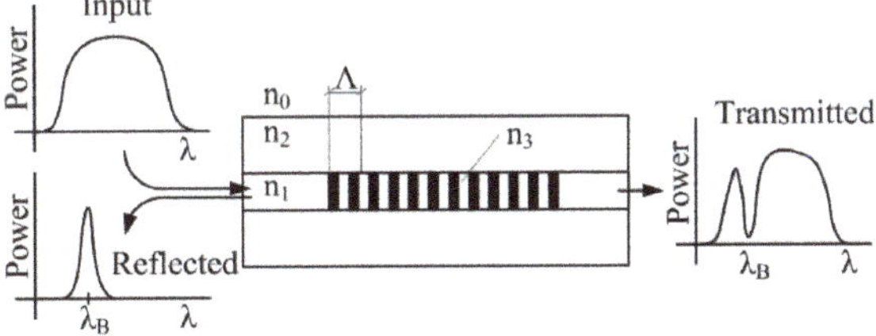

Figure 2. The Bragg grating principle.

The central wavelength of the reflected spectral portion is called the Bragg wavelength λ_B and is defined by the optical and geometric properties of the structure according to (2):

$$\lambda_B = 2n_{eff}\Lambda \quad (2)$$

where n_{eff} is the effective refractive index of the periodic structure and Λ is the distance between the periodic changes in the refractive index. The external effects of the temperature and the deformation influence the optical and geometric properties and thus, the spectral position of the Bragg wavelength. Thanks to this feature, Bragg gratings are used in sensory applications. The dependence of the Bragg wavelength on the deformation and the temperature is expressed by (3):

$$\frac{\Delta\lambda_B}{\lambda_B} = k\varepsilon + (\alpha_\Lambda + \alpha_n)\Delta T \quad (3)$$

where k is the deformation coefficient, ε is the optical fibre deformation caused by measurement, α_Λ is the coefficient of thermal expansion, α_n is the thermo-optic coefficient and ΔT is the change in the operating temperature [17].

Bragg gratings in a standard optical fibre with a central wavelength of 1550 nm show a deformation sensitivity of 1.1 pm/µstrain and a temperature sensitivity of 10.3 pm/°C. By using a suitable encapsulation, it is possible to implement a sensor of almost any physical quantity. FBG sensors are used in automobile [18] and railway transport [19], the construction industry [20], power engineering, biomedical [21] or perimetric applications [22], etc.

Bragg gratings are single-point sensors. By using a wavelength or time multiplex, it is possible to connect tens or hundreds of these sensors in a single optical fibre in order to achieve a quasi-distributed sensory system [23].

2.1.1. Fiberglass Bragg Sensors

One of the most widespread applications of Bragg gratings includes deformation and compression measurements. Depending on the type of application, Bragg gratings can be encapsulated in many ways. The principle of the encapsulation is to protect the fragile glass fibre, to enhance the sensitivity to the desired quantity and to suppress the surrounding interference. Bragg gratings can be encapsulated in polymers [24] of fibreglass or composite materials [25,26], and special steel jigs [27], which are mechanically attached to the structure that is to be measured, etc. Based on the advantages of using Bragg gratings—such as their reliable and very accurate measurement—these types of sensors were used for the reference measurement of the occupancy recognition of room R104 in the SH.

Because the grating sensors were installed on a wooden staircase, the encapsulation of Bragg gratings in fibreglass strips was used. This method enables the implementation of a very thin sensor that transmits deformations from the step (passage of persons) to the optical fibre itself.

Two Bragg gratings in a single-mode optical fibre with primary acrylate protection were used for implementing the sensors. Bragg gratings A and B had the following parameters: The Bragg wavelengths were 1547.510 nm and 1552.369 nm, respectively; the reflection spectrum width was 256 pm and 227 pm, respectively; the reflectivity was 91.2% and 91.3%, respectively. Each Bragg grating was placed between the glass fabrics (2 layers below and 2 layers above the optical fibre). The glass fabric was then coated with a polymer resin. The actual curing caused a Bragg wavelength shift of 23 µm for Sensor A and of 19 µm for Sensor B to lower wavelengths (Figure 3). The Bragg grating is located in the middle of the fibreglass strip, marked in red.

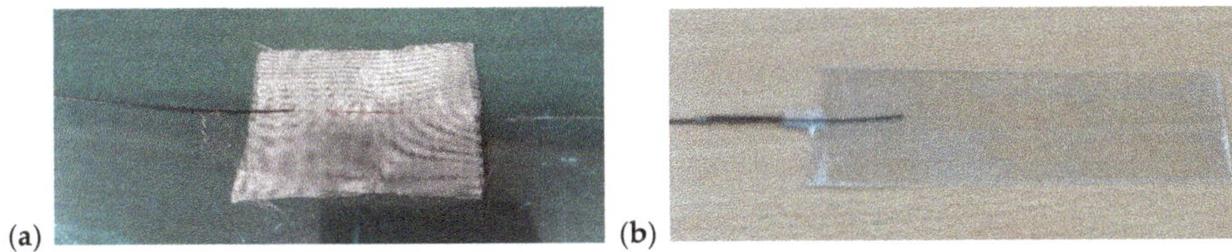

Figure 3. Implementation of the sensor by encapsulating the Bragg grating in fibreglass (a); the resulting fibre Bragg grating (FBG) fibreglass sensor (b).

2.1.2. Implementation of FBG Sensors

FBG sensors were implemented on the staircase leading from the ground floor to the first floor (Figure 4). The sensors were glued with cyanoacrylate adhesive to the bottom of the second step (FBG A) and the third step (FBG B).

Figure 4. Placement of FBG sensors on the staircase in the smart home (SH), room R104.

2.2. Use of a CO_2 Sensor Network for Monitoring SH Space Occupancy

Common CO_2 sensors can be used to detect the occupancy of individual SH spaces. In room R104, the BT 12.09 sensor (Figure 5), in room R203 the BT 12.10 sensor and in room R204 the BT 12.10 sensor (Figure 6) were used for measuring the CO_2 concentration in the framework of forced Air Condition (AC) control (Figure 7). The BACnet technology is used in the SH to control HVAC (Figure 8). The presence of occupants in the SH can be detected by measuring the CO_2 concentration. Figure 5 shows the ground plan of the ground floor of the SH with the location of the individual sensors for CO_2 measurement.

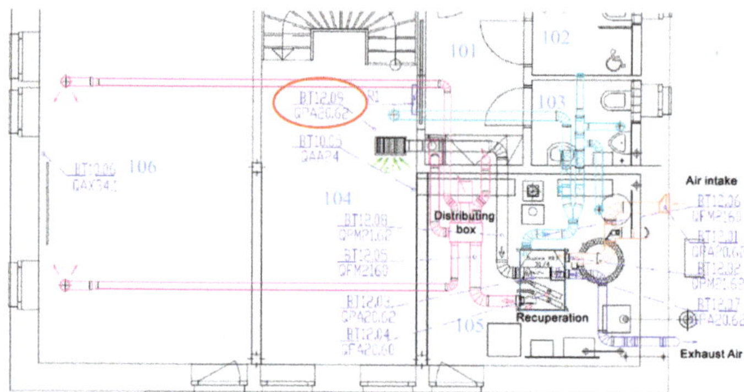

Figure 5. The ground plan of the ground floor of the SH with the location of the sensors for CO_2 measurement.

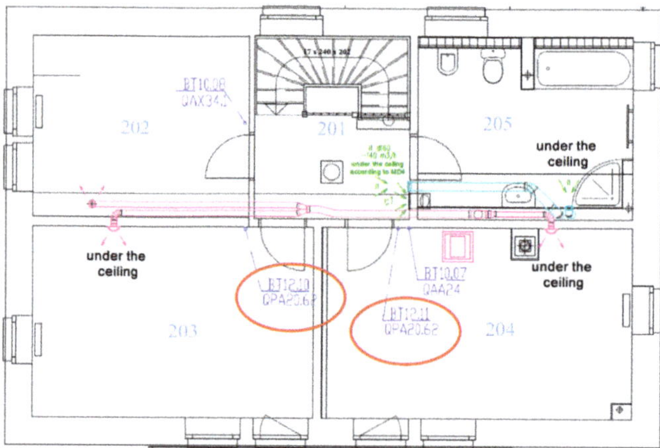

Figure 6. The ground plan of the first floor of the SH with the location of the sensors for CO_2 measurement.

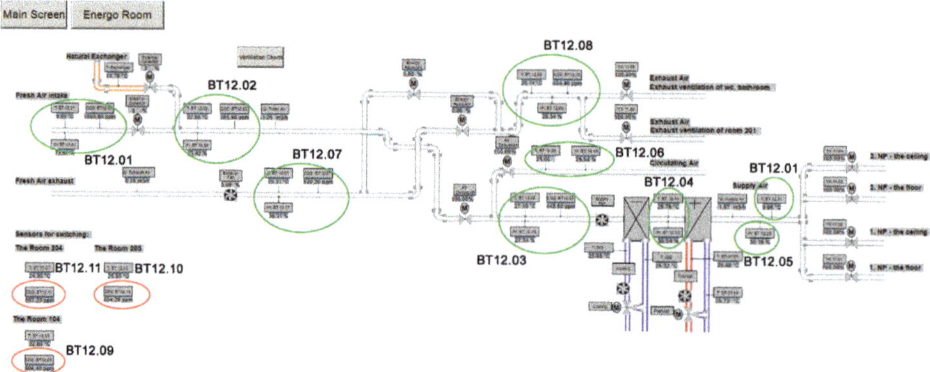

Figure 7. The ventilation distribution technology with the location of the individual sensors for CO_2 measurement in an SH.

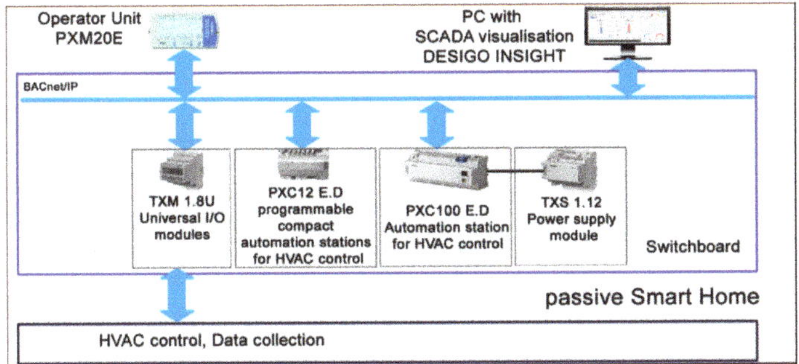

Figure 8. Block diagram of the BACnet technology used in an SH for HVAC control.

The individual rooms on the ground floor of the SH are marked as follows (Figure 5):

- R101—door space, entrance hall,
- R102—toilet 1,
- R103—toilet 2,
- R104—entrance room; FBG sensor is placed on the staircase,
- R105—utility room, there are heating sources,
- R106—classroom.

Figure 6 shows the ground plan of the first floor of the SH with the location of the individual sensors for CO_2 measurement.

The individual rooms on the SH first floor are marked as follows (Figure 6):

- R201—staircase,
- R202—control room,
- R203—classroom (office),
- R204—classroom (office),
- R205—toilet and bathroom.

The list (legend) of the individual sensors used (Figures 5–7):

- BT 12.01—Measurement at the fresh outdoor air inlet into QPA 2062 SH.
- BT 12.02—Measurement at the recirculation air inlet from SH spaces into QPM 2162 heat recovery unit.
- BT 12.03—Measurement at the recirculation air inlet from SH spaces into QPA 2062 heat recovery unit.
- BT 12.04—Measurement in QFA 2060 heat recovery unit.
- BT 12.05—Measurement at the recirculation and fresh air outlet from the heat recovery unit into QFM 2160 SH.
- BT 12.06—Measurement at the exhaust air inlet into QFM 2160 recuperation unit.
- BT 12.07—Measurement at the exhaust air outlet from QPA 2062 recuperation unit.
- BT 12.08—Measurement at the recirculation air inlet from SH spaces into QPM 2162 heat recovery unit.
- BT 12.09—sensor located in room R104, QPA 2062.
- BT 12.10—sensor located in room R203, QPA 2062.
- BT 12.11—sensor located in room R204, QPA 2062.

The technical specification of the individual sensors used:

- QPA 20.62 room sensor for measuring the air quality—CO_2, relative humidity and temperature—with a measurement accuracy: (50 ppm + 2% of the value measured, long-term drift: 5% of the measuring range/5 years (typically). The CO_2 sensor principle is based on non-dispersive infrared absorption (NDIR) measurement.
- QPM 21.62 channel sensors for air quality—CO_2, relative humidity, temperature. Measurement accuracy: (50 ppm + 2% of the value measured), long-term drift: 5% of the measuring range/5 years (typically). The CO_2 sensor is based on non-dispersive infrared absorption (NDIR) measurement.
- QFA 20.60 room sensor for temperature and relative humidity. Measurement accuracy ± 3% rH_{in} within the comfort range. Application range −15 ... +50 °C/0 ... 95% rH_{in} (no condensation).
- QFM 21.60 channel sensor for relative humidity and temperature. Measurement accuracy ± 3% rH_{in} within the comfortable range. Application range −15 ... +60 °C/0 ... 95% rH_{in} (no condensation).

Figure 7 shows the ventilation distribution technology with the location of the individual CO_2 sensors.

A block diagram containing a description of the individual components and function blocks within the BACnet technology in the SH for HVAC control is shown in Figure 8.

2.3. The Design of the New Method for CO_2 Prediction

The newly devised method for CO_2 prediction from the temperature indoor and relative humidity indoor values measured by means of ANN SCG (multiple one-day measurements) was used for the location of an occupant (in rooms R104, R203 a R204) in SH with the highest possible accuracy. Block diagram of processing the quantities measured in SH for multiple one-day measurements in the period from 25 June 2018, to 28 June 2018, using the method devised for CO_2 prediction by means of ANN SCG is shown in Figure 9.

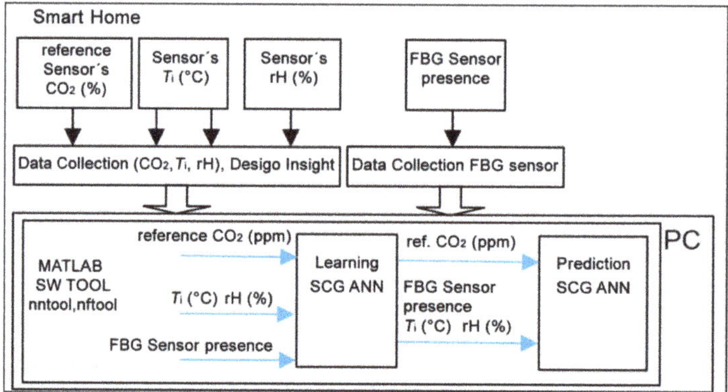

Figure 9. Block diagram describing processing of the data measured by means of a scaled conjugate gradient artificial neural network (ANN SCG) within the method devised for CO_2 prediction.

The measured values that were used in these experiments are the indoor CO_2 concentration, indoor relative humidity and indoor temperature. The data were pre-processed to improve the efficiency of neural network training. That means that data were normalized so that all the values are between 0 and 1 (Figure 10, Step 1).

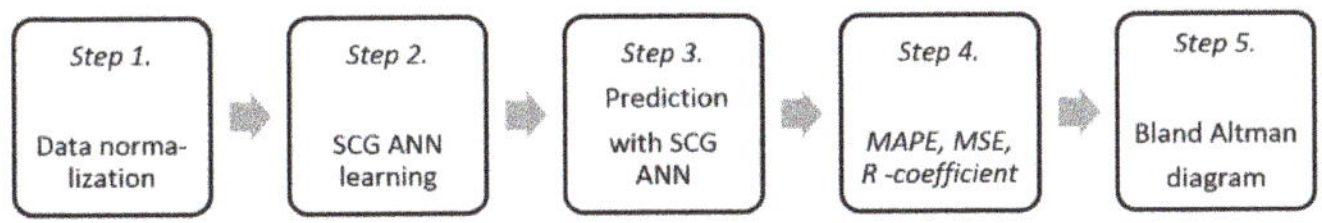

Figure 10. Block scheme summarizing the experiment steps.

In Matlab, the function nftool (neural fitting) was used with neurons varying from 10 to 100 and the three methods previously mentioned. The data samples were divided into 3 sets: training (used to teach the network), validation and testing (provides an independent measure of the network training) (Figure 10, Step 2). After training the networks with data measured on 25 July for room 203, the 90 networks were used to predict data for the rest of the dates from the same room, as well as the two others. Since the learning date was different from the prediction dates, this was called "cross-validation". In this step, the function used in Matlab was 'nntool' (Figure 10, Step 3). Once the results were ready, the next step was to calculate some parameters that would allow us to quantify the precision of the results, and, therefore, compare the prediction quality between the different ANN SCG (Figure 10, Step 4).

The three parameters we relied on for our experiments are as follows.

R (correlation coefficient) is a statistical measure that calculates the strength of the relationship between the relative movements of two variables and is calculated with the formula (4). The values range between −1 and 1. A value of exactly 1.0 means there is a perfect positive relationship between the two variables. For a positive increase in one variable, there is also a positive increase in the second variable. A value of −1.0 means there is a perfect negative relationship between the two variables. This shows that the variables move in opposite directions—for a positive increase in one variable, there is a decrease in the second variable. If the correlation is 0, there is no relationship between the two variables [28].

$$R = \frac{\sum(x-\overline{x})(y-\overline{y})}{\sqrt{\sum(x-\overline{x})^2 \sum(y-\overline{y})^2}} \quad (4)$$

The MSE (mean squared error) parameter describes how close a regression line is to a set of points and is calculated with formula (5). It does this by taking the distances from the points to the regression line (these distances are the "errors") and squaring them. The squaring is necessary to remove any negative signs. It also gives more weight to larger differences. It is called the mean squared error as you are finding the average of a set of errors [29]:

$$MSE = \frac{1}{n}\sum_{i=1}^{n}\left(y_i - y_i^*\right) \quad (5)$$

MAPE (average absolute percentage error) is a statistical measurement parameter of how accurate a forecast system is. It measures this accuracy as a percentage, and it can be calculated as the average absolute percent error for each time period minus actual values divided by actual values which are given by (6) [30]:

$$MAPE = \frac{1}{n}\sum_{i=1}^{n}\frac{\left|y_i - y_i^*\right|}{y_i^*} \quad (6)$$

y_i: reference value,
y_i^*: predicted value,
n: total number of values,
$\overline{x}, \overline{y}$: mean of x, y.

After calculating the MAPE, MSE, and R correlation parameters, we plot two figures. The first one has a reference and predicted CO_2 over time (Figures 18, 20 and 22) and the second one is a

Bland–Altmann plot (Figures 19, 21 and 23). The Bland–Altman technique allows us to make a comparison between two measurement methods of the same sample. It was established by J. Martin Bland and Douglas G. Altman. Essentially, it quantifies the difference between measurements using a graphical method. It consists of a scatterplot with the average and the difference represented on the X-axis and the Y-axis, respectively. The plot also has horizontal lines drawn at the mean difference and at the limits of agreement, which are defined as the mean difference plus and minus 1.96 times the standard deviation of the differences (Figure 10, Step 5). The final step is to compare the results and the plots, so as to decide which conditions and which methods are optimal for future predictions.

The procedure used to carry out the learning process in the neural networks is called optimization algorithms. In the following experiments, a ANN SCG type of mathematical algorithm was used to train the ANNs.

Scaled Conjugate Gradient Algorithm:

SCG is a supervised learning algorithm for feedforward neural networks and is a member of the class of conjugate gradient methods (CGMs). Let $p = \exp\left(\frac{-\Delta E}{T}\right)$ be a vector, N the sum of the number of weights and of the number of biases of the network, and E the error function we want to minimize. SCG differs from other conjugate gradient methods in two ways [31]:

Each iteration k of a CGM computes ω_i, where R^N is a new conjugate direction, and $\omega_{k+1} = \omega_k + \alpha_k.p_k$ is the size of the step in this direction. In fact, p_k is a function of α_k, the Hessian matrix of the error function, namely the matrix of the second derivatives. In contrast to other CGMs that avoid the complex computation of the Hessian and approximate α_k with a time-consuming line search procedure, SCG makes the following simple approximation of the term $E''(\omega_k)$, a key component of the computation of α_k: s_k [32] as the Hessian is not always positive, which prevents the algorithm from achieving good performance; SCG uses a scalar α_k which is supposed to regulate the indefiniteness of the Hessian. This resembles the Levenberg–Marquardt method, and is performed by using the following equation (7) [33]:

$$s_k = E''(\omega_k).p_k \approx \frac{E'(\omega_k + \alpha_k.p_k) - E'(\omega_k)}{\alpha_k},\ 0 < \alpha_k \ll 1 \tag{7}$$

and adjusting λ_k at each iteration.

The final step is to compare the results and the plots, so as to decide which conditions and which methods are optimal for future predictions [34].

2.4. The Signed–Regressor LMS Adaptive Filter

The signed–regressor LMS adaptive filter was used to filter the predicted course in order to determine the occupancy of the monitored areas more precisely (Figures 19, 21, 23, 26, 28 and 30).

2.4.1. The Conventional LMS Algorithm

The LMS algorithm is a linear adaptive filtering algorithm, which consists of two basic processes:

(a) a filtering process, which involves computing the output $y(n)$ of the linear filter in response to an input signal $x(n)$ (8), generating an estimation error $e(n)$ by comparing this output $y(n)$ with the desired response $d(n)$ (9),

(b) an adaptive process (10), which involves the automatic adjustment of the parameters $w(n+1)$ of the filter in accordance with the estimation error $e(n)$.

$$y(n) = \sum_{i=0}^{M-1} w_i(n)x(n-i) \tag{8}$$

$$e(n) = d(n) - y(n) \tag{9}$$

$$\mathbf{w}(n+1) = \mathbf{w}(n) + 2\mu e(n)\mathbf{x}(n) \tag{10}$$

where $\mathbf{w}(n)$ is M tap—weight vector, $\mathbf{w}(n+1)$ is M tap—weight vector update [35,36].

2.4.2. The Signed–Regressor LMS Algorithm

The signed–regressor algorithm is obtained from the conventional recursion (10) by replacing the tap-input vector $\mathbf{x}(n)$ with the vector sign($\mathbf{x}(n)$)), where the sign function is applied to the vector $\mathbf{x}(n)$ on an element-by-element basis. The signed–regressor recursion is then [35]:

$$\mathbf{w}(n+1) = \mathbf{w}(n) + 2\mu\, e(n)\text{sign}(\mathbf{x}(n)) \tag{11}$$

3. Experiments and Results

3.1. Using Fiber Bragg Grating Sensor for Recognition of Number of Occupants in SH Room R104

The results of the experiments to detect the number of occupants in the monitored space of room R104 are described below. On 25 June 2018, we installed the individual sensors (FBG A and FBG B) on the staircase in room R104 (Figure 4). The actual measurement to unambiguously detect the number of occupants in the monitored space took place from 25 to 28 June 2018. The record of the continuous measurement conducted in the period from 25 to 28 June 2018, is shown in Figure 11. These are signals from the Bragg sensors where the individual peaks represent persons treading on a particular step. Blue shows the waveform of the FBG A sensor on the second step and red shows the signal from FBG B sensor on the third step.

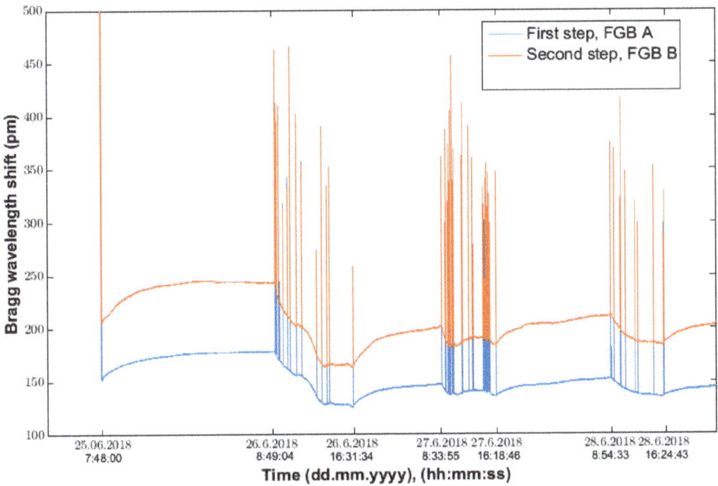

Figure 11. The waveform of signals from FBG A and FBG B sensors during the 24-h measurement of recognition of the number occupants in room R104 in the period from 25 June 2018 (7:48:00), to 28 June 2018 (23:59:00).

To detect the number of occupants on the first floor, an algorithm was implemented in the MATLAB (Matrix Laboratory) computational software environment. The algorithm is based on the detection of peaks (treading on the step) from both FBG sensors. The time-corresponding peaks were compared over time, while the direction of the passage (up or down) was detected based on earlier treading on the first or second step. The number of occupants on the first floor was detected using a counter. The occupancy waveform of the first floor is shown in Figure 12.

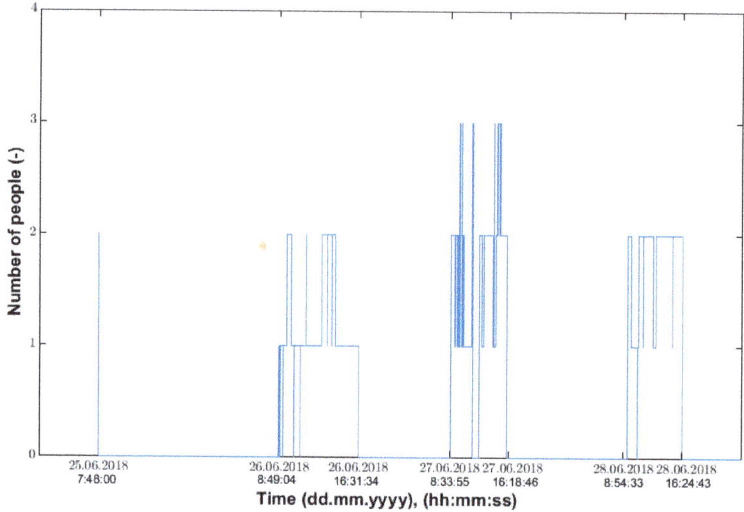

Figure 12. First floor of the recognition of the number occupants obtained from the measurement conducted in the period from 25 June 2018 (7:48:00), to 28 June 2018 (23:59:00).

The number of occupants can be detected from the following Table 1:

Table 1. Unambiguous determination of the number of occupants on the staircase in room R104 in the SH.

Date	Time	Number of Occupants	Date	Time	Number of Occupants	Date	Time	Number of Occupants
(dd.mm.yyyy)	(hh:mm:ss)		(dd.mm.yyyy)	(hh:mm:ss)		(dd.mm.yyyy)	(hh:mm:ss)	
25 June 2018	7:48:07	1	27 June 2018	9:40:24	2	27 June 2018	14:52:53	2
25 June 2018	7:53:11	1	27 June 2018	9:44:09	3	27 June 2018	14:53:13	3
25 June 2018	7:53:13	2	27 June 2018	9:47:55	2	27 June 2018	14:56:13	2
25 June 2018	7:53:23	1	27 June 2018	9:51:23	1	27 June 2018	14:56:49	3
26 June 2018	8:49:04	1	27 June 2018	9:52:31	2	27 June 2018	15:01:32	2
26 June 2018	8:59:39	1	27 June 2018	10:04:16	3	27 June 2018	15:02:11	3
26 June 2018	9:21:15	1	27 June 2018	10:06:44	2	27 June 2018	15:12:12	2
26 June 2018	9:55:43	2	27 June 2018	10:08:47	1	27 June 2018	15:13:37	2
26 June 2018	10:29:44	1	27 June 2018	10:09:10	2	27 June 2018	15:13:41	3
26 June 2018	10:54:16	1	27 June 2018	10:21:31	1	27 June 2018	15:14:28	2
26 June 2018	11:49:18	1	27 June 2018	11:25:30	2	27 June 2018	15:14:50	3
26 June 2018	12:34:02	2	27 June 2018	11:29:22	3	27 June 2018	15:15:49	2
26 June 2018	12:35:40	1	27 June 2018	11:33:34	2	27 June 2018	15:22:19	3
26 June 2018	14:38:27	2	27 June 2018	11:36:08	1	27 June 2018	15:23:10	2
26 June 2018	15:24:31	1	27 June 2018	11:36:48	0	27 June 2018	16:18:22	1
26 June 2018	15:24:52	2	27 June 2018	12:28:48	1	27 June 2018	16:18:46	0
26 June 2018	16:07:57	1	27 June 2018	12:28:57	2	28 June 2018	8:54:33	1
26 June 2018	16:08:25	2	27 June 2018	12:53:14	1	28 June 2018	8:54:35	2
26 June 2018	16:31:34	1	27 June 2018	13:03:53	2	28 June 2018	9:22:11	1
27 June 2018	8:33:55	1	27 June 2018	14:25:57	1	28 June 2018	10:19:19	1
27 June 2018	8:35:22	2	27 June 2018	14:26:27	2	28 June 2018	10:20:05	2
27 June 2018	9:07:51	1	27 June 2018	14:29:00	3	28 June 2018	10:59:54	1
27 June 2018	9:09:36	2	27 June 2018	14:30:04	2	28 June 2018	11:00:07	2

Table 1. *Cont.*

Date	Time	Number of Occupants	Date	Time	Number of Occupants	Date	Time	Number of Occupants
(dd.mm.yyyy)	(hh:mm:ss)		(dd.mm.yyyy)	(hh:mm:ss)		(dd.mm.yyyy)	(hh:mm:ss)	
27 June 2018	9:26:26	1	27 June 2018	14:33:03	1	28 June 2018	12:23:23	1
27 June 2018	9:30:32	2	27 June 2018	14:44:50	2	28 June 2018	12:45:23	2
27 June 2018	9:34:57	1	27 June 2018	14:45:09	1	28 June 2018	15:01:41	1
27 June 2018	9:35:50	2	27 June 2018	14:46:08	2	28 June 2018	15:01:57	2
27 June 2018	9:39:48	1	27 June 2018	14:51:29	3	28 June 2018	16:22:41	1

Discussion—Experiment 3.1:

On the basis of measuring the movement of occupants on the staircase in room R104 using FBG sensors in the period from 26 to 29 June 2019, it is possible to clearly identify the number of occupants (Figure 12) in the monitored SH space (Figure 4)—specifically in room R104 (Figure 5)—with an accuracy of 1 s. Within the initial testing of the experiment conducted, two persons walking up the stairs to the first floor of the SH were detected on 25 June 2018, when installing the FBG sensors in the SH (room R104) (Figure 12), (Table 1). On 26 June 2018, movement of one and two persons on the staircase from the ground floor to the first floor was detected during the day (Figure 12), (Table 1). On 27 June 2018, and 29 June 2018, movement of one, two and three persons was detected during the day. On 28 June the movement of one and two persons was detected during the day (Figure 12), (Table 1). The drawback of this method of using the FBG sensor to detect the number of occupants in the monitored space is the lack of information on the occupancy of the monitored space of R104 in the SH. This drawback has been eliminated by adding CO_2 sensors to rooms R104, R203 and R204 in experiment 2, which is described below.

3.2. Use of CO_2 Sensors for Monitoring SH Space Occupancy

CO_2 sensors (R104 (BT12.09), R203 (BT12.10), R204 (BT12.11)) can be used to detect occupancy of the monitored spaces (rooms R104, R203, R204) in the SH (Figures 5 and 6); these sensors are used to control the quality of the indoor environment in these rooms using BACnet technology in the SH (Figure 7). To control HVAC in SH (Figure 7), other CO_2 sensors which provide "measurement at the fresh outdoor air inlet into QPA 2062 SH (BT 12.01) were used; measurement at the recirculation air inlet from SH spaces into the QPM 2162 heat recovery unit (BT 12.02); measurement at the recirculation air inlet from SH spaces into the QPA 2062 heat recovery unit (BT 12.03); measurement in the QFA 2060 heat recovery unit (BT 12.04); measurement at the recirculation and fresh air outlet from the heat recovery unit into the QFM 2160 SH (BT 12.05); measurement at the exhaust air inlet into the QFM 2160 recuperation unit (BT 12.06); measurement at the exhaust air outlet from the QPA 2062 recuperation unit (BT 12.07); measurement at the recirculation air inlet from SH spaces into the QPM 2162 heat recovery unit (BT 12.08)".

Discussion—Experiment 3.2:

Using the information from the FBG sensor (Figure 13), it is possible to unambiguously detect the number of occupants present in the monitored space. Based on the measured values (CO_2 concentration), it is possible to detect the occupancy of the monitored SH spaces, the arrival of a person into the monitored room or the exit from the monitored space, or the length of stay in the monitored space (Figures 14–16). The aforementioned procedure enables the unambiguous determination of the occupancy rate of the monitored SH spaces, indirectly by measuring common non-electrical quantities (CO_2) within the operational–technical function control in the SH.

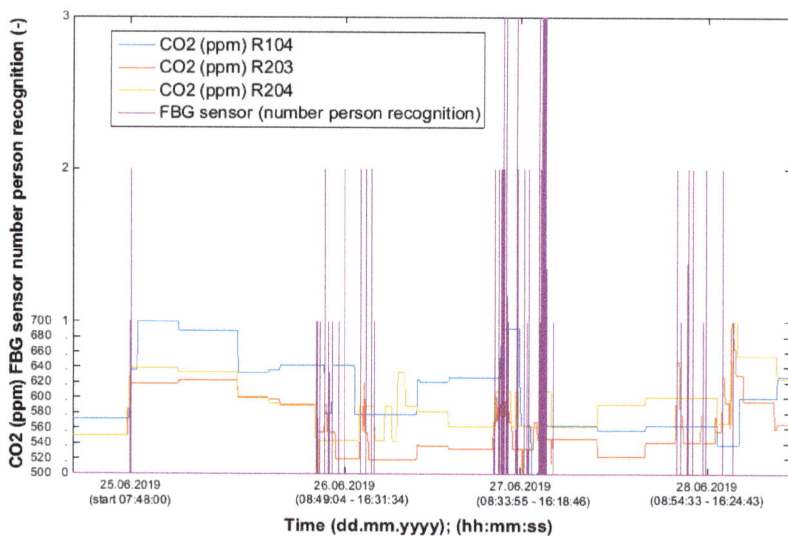

Figure 13. Waveforms of the CO_2 concentration values measured in SH rooms R104 (BT 12.09), R203 (BT 12.10), R204 (BT 12.11) in order to detect the occupancy of the monitored spaces with the representation of the recognition of the number of occupants in room R104 using the FBG sensor.

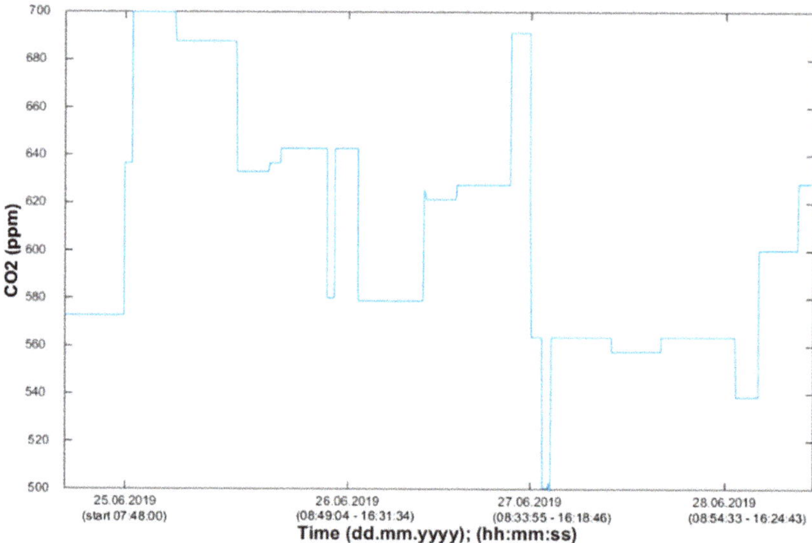

Figure 14. Waveforms of the CO_2 concentration values measured in SH room R104 (BT 12.09).

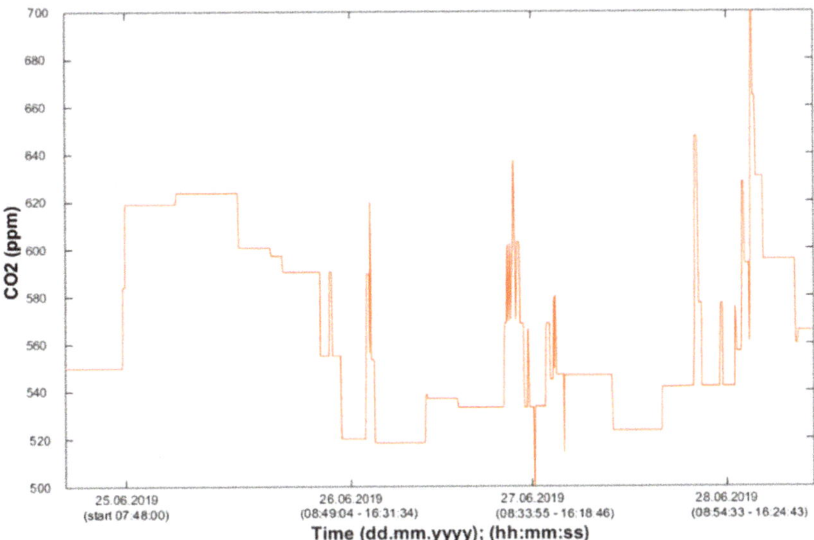

Figure 15. Waveforms of the CO_2 concentration values measured in SH room R203 (BT 12.10).

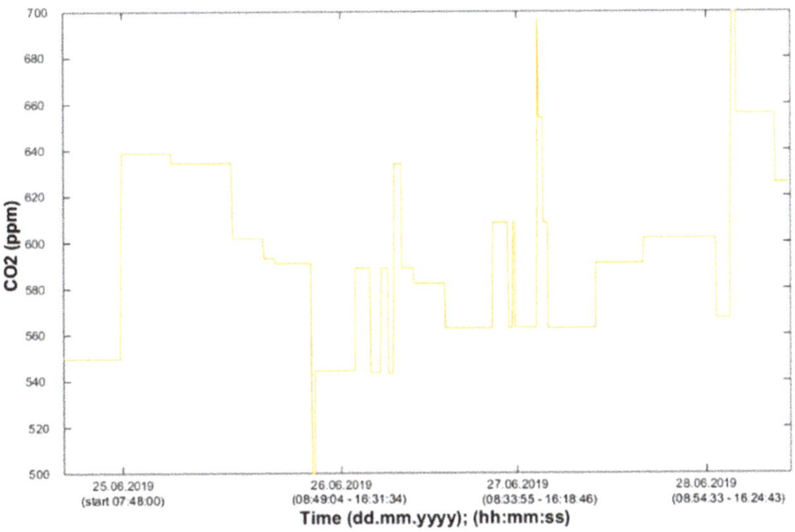

Figure 16. Waveforms of the CO_2 concentration values measured in SH room R204 (BT 12.11).

However, if the building is of an administrative type, the acquisition of CO_2 sensors for dozens of rooms is a major investment. In market research, we ascertained that CO_2 sensors are two to three times (in some cases, greater) more expensive than temperature and humidity sensors, which are, moreover, a common part of individual rooms in administrative buildings in the Czech Republic.

Due to the higher costs of acquiring CO_2 sensors, we proposed the possibility of lowering the initial investment costs for larger administrative buildings by providing information about the occupancy of the individual rooms within the newly devised method of CO_2 prediction. The mathematical method of ANN SCG was used to predict the waveform of CO_2 concentration using the values measured from indoor temperature (T_{in}) and indoor relative humidity (rH_{in}) sensors. The experiments performed within the newly devised method are presented in the following text.

3.3. Experimental Verification of the Method with the Devised Artificial Neural Network (ANN) Scaled Conjugate Gradient (SCG)

The conditions of experiments 3.3a and 3.3b were as follows. The experiments were performed for the waveforms of temperature (T_{in}), relative humidity (rH_{in}) and CO_2 concentration measured on 26 June 2018, 27 June 2018, and 28 June 2018, (step 1) for rooms R104, R203 and R204 (Figure 17).

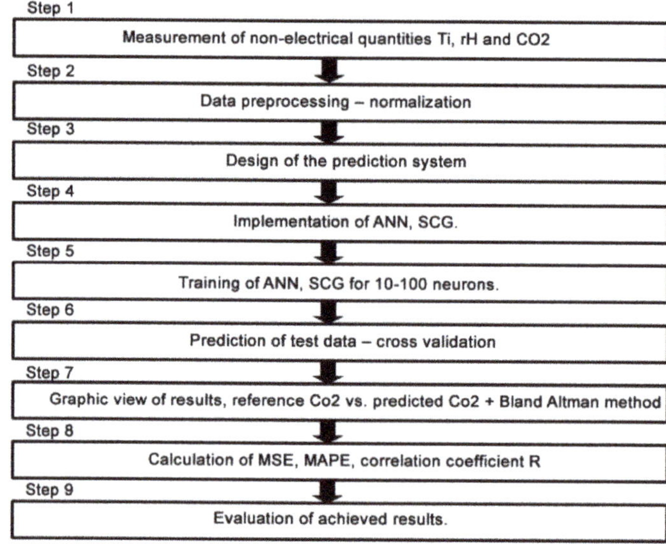

Figure 17. The waveform of the experiments performed on 26 June 2018, 27 June 2018, and 28 June 2018 for rooms R104, R203 and R204.

Furthermore, the measured data were pre-processed (step 2) (Figure 17). The design of the prediction system structure (step 3), the implementation (step 4) and ANN SCG training (step 5) for CO_2 prediction using two input values of temperature (T_{in}) and relative humidity (rH_{in}) were carried out. The ANN SCG structure designed (Figure 18) for experiment 3.3a was trained (step 5) for the temperature (T_{in}), relative humidity (rH_{in}) and CO_2 concentration values measured in room R203. The actual CO_2 prediction (step 6) was implemented for the data measured in rooms R104, R203 and R204. To increase the accuracy of the newly devised method, an additional quantity from the FBG sensor containing the number of occupants in the monitored space of R104 was added to the original two input quantities. The ANN SCG structure designed for experiment 3.3b for CO_2 prediction using three input quantities—temperature (T_{in}), relative humidity (rH_{in}) and the number of occupants (FBG sensor). The procedure for performing the experiments was the same as that described in Figure 17.

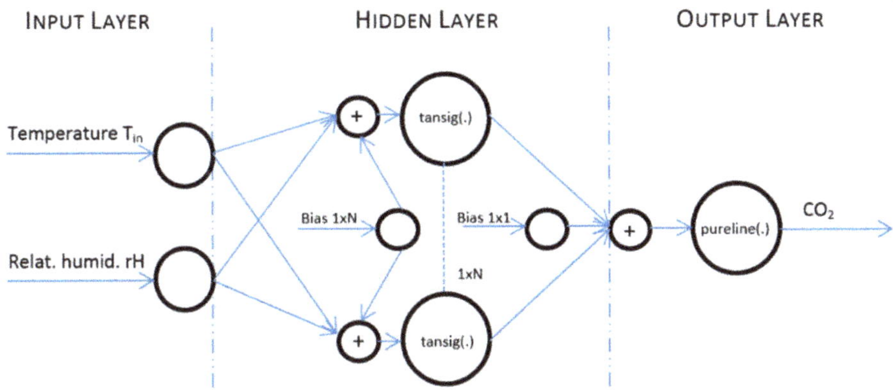

Figure 18. The architecture of the designed ANN SCG on test data measured in R203 from 25 June 2018 for two inputs T_{in} and rH_{in} without and FBG sensor for person presence measuring (PPM).

Experiment 3.3a:

Input values, T_{in} and rH_{in}, to ANN SCG for prediction of CO_2 in rooms R203, R204, and R104 (Figure 18). Prediction of CO_2 in rooms R203, R204, R104 for the dates 26, 27 and 28 June 2018 with inputs T_{in} and rH_{in}, using learned ANN SCG (Figure 18) from 25 June 2018. Tables 2–4 show R, MSE and MAPE parameter values, followed by plots of reference and predicted CO_2, as well as Bland–Altmann plots in rooms 203 (Figures 19 and 20), 204 (Figures 21 and 22) and 104 (Figures 23 and 24).

Table 2. Learned ANN SCG from 25 June 2018 in R203 (with T_{in} and rH_{in}) and prediction with cross-validation for 26 June 2018 in R203 (with T_{in} and rH_{in}), 27 June 2018 in R203 (with T_{in} and rH_{in}), 28 June 2018 in R203 (with T_{in} and rH_{in}).

Number of Neurons ANN SCG	26 June 2018 in R203			27 June 2018 in R203			28 June 2018 in R203		
	MSE	R	MAPE	MSE	R	MAPE	MSE	R	MAPE
	(-)	(-)	(-)	(-)	(-)	(-)	(-)	(-)	(-)
10	0.0062	0.8867	0.2858	0.006	0.5094	0.198	0.0206	0.6398	0.3275
20	0.0063	0.8846	0.289	0.006	0.5098	0.1992	0.0204	0.6449	0.3238
30	0.0058	0.8943	0.2872	0.0058	0.5355	0.1794	0.0198	0.6586	0.3265
40	0.0071	0.8728	0.3237	0.0056	0.5555	0.2036	0.0193	0.6687	0.3186
50	0.0047	0.9151	0.2171	0.0056	0.5591	0.1633	0.0138	0.7784	0.212
60	0.006	0.8918	0.3298	0.0061	0.5084	0.1906	0.0103	0.8401	0.1818
70	0.0059	0.8957	0.3025	0.0054	0.5824	0.1861	0.0127	0.7983	0.2091
80	0.0048	0.9145	0.2475	0.0059	0.5261	0.1845	0.0127	0.7988	0.2558
90	0.0069	0.877	0.1623	0.0065	0.4645	0.2097	0.0106	0.8346	0.2221
100	0.0047	0.916	0.2559	0.0047	0.6489	0.1614	0.0111	0.8266	0.1884

Table 3. Learned ANN SCG from 25 June 2018 in R203 (with T_{in} and rH_{in}) and prediction with cross-validation for 26 June 2018 in R204 (with T_{in} and rH_{in}), 27 June 2018 in R204 (with T_{in} and rH_{in}), and 28 June 2018 in R204 (with T_{in} and rH_{in}).

	26 June 2018 in R204			27 June 2018 in R204			28 June 2018 in R204		
Number of Neurons ANN SCG	MSE	R	MAPE	MSE	R	MAPE	MSE	R	MAPE
	(-)	(-)	(-)	(-)	(-)	(-)	(-)	(-)	(-)
10	0.0107	0.6276	0.2117	0.0112	0.2603	0.209	0.0187	0.4399	0.1699
20	0.0117	0.5816	0.2634	0.0057	0.7272	0.156	0.0087	0.8007	0.1154
30	0.0116	0.5914	0.2917	0.0049	0.7726	0.1232	0.0186	0.4429	0.1678
40	0.0104	0.6402	0.2231	0.0058	0.7213	0.1535	0.0074	0.825	0.1015
50	0.0103	0.6482	0.2425	0.0061	0.7017	0.1387	0.0186	0.4407	0.1692
60	0.0122	0.5606	0.2589	0.0054	0.7425	0.1463	0.0077	0.817	0.11
70	0.0111	0.6125	0.2503	0.0073	0.6261	0.1668	0.0077	0.8159	0.1065
80	0.0106	0.6354	0.2188	0.0052	0.7551	0.1133	0.0085	0.7971	0.1174
90	0.0111	0.6228	0.1671	0.0052	0.7555	0.1458	0.0077	0.8178	0.1078
100	0.0109	0.6232	0.2742	0.0048	0.7755	0.1404	0.0104	0.7416	0.1202

Table 4. Learned ANN SCG from 25 June 2018 in R203 (with T_{in} and rH_{in}) and prediction with cross-validation for 26 June 2018 in R104 (with T_{in} and rH_{in}), 27 June 2018 in R104 (with T_{in} and rH_{in}), and 28 June 2018 in R104 (with T_{in} and rH_{in}).

	26 June 2018 in R104			27 June 2018 in R104			28 June 2018 in R104		
Number of Neurons ANN SCG	MSE	R	MAPE	MSE	R	MAPE	MSE	R	MAPE
	(-)	(-)	(-)	(-)	(-)	(-)	(-)	(-)	(-)
10	0.0097	0.6714	0.3365	0.0121	0.2188	0.2147	0.0187	0.4399	0.1694
20	0.0102	0.6516	0.2259	0.0052	0.7541	0.1466	0.0186	0.4418	0.1713
30	0.0105	0.6398	0.2233	0.0061	0.7052	0.1395	0.0187	0.4396	0.1699
40	0.0099	0.6643	0.2074	0.0055	0.7389	0.144	0.0128	0.6684	0.1197
50	0.0098	0.6685	0.2099	0.0059	0.7136	0.1319	0.0186	0.4411	0.1692
60	0.0098	0.6697	0.2044	0.0055	0.7387	0.1544	0.0025	0.9135	0.0905
70	0.0099	0.6626	0.1937	0.0056	0.7345	0.1464	0.0082	0.8031	0.1089
80	0.0101	0.6542	0.2063	0.0051	0.7567	0.1134	0.0096	0.7676	0.1121
90	0.0095	0.6822	0.1859	0.0051	0.7613	0.1429	0.0085	0.7959	0.1216
100	0.01	0.6587	0.2005	0.0043	0.8027	0.1202	0.0077	0.8175	0.1041

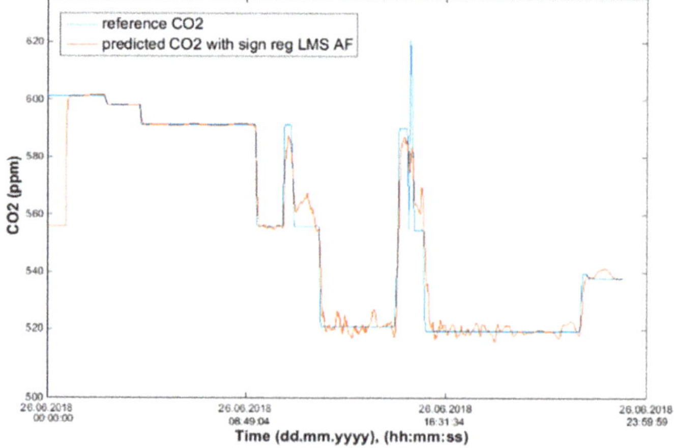

Figure 19. Comparison of the reference CO_2 concentration waveform and predicted CO_2 waveforms (SRLMS AF) from 26 June 2018 in R203 with an ANN with 100 neurons and SCG method trained with data from 25 June 2018 in R203.

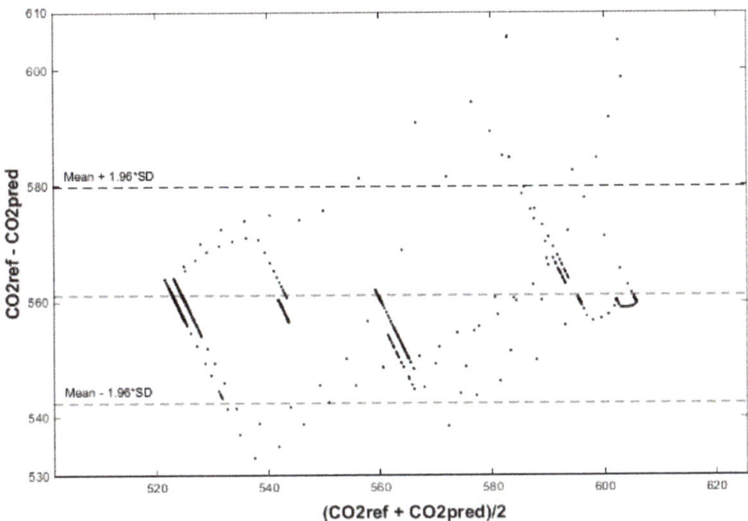

Figure 20. Bland–Altman plot for the reference and predicted CO_2 waveforms (SRLMS AF) from 26 June 2018 in R203 with an ANN with 100 neurons and SCG method trained with data from 25 June 2018 in R203.

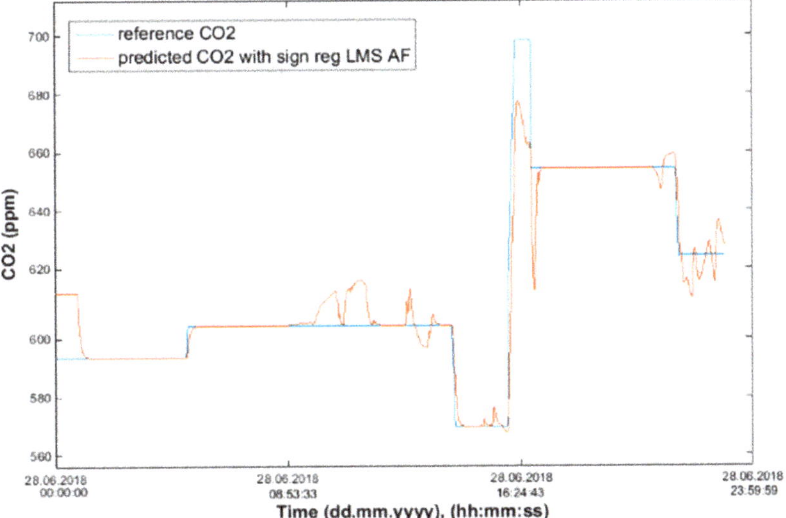

Figure 21. Comparison of the reference and predicted CO_2 waveforms (SRLMS AF) from 28 June 2018 in R204 with an ANN with 40 neurons and SCG method trained with data from 25 June 2018 in R203.

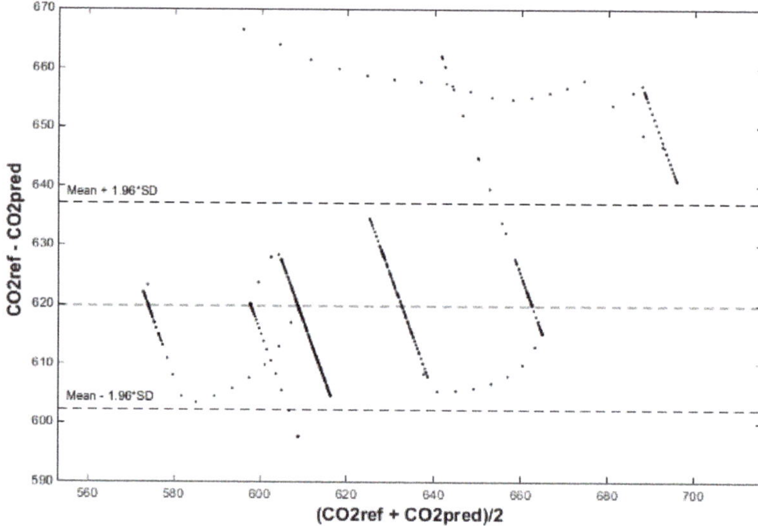

Figure 22. Bland–Altman plot for the reference and predicted CO_2 waveforms (SRLMS AF) from 28 June 2018 in R204 with an ANN with 40 neurons and SCG method trained with data from 25 June 2018 in R203.

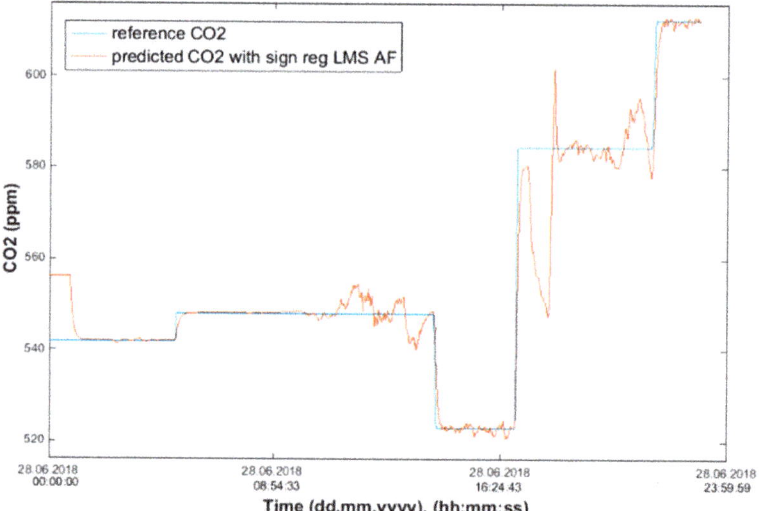

Figure 23. Comparison of the reference and predicted CO_2 waveforms (SRLMS AF) from 28 June 2018 in R104 with an ANN with 60 neurons and SCG method trained with data from 25 June 2018 in R203.

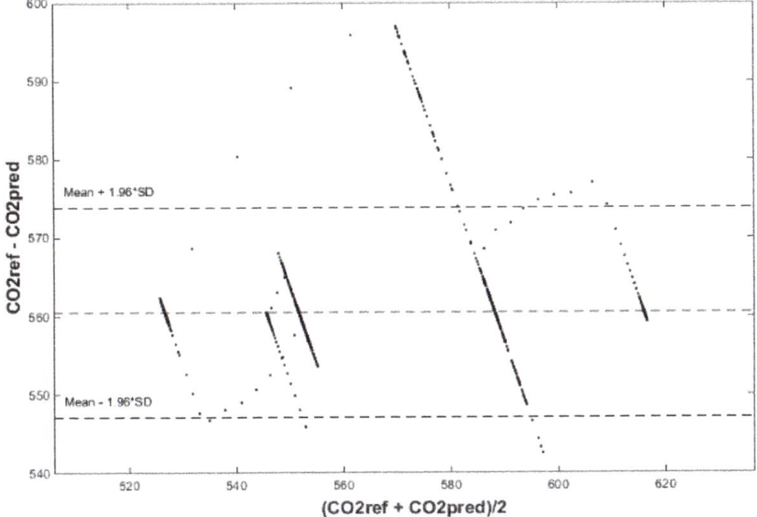

Figure 24. Bland–Altman plot for the reference and predicted CO_2 waveforms (SRLMS AF) from 28 June 2018 in R104 with an ANN with 60 neurons and SCG method trained with data from 25 June 2018 in R203.

Experiment 3b:

Our next experiment was performed with input values of T_{in} and rH_{in} and presence values from the FBG sensor input into ANN SCG for the prediction of CO_2 in rooms R203, R204, R104 (Figure 25). Prediction of CO_2 in rooms R203, R204, R104 for the dates 26, 27 and 28 June 2018 with input values from sensors T_{in}, rH_{in} and FBG using learned ANN SCG (Figure 25) from 25 June 2018. Tables 5–7 show R, MSE and MAPE parameter values, followed by plots of reference and predicted CO_2 as well as Bland–Altmann plots in rooms 203 (Figures 26 and 27), 204 (Figures 28 and 29) and 104 (Figures 30 and 31).

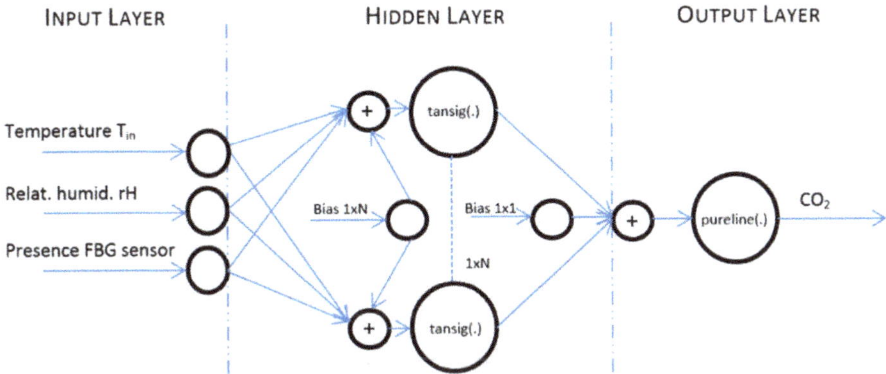

Figure 25. The architecture of designed ANN SCG on test data measured in R203 from 25 June 2018 for three inputs T_{in}, rH_{in} and those of an FBG sensor for PPM.

Table 5. Learned ANN SCG from 25 June 2018 in R203 (with $T_{in,}$ rH_{in} and an FBG sensor for PPM) and prediction with cross-validation for 26 June 2018 in R203 (with $T_{in,}$ rH_{in} and an FBG sensor for PPM), 27 June 2018 in R203 (with $T_{in,}$ rH_{in} and an FBG sensor for PPM), 28 June 2018 in R203 (with $T_{in,}$ rH_{in} and an FBG sensor for PPM).

	26 June 2018 in R203			27 June 2018 in R203			28 June 2018 in R203		
Number of Neurons ANN SCG	MSE (-)	R (-)	MAPE (-)	MSE (-)	R (-)	MAPE (-)	MSE (-)	R (-)	MAPE (-)
10	0.0048	0.9139	0.2285	0.006	0.5094	0.198	0.0201	0.6521	0.2644
20	0.0055	0.9013	0.2814	0.006	0.5098	0.1992	0.0176	0.7052	0.2928
30	0.0057	0.8976	0.301	0.0058	0.5355	0.1794	0.0144	0.7684	0.2296
40	0.0049	0.9117	0.2189	0.0056	0.5555	0.2036	0.0206	0.64	0.3237
50	0.0064	0.8835	0.2802	0.0056	0.5591	0.1633	0.0141	0.7729	0.2079
60	0.004	0.9281	0.2255	0.0061	0.5084	0.1906	0.0132	0.7894	0.2
70	0.0048	0.9135	0.2377	0.0054	0.5824	0.1861	0.0142	0.7704	0.2353
80	0.0052	0.906	0.2541	0.0059	0.5261	0.1845	0.0124	0.8038	0.2117
90	0.0074	0.8684	0.2987	0.0065	0.4645	0.2097	0.0203	0.6467	0.3084
100	0.0044	0.9218	0.1693	0.0047	0.6489	0.1614	0.0203	0.6465	0.3002

Table 6. Learned ANN SCG from 25 June 2018 in R203 (with $T_{in,}$ rH_{in} and an FBG sensor for PPM) and prediction with cross-validation for 26 June 2018 in R204 (with $T_{in,}$ rH_{in} and an FBG sensor for PPM), 27 June 2018 in R204 (with $T_{in,}$ rH_{in} and an FBG sensor for PPM), 28 June 2018 in R204 (with $T_{in,}$ rH_{in} and an FBG sensor for PPM).

	26 June 2018 in R204			27 June 2018 in R204			28 June 2018 in R204		
Number of Neurons ANN SCG	MSE (-)	R (-)	MAPE (-)	MSE (-)	R (-)	MAPE (-)	MSE (-)	R (-)	MAPE (-)
10	0.0111	0.6096	0.0312	0.0111	0.2745	0.2039	0.0187	0.4407	0.172
20	0.0115	0.5939	0.2684	0.0111	0.302	0.2044	0.0053	0.8785	0.1011
30	0.0106	0.6343	0.2093	0.0063	0.6936	0.1521	0.0186	0.4421	0.1699
40	0.0113	0.6022	0.2116	0.0061	0.702	0.1549	0.0186	0.4422	0.1709
50	0.0107	0.6311	0.2047	0.0058	0.7221	0.1323	0.0082	0.8024	0.1139
60	0.01	0.6595	0.1621	0.0037	0.8352	0.1111	0.0078	0.8139	0.101
70	0.0098	0.6679	0.2093	0.0062	0.6994	0.1459	0.0084	0.7997	0.109
80	0.013	0.5346	0.2705	0.0045	0.7912	0.1289	0.0186	0.4409	0.1695
90	0.0109	0.6201	0.2235	0.0052	0.756	0.1452	0.0048	0.8912	0.0891
100	0.0095	0.6812	0.1807	0.0038	0.8289	0.1018	0.0068	0.8397	0.0948

Table 7. Learned ANN SCG from 25 June 2018 in R203 (with $T_{in,}$ rH_{in} and an FBG sensor for PPM) and prediction with cross-validation for 26 June 2018 in R104 (with $T_{in,}$ rH_{in} and an FBG sensor for PPM), 27 June 2018 in R104 (with $T_{in,}$ rH_{in} and an FBG sensor for PPM), 28 June 2018 in R104 (with $T_{in,}$ rH_{in} and an FBG sensor for PPM).

	26 June 2018 in R104			27 June 2018 in R104			28 June 2018 in R104		
Number of Neurons ANN SCG	MSE (-)	R (-)	MAPE (-)	MSE (-)	R (-)	MAPE (-)	MSE (-)	R (-)	MAPE (-)
10	0.0104	0.6432	0.2095	0.0112	0.2735	0.2028	0.0231	0.3505	0.2229
20	0.0117	0.5828	0.0638	0.0109	0.3042	0.2001	0.0195	0.4006	0.2088
30	0.0115	0.5987	0.2458	0.006	0.7079	0.151	0.0187	0.4416	0.1723
40	0.0103	0.6482	0.2155	0.0065	0.6818	0.1532	0.0187	0.4404	0.173
50	0.01	0.6612	0.2053	0.0051	0.7592	0.1147	0.0082	0.8034	0.1133
60	0.01	0.6613	0.2	0.0033	0.8534	0.0968	0.0074	0.8245	0.0953
70	0.0114	0.6135	0.2786	0.0066	0.6759	0.1525	0.0083	0.8015	0.108
80	0.0105	0.6379	0.2151	0.0042	0.8063	0.0555	0.0186	0.443	0.1678
90	0.0095	0.6799	0.1926	0.0048	0.7748	0.1293	0.0082	0.8052	0.1112
100	0.0089	0.7067	0.1906	0.0043	0.8026	0.103	0.0076	0.8194	0.1059

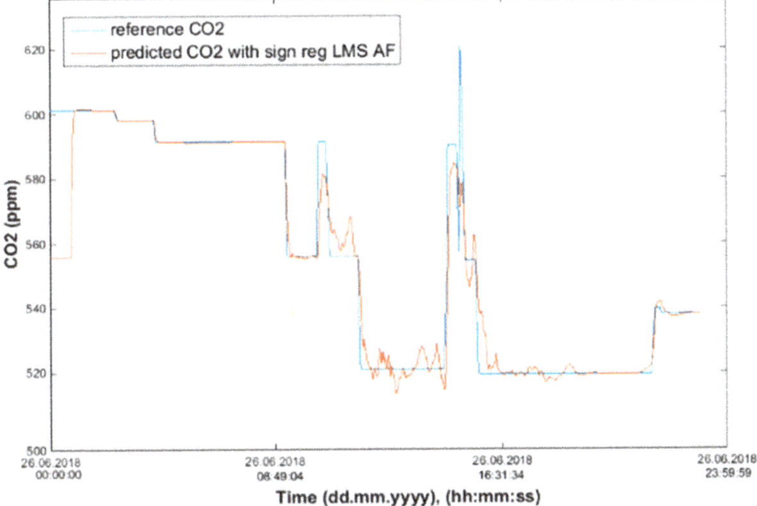

Figure 26. Comparison of the reference and predicted CO_2 waveforms (SRLMS AF) from 26 June 2018 in R203 with an ANN with 60 neurons and SCG method trained with data from 25 June 2018 in R203.

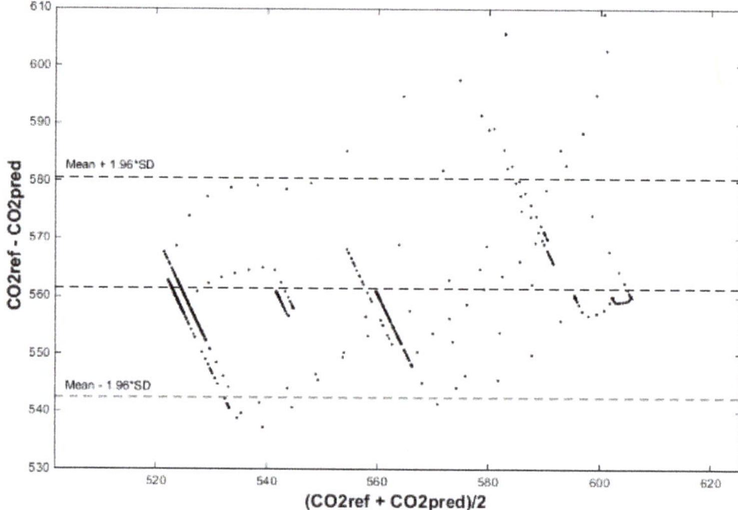

Figure 27. Bland–Altman for the reference and predicted CO_2 waveforms (SRLMS AF) from 26 June 2018 in R203 with an ANN with 60 neurons and SCG method trained with data from 25 June 2018 in R203.

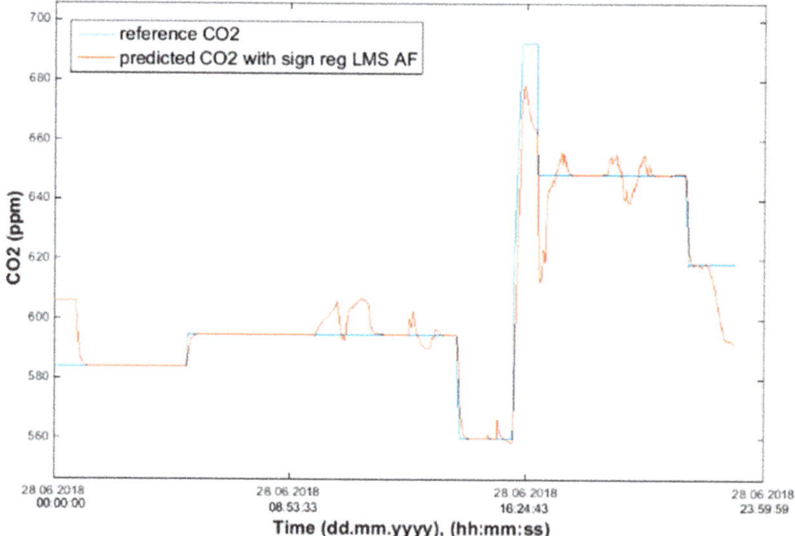

Figure 28. Comparison of the reference and predicted CO_2 waveforms (SRLMS AF) from 28 June 2018 in R204 with an ANN with 90 neurons and SCG method trained with data from 25 June 2018 in R203.

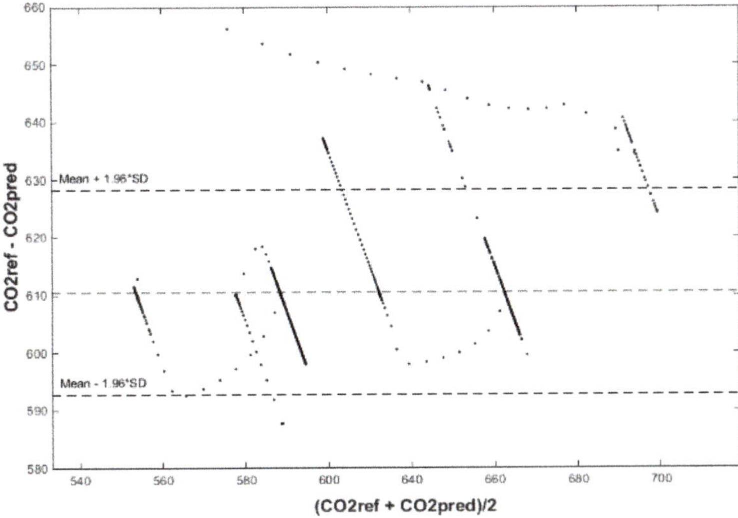

Figure 29. Bland–Altman for the reference and predicted CO_2 waveforms (SRLMS AF) from 28 June 2018 in R204 with an ANN with 90 neurons and SCG method trained with data from 25 June 2018 in R203.

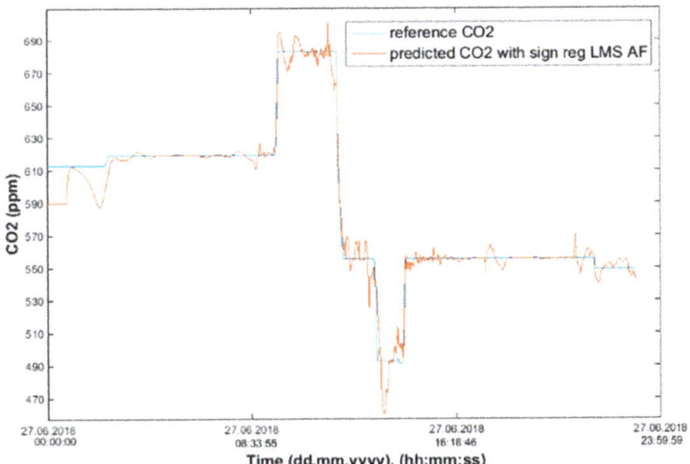

Figure 30. Comparison of the reference and predicted CO_2 waveforms (SRLMS AF) from 27 June 2018 in R104 with an ANN with 60 neurons and SCG method trained with data from 25 June 2018 in R203.

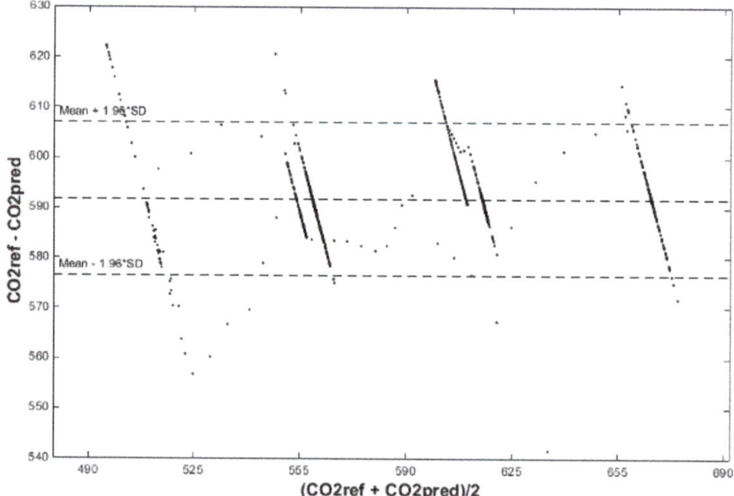

Figure 31. Bland–Altman for the reference and predicted CO_2 waveforms (SRLMS AF) from 27 June 2018 in R104 with an ANN with 60 neurons and SCG method trained with data from 25 June 2018 in R203.

Discussion—Experiments 3.3a and 3.3b:

The best result within the designed three-day (26 June to 28 June 2018) experiments for the learned ANN SCG from 25 June 2018, in R203 (with T_{in} and rH_{in}) and the prediction with cross-validation for 26 June 2018, in R203, R204, R104 (with T_{in} and rH_{in}), 27 June 2018, in R203, R204, R104 (with T_{in} and rH_{in}), 28 June 2018, in R203, R204, R104 (with T_{in} and rH_{in}), without using the FBG sensor measurements, was detected for ANN SCG in room R203 (Table 2) with the following values: R = 91.6 (%), MSE = 0.47×10^{-2}, MAPE = 25.59×10^{-2}, for 100 ANN SCG neurons (Figures 19 and 20). This is because ANN SCG learned for the data measured on 25 June 2018, in room R203 was used for CO_2 prediction. By contrast, the worst calculated result was detected for ANN SCG in room R104 (Table 4), where R = 21.88 (%), MSE = 1.21×10^{-2} and MAPE = 21.47×10^{-2} for a 10 ANN SCG neurons. This is because room R104 was the furthest away from room R203 for which ANN SCG was learned.

As for the CO_2 prediction, the best result within the designed three-day experiments using the FBG sensor for learned ANN SCG from 25 June 2018, in R203 (with T_{in}, rH_{in} and the FBG sensor for PPM) and prediction with cross-validation for 26 June 2018, in R203, R204, R104 (with T_{in}, rH_{in} and the FBG sensor for PPM), 27 June 2018, in R203, R204, R104 (with T_{in}, rH_{in} and the FBG sensor for PPM), 28 June 2018, in R203, R204, R104 (with T_{in}, rH_{in} and the FBG sensor for PPM) was detected for ANN SCG—number of neurons = 60 in room R203 (Figures 26 and 27), (Table 5), where, on 26 June 2018, R = 92.81 (%), MSE = 0.40×10^{-2}, and MAPE = 21.55×10^{-2}. This is because ANN SCG learned for the data measured on 25 June 2018, in room R203 was used for CO_2 prediction. By contrast, the worst calculated result was detected for ANN SCG in room R204 (Table 6), where R = 21.45 (%), MSE = 1.11×10^{-2}, and MAPE = 20.39×10^{-2}, for 10 ANN SCG neurons. This is because there was no FBG sensor in room R204 to detect the presence of persons.

Based on the graphs illustrated in Figures 32a–c and 33a–c, it can be stated that one universal learned ANN SCG cannot be used for the most accurate calculation of CO_2 prediction for all SH spaces. In our experiments, ANN SCG was learned for the data measured in room R203 on 25 June 2018. The calculated values of the verification parameters measured (R coefficient, MSE and MAPE) in Tables 2 and 5 (Figures 19, 20, 26 and 27) indicate that, in order to achieve the greatest accuracy of CO_2 prediction, it is necessary that ANN SCG is always learned for a specific room, for a specific monitored space in the SH. The expectations related to the use of the FBG sensor to increase the accuracy of CO_2 prediction by means of ANN SCG were not fulfilled in the above-mentioned experiments (Figures 32 and 33). However, the use of the FBG sensor in the SH is very useful from the point of view of a robust and stable solution for recognizing the number of occupants in the monitored SH spaces or intelligent administrative buildings. The mathematical method with ANN SCG is not the most suitable for the devised CO_2 prediction method. For greater accuracy, it is necessary to implement filter algorithms [28] to remove additive noise from the predicted CO_2 concentration waveform. Based on the results achieved and described above, further experiments on CO_2 prediction will be conducted using more precise mathematical methods [29–34,37–39] within the IoT platform [40].

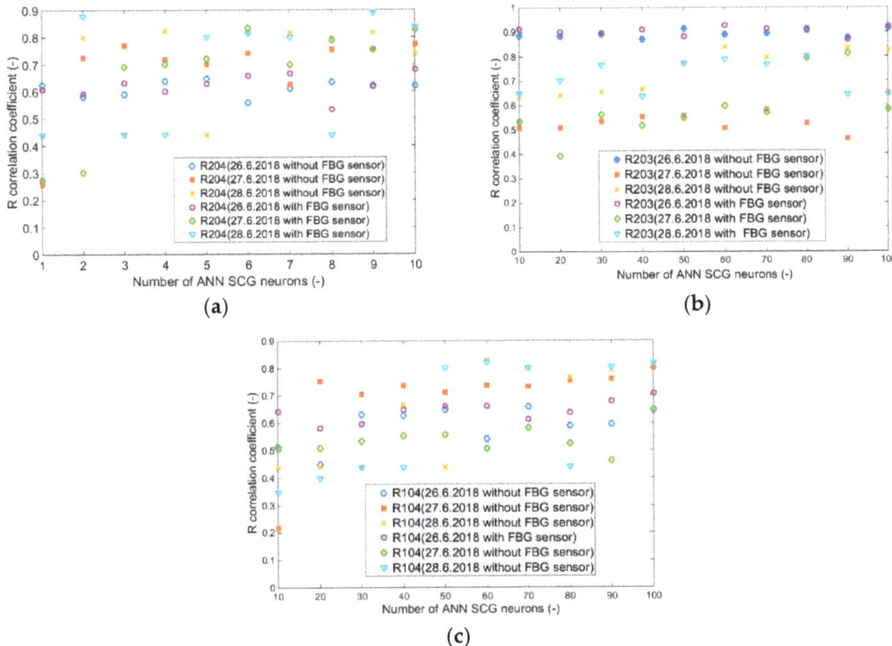

Figure 32. R-correlation coefficients for (**a**) room R203 with and without FBG, (**b**) room R204 with and without FBG, (**c**) room R104 with and without FBG.

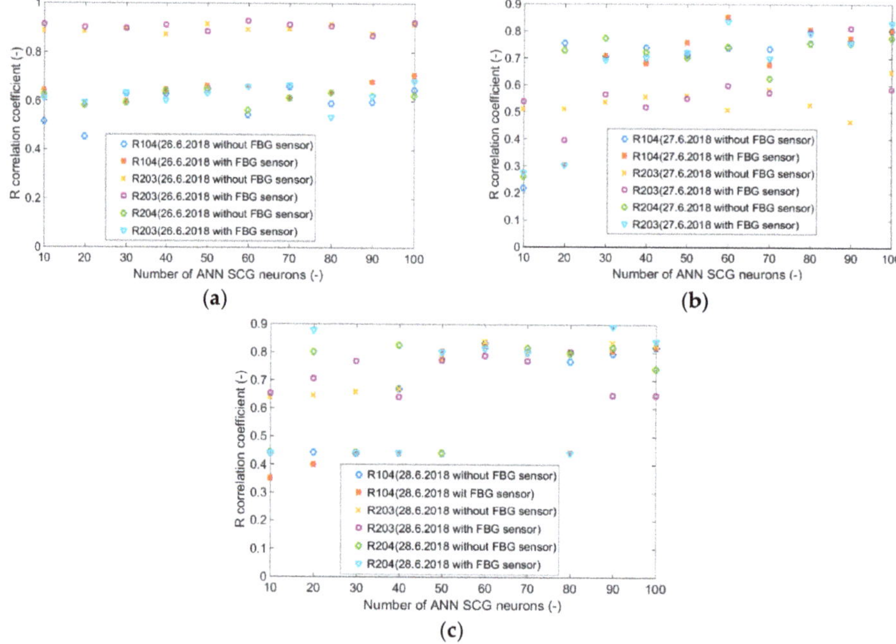

Figure 33. R-correlation coefficients for rooms R104, R203, R204 with and without FBG (**a**) 26 June 2018, (**b**) 27 June 2018, (**c**) 28 June 2018.

4. Conclusions

In order to detect the occupancy of the individual SH spaces by indirect methods (without using cameras), common operational and technical sensors were used in the experiments conducted to measure the indoor temperature, the indoor relative humidity and the indoor CO_2 concentration within the BACnet technology for HVAC control in the SH.

Based on the experiments conducted in rooms R104, R203 and R204, it is possible to unambiguously detect the suitability of the sensors used for measuring the CO_2 concentration in order to detect the occupancy and the length of stay in the monitored SH space. Moreover, the CO_2 sensors used in the technology that were exclusively for HVAC control in the SH can be used to detect the occupancy and the length of stay in the monitored spaces, which will make the required information more accurate.

From the perspective of cost reduction, a method for predicting the CO_2 concentration waveform by means of ANN SCG from the indoor temperature and indoor relative humidity values measured were devised and verified for the space location of an occupant (in rooms R104, R203, and R204) in the SH with the highest possible accuracy. The ANN SCG mathematical method used does not achieve such accuracy compared to the methods used in [28–30] and [34,37–40]. For selected experiments, nevertheless, the correlation coefficient in this article was greater than 90%. An increase in the correlation coefficient value can be achieved by using filter methods for suppressing additive noise from the predicted CO_2 concentration waveform [28]. In the experiments conducted, we verified that, for each monitored space in the SH, the ANN SCG should be learned so that the CO_2 prediction was as accurate as possible.

The FBG sensor was used to unambiguously detect the number of occupants in the monitored space of room R104. The experiments confirmed the suitability of using FBG sensors for reasons of robustness, accuracy and high reliability in the detection of the number of occupants moving in the monitored SH or Intelligent Building space without the need to deploy cameras. The experiments did not confirm the assumption that the input value of the presence of occupants added from the FBG sensor to the ANN SCG will make the prediction of the CO_2 concentration using the ANN SCG method more accurate.

The application of FBG sensors in a passive SH is described. These sensors are electrically passive, while allowing the evaluation of multiple quantities at the same time, and with suitable encapsulation of FBG, the sensor is immune to EMI (electromagnetic interference). Due to its small size and weight, implementation in an SH is both straightforward and cost-effective. Using a spectral approach, tens of FBG sensors can be evaluated using a single evaluation unit [21].

In the next work, the authors will focus on verifying the practical implementation of the method devised for CO_2 prediction in an SH for monitoring the occupancy of the SH within the IoT platform [40].

Author Contributions: J.V., methodology; J.V., J.N., M.F. software; J.V., J.N., M.F. validation; J.V., formal analysis; J.V. investigation; J.V., J.N., M.F. resources; J.V., J.N., M.F., data creation; J.V., J.N., M.F. writing—original draft preparation; J.V., J.N., M.F. writing—review and editing; J.V., visualization; J.V., supervision; R.M., project administration; and R.M., funding acquisition. All authors have read and agreed to the published version of the manuscript.

Funding: This work was supported by the European Regional Development Fund in the Research Centre of Advanced Mechatronic Systems project, project number CZ.02.1.01/0.0/0.0/16_019/0000867 within the Operational Programme Research, Development and Education.

Acknowledgments: This work was supported by the Student Grant System of VSB Technical University of Ostrava, grant number SP2019/118. This work was supported by the European Regional Development Fund in the Research Centre of Advanced Mechatronic Systems project, project number CZ.02.1.01/0.0/0.0/16_019/0000867 within the Operational Programme Research, Development and Education.

Conflicts of Interest: The authors declare no conflict of interest.

Abbreviations

The following abbreviations are used in this manuscript:

(ANN)	artificial neural network
(FBG)	fiber Bragg grating
(SCG)	scaled conjugate gradient
(LMS)	least mean squares
(SRLMS)	signed—regressor least mean squares algorithm
(AF)	adaptive filter
(SH)	smart home
(SW)	software
(CO2)	carbon dioxide
(rH_{in})	relative humidity indoor
(T_{in})	temperature indoor
R104	room 104 in the smart home
R203	room 203 in the smart home
R204	room 204 in the smart home
(HVAC)	heating, ventilation and air conditioning
(IoT)	internet of things
(MSE)	mean squared error
(MAPE)	average absolute percentage error
(R)	correlation coefficient
(BACnet)	data communication protocol for building automation and control networks
(PPM)	person presence measuring

References

1. Azghandi, M.V.; Nikolaidis, I.; Stroulia, E. Multi–occupant movement tracking in smart home environments. In Proceedings of the International Conference on Smart Homes and Health Telematics, Geneva, Switzerland, 10–12 June 2015; Volume 9102, pp. 319–324.

2. Benmansour, A.; Bouchachia, A.; Feham, M. Multioccupant activity recognition in pervasive smart home environments. *ACM Comput. Surv.* **2015**, *48*, 34. [CrossRef]
3. Braun, A.; Majewski, M.; Wichert, R.; Kuijper, A. Investigating low-cost wireless occupancy sensors for beds. In Proceedings of the 2016 International Conference on Distributed, Ambient, and Pervasive Interactions, Toronto, ON, Canada, 17–22 July 2016; Volume 9749, pp. 26–34.
4. Chan, Z.Y.; Shum, P. *Smart Office—A Voice-Controlled Workplace for Everyone*; Nanyang Technological University: Singapore, 2018.
5. Chen, Z.; Wang, Y.; Liu, H. Unobtrusive sensor-based occupancy facing direction detection and tracking using advanced machine learning algorithms. *IEEE Sens. J.* **2018**, *18*, 6360–6368. [CrossRef]
6. Khan, M.A.A.H.; Kukkapalli, R.; Waradpande, P.; Kulandaivel, S.; Banerjee, N.; Roy, N.; Robucci, R. RAM: Radar-Based Activity Monitor. In Proceedings of the IEEE INFOCOM 2016—The 35th Annual IEEE International Conference on Computer Communications, San Francisco, CA, USA, 10–14 April 2016.
7. Lee, H.; Wu, C.; Aghajan, H. Vision-based user-centric light control for smart environments. *Pervasive Mob. Comput.* **2011**, *7*, 223–240. [CrossRef]
8. Mokhtari, G.; Zhang, Q.; Nourbakhsh, G.; Ball, S.; Karunanithi, M. BLUESOUND: A New Resident Identification Sensor—Using Ultrasound Array and BLE Technology for Smart Home Platform. *IEEE Sens. J.* **2017**, *17*, 1503–1512. [CrossRef]
9. Yan, D.; Jin, Y.; Sun, H.; Dong, B.; Ye, Z.; Li, Z.; Yuan, Y. Household appliance recognition through a Bayes classification model. *Sustain. Cities Soc.* **2019**, *46*, 101393. [CrossRef]
10. Yang, J.; Zou, H.; Jiang, H.; Xie, L. Device-Free Occupant Activity Sensing Using WiFi-Enabled IoT Devices for Smart Homes. *IEEE Internet Things J.* **2018**, *5*, 3991–4002. [CrossRef]
11. Feng, X.; Yan, D.; Hong, T. Simulation of occupancy in buildings. *Energy Build.* **2015**, *87*, 348–359. [CrossRef]
12. Yin, J.; Fang, M.; Mokhtari, G.; Zhang, Q. Multi-resident location tracking in smart home through non-wearable unobtrusive sensors. In Proceedings of the 14th International Conference on Smart Homes and Health Telematics (ICOST 2016), Wuhan, China, 25–27 May 2016; Volume 9677, pp. 3–13.
13. Hong, T.; Sun, H.; Chen, Y.; Taylor-Lange, S.C.; Yan, D. An occupant behavior modeling tool for co-simulation. *Energy Build.* **2016**, *117*, 272–281. [CrossRef]
14. Vanus, J.; Martinek, R.; Bilik, P.; Zidek, J.; Skotnicova, I. Evaluation of thermal comfort of the internal environment in smart home using objective and subjective factors. In Proceedings of the 17th International Scientific Conference on Electric Power Engineering, Prague, Czech Republic, 16–18 May 2016.
15. Polednik, J.; Pavlik, J.; Skotnicova, I. Research and Innovation Centre. Available online: http://www.vyzkumneinovacnicentrum.cz/wp-content/uploads/2014/10/Vyzkumne-inovacni-centrum.pdf (accessed on 15 August 2019).
16. Hill, K.O.; Malo, B.; Bilodeau, F.; Johnson, D.C.; Albert, J. Bragg gratings fabricated in monomode photosensitive optical fiber by UV exposure through a phase mask. *Appl. Phys. Lett.* **1993**, *62*, 1035–1037. [CrossRef]
17. Othonos, A. Fiber bragg gratings. *Rev. Sci. Instrum.* **1997**, *68*, 4309–4341. [CrossRef]
18. Kunzler, M.; Udd, E.; Taylor, T.; Kunzler, W. Traffic Monitoring Using Fiber Optic Grating Sensors on the I-84 Freeway & Future Uses in WIM. In Proceedings of the Sixth Pacific Northwest Fiber Optic Sensor Workshop, Troutdale, OR, USA, 14 May 2003; Volume 5278, pp. 122–127.
19. Wei, C.-L.; Lai, C.-C.; Liu, S.-Y.; Chung, W.H.; Ho, T.K.; Tam, H.-Y.; Ho, S.L.; McCusker, A.; Kam, J.; Lee, K.Y. A fiber Bragg grating sensor system for train axle counting. *IEEE Sens. J.* **2010**, *10*, 1905–1912.
20. Moorman, W.; Taerwe, L.; De Waele, W.; Degrieck, J.; Himpe, J. Measuring ground anchor forces of a quay wall with Bragg sensors. *J. Struct. Eng.* **2005**, *131*, 322–328. [CrossRef]
21. Fajkus, M.; Nedoma, J.; Martinek, R.; Vasinek, V.; Nazeran, H.; Siska, P. A non-invasive multichannel hybrid fiber-optic sensor system for vital sign monitoring. *Sensors* **2017**, *17*, 111. [CrossRef] [PubMed]
22. Fajkus, M.; Nedoma, J.; Siska, P.; Bednarek, L.; Zabka, S.; Vasinek, V. Perimeter system based on a combination of a Mach-Zehnder interferometer and the bragg gratings. *Adv. Electr. Electron. Eng.* **2016**, *14*, 318–324. [CrossRef]
23. Fajkus, M.; Navruz, I.; Kepak, S.; Davidson, A.; Siska, P.; Cubik, J.; Vasinek, V. Capacity of wavelength and time division multiplexing for quasi-distributed measurement using fiber bragg gratings. *Adv. Electr. Electron. Eng.* **2015**, *13*, 575–582. [CrossRef]

24. Nedoma, J.; Fajkus, M.; Bednarek, L.; Frnda, J.; Zavadil, J.; Vasinek, V. Encapsulation of FBG sensor into the PDMS and its effect on spectral and temperature characteristics. *Adv. Electr. Electron. Eng.* **2016**, *14*, 460–466. [CrossRef]
25. Zhang, Y.; Chang, X.; Zhang, X.; He, X. The Packaging Technology Study on Smart Composite Structure Based on the Embedded FBG Sensor. In *IOP Conference Series: Materials Science and Engineering*; IOP Publishing Ltd.: Bristol, UK, 2018; Volume 322.
26. Anoshkin, A.N.; Shipunov, G.S.; Voronkov, A.A.; Shardakov, I.N. Effect of temperature on the spectrum of fiber Bragg grating sensors embedded in polymer composite. In *AIP Conference Proceedings*; AIP Publishing: Tomsk, Russia, 2017; Volume 1909.
27. Wang, T.; He, D.; Yang, F.; Wang, Y. Fiber Bragg grating sensors for strain monitoring of steelwork. In Proceedings of the 2009 International Conference on Optical Instruments and Technology, Shanghai, China, 19–22 October 2009; Volume 7508.
28. Vanus, J.; Belesova, J.; Martinek, R.; Nedoma, J.; Fajkus, M.; Bilik, P.; Zidek, J. Monitoring of the daily living activities in smart home care. *Human Cent. Comput. Inf. Sci.* **2017**, *7*, 30. [CrossRef]
29. Vanus, J.; Kubicek, J.; Gorjani, O.M.; Koziorek, J. Using the IBM SPSS SW Tool with Wavelet Transformation for CO_2 Prediction within IoT in Smart Home Care. *Sensors* **2019**, *19*, 1407. [CrossRef]
30. Vanus, J.; Machac, J.; Martinek, R.; Bilik, P.; Zidek, J.; Nedoma, J.; Fajkus, M. The design of an indirect method for the human presence monitoring in the intelligent building. *Human Cent. Comput. Inf. Sci.* **2018**, *8*, 28. [CrossRef]
31. Moller, M.F. A scaled conjugate gradient algorithm for fast supervised learning. *Neural Netw.* **1993**, *6*, 525–533. [CrossRef]
32. Steihaug, T. The conjugate gradient method and trust regions in large scale optimization. *SIAM J. Numer. Anal.* **1983**, *20*, 626–637. [CrossRef]
33. Branch, M.A.; Coleman, T.F.; Li, Y. Subspace, interior, and conjugate gradient method for large-scale bound-constrained minimization problems. *SIAM J. Sci. Comput.* **1999**, *21*, 1–23. [CrossRef]
34. Vanus, J.; Martinek, R.; Bilik, P.; Zidek, J.; Dohnalek, P.; Gajdos, P. New method for accurate prediction of CO2 in the Smart Home. In Proceedings of the 2016 IEEE International Instrumentation and Measurement Technology Conference Proceedings, Taipei, Taiwan, 23–26 May 2016.
35. Farhang-Boroujeny, B. *Adaptive Filters: Theory and Applications*; John Wiley & Sons: Chichester, UK, 1998.
36. Poularikas, A.D.; Ramadan, Z.M. *Adaptive Filtering Primer with MATLAB*; CRC Press: Boca Raton, FL, USA, 2017.
37. Vanus, J.; Martinek, R.; Nedoma, J.; Fajkus, M.; Cvejn, D.; Valicek, P.; Novak, T. Utilization of the LMS Algorithm to Filter the Predicted Course by Means of Neural Networks for Monitoring the Occupancy of Rooms in an Intelligent Administrative Building. *IFAC-PapersOnLine* **2018**, *51*, 378–383. [CrossRef]
38. Vanus, J.; Martinek, R.; Nedoma, J.; Fajkus, M.; Valicek, P.; Novak, T. Utilization of Interoperability between the BACnet and KNX Technologies for Monitoring of Operational-Technical Functions in Intelligent Buildings by Means of the PI System SW Tool. *IFAC-PapersOnLine* **2018**, *51*, 372–377. [CrossRef]
39. Vanus, J.; Sykora, J.; Martinek, R.; Bilik, P.; Koval, L.; Zidek, J.; Fajkus, M.; Nedoma, J. Use of the software PI system within the concept of smart cities. In Proceedings of the 9th International Scientific Symposium on Electrical Power Engineering (ELEKTROENERGETIKA 2017), Stará Lesná, Slovakia, 12–14 September 2017.
40. Petnik, J.; Vanus, J. Design of Smart Home Implementation within IoT with Natural Language Interface. *IFAC-PapersOnLine* **2018**, *51*, 174–179. [CrossRef]

Article

Wavelet-Based Filtration Procedure for Denoising the Predicted CO_2 Waveforms in Smart Home within the Internet of Things

Jan Vanus *, Klara Fiedorova, Jan Kubicek, Ojan Majidzadeh Gorjani and Martin Augustynek

Department of Cybernetics and Biomedical Engineering, VSB-Technical University of Ostrava, FEECS, 708 Ostrava-Poruba, Czech Republic; klara.fiedorova@vsb.cz (K.F.); jan.kubicek@vsb.cz (J.K.); ojan.majidzadeh.gorjani@vsb.cz (O.M.G.); martin.augustynek@vsb.cz (M.A.)

* Correspondence: jan.vanus@vsb.cz

Received: 17 December 2019; Accepted: 17 January 2020; Published: 22 January 2020

Abstract: The operating cost minimization of smart homes can be achieved with the optimization of the management of the building's technical functions by determination of the current occupancy status of the individual monitored spaces of a smart home. To respect the privacy of the smart home residents, indirect methods (without using cameras and microphones) are possible for occupancy recognition of space in smart homes. This article describes a newly proposed indirect method to increase the accuracy of the occupancy recognition of monitored spaces of smart homes. The proposed procedure uses the prediction of the course of CO_2 concentration from operationally measured quantities (temperature indoor and relative humidity indoor) using artificial neural networks with a multilayer perceptron algorithm. The mathematical wavelet transformation method is used for additive noise canceling from the predicted course of the CO_2 concentration signal with an objective increase accuracy of the prediction. The calculated accuracy of CO_2 concentration waveform prediction in the additive noise-canceling application was higher than 98% in selected experiments.

Keywords: intelligent buildings; wavelet transformation; prediction; artificial neural network; multilayer perceptron; cloud computing; Internet of Things; smart home

---

1. Introduction

In the field of intelligent building (IB) automation and in the context of optimized management of operational-technical functions to reduce operating costs, increase control, and comfort, the European Union has published the directive "Directive (European Union) 2018/844 of the European Parliament and of the Council of 30 May 2018" emphasizes monitoring and processing of measured data in real time. Building automation and electronic monitoring of building technical systems offer considerable potential for cost-effective and significant energy savings for both the consumers and businesses [1].

In this article, the authors describe the implementation of KNX (Konnex (standard EN 50090, ISO/IEC 14543) technology with IBM Internet of Things (IoT) platform connectivity for monitoring and processing of measured data in real time within IB automation. On the basis of the measured values of carbon dioxide (CO_2) concentration, it is possible to detect the occupancy of the monitored smart home (SH) spaces, the arrival of a person into the monitored room, or the exit from the monitored space, or the length of stay in the monitored space. The aforementioned procedure enables the unambiguous determination of the occupancy rate of the monitored SH spaces, indirectly by measuring common nonelectrical quantities of CO_2 within the operational–technical function control in the SH. In market research, we ascertained that CO_2 sensors are two to three times (in some cases, greater) more expensive than temperature and humidity sensors, which are, moreover, a common part of IB in the Czech Republic.

Due to the higher costs of acquiring CO_2 sensors, we proposed the possibility of lowering the initial investment costs for IB by providing information about the occupancy of the individual rooms within a novel indirect method for monitoring IB spaces presence occupancy using a temperature indoor sensor and relative humidity sensor instead of a CO_2 indoor sensor. This method uses predictive modeling (using Statistical Package for the Social Sciences from the company IBM (SPSS) Modeler 18) with the application of an artificial neural network (ANN) for the prediction of CO_2 concentration using the measured values of indoor temperature and indoor relative humidity. To increase the accuracy of this method, the additive noise canceling method was used with wavelet transformation. This article emphasizes the adjustment and optimization of individual parameters of wavelet transformation (mathematical filtration method) for additive noise canceling and an increase in CO_2 prediction accuracy. Additionally, this article illustrates the data collection and the IoT platform connectivity within KNX technology as a practical part of (SH) control simulation. The main goal of the authors is to find the optimal setting of individual parameters of the wavelet transform in the additive noise suppression application with an emphasis on increasing the accuracy of predictions of CO_2 concentration from measured values of indoor temperature and indoor relative humidity for monitoring and recognition of IB space occupancy using common operation sensors.

Related Works

The monitoring of technical systems can be utilized using android mobile visualization applications [2] (the prediction was performed by the ANN Bayesian regulation method (BRM) with least mean square (LMS) adaptive filtration (AF) additive noise canceling, best accuracy was better than 90%), supervisory control and data acquisition (SCADA) visualization systems, or robust software (SW) tools for collecting and archiving measured data in smart home care (SHC) [3] (the prediction was performed by the ANN-based on the Levenberg–Marquardt algorithm (LMA), experimental results verified high method accuracy > 95%). Similarly, the IoT platform can also be employed to monitor and visualize technical systems in IB [4]. KNX is one of the many technologies that are widely used to control the technical and operational functions of IB worldwide [5]. Petnik et al. describes the implementation of KNX technology for controlling and monitoring operational and technical functions in SHs within the IoT cloud platform [6]. The SHC platform and health care platform [7,8] are being prepared to use the IoT concepts within the fifth generation of the mobile network standard [9].

Measured values of nonelectrical and electrical quantities in real time using implemented KNX technology in SH (presence of persons, power consumption, temperature, relative humidity, or CO_2 concentration) need to be preprocessed and adjusted for subsequent calculations using appropriate mathematical methods (classification [10,11], recognition [12–14], and prediction [15]) (the prediction was performed by the ANN-based on the scaled conjugate gradient (SCG), experimental results verified high method accuracy > 90%), [16] (the prediction was performed by the ANN Bayesian regulation method (BRM) with LMS AF additive noise canceling, best accuracy was better than 95%) [17] (the prediction was performed by decision tree regression method with the accuracy of 46.25 ppm). An important area of the described chain is the suppression of additive noise from the measured and calculated waveforms of monitored quantities [18,19]. The disadvantage of using an LMS adaptive filter in an additive noise suppression application is the slow startup of the filtering process in the initial phase, depending on the step size parameter set and the adaptive filter order.

Therefore, we decided to use the wavelet transform in an application to suppress additive noise from the predicted CO_2 signal in order to increase the accuracy of CO_2 prediction.

The signal noise represents a significant problem for signal representation and further processing [20]. Each signal is represented by the trend component, determining the signal evolution over time [21]. The further signal components, including periodic and aperiodic parts, are superimposed on the signal trend [22]. These components standard represent an additive part having the noise character, thus, deteriorate the visual quality and features of the CO_2 signals [23]. Therefore, we aimed to extract the trend component of the CO_2 signal, while the additive noise is supposed to be suppressed. For noise suppression, many methods have been developed. Mostly, these methods utilize a sliding window, passing through the signal area and approximate local features of the signals by statistical parameters, such as the average or median filters, adaptive filters, or fitting procedures, including the Savitzky–Golay filter [24,25]. Nevertheless, such methods are not capable of adjusting to the local frequency content, except for the adaptive filters which are time-consuming and require a reference signal, which can be a complication. An approximation of the local frequencies by an adjustable window function is crucial for the nonstationary signals, where we observe time-varying frequency content over time [26–28]. For this reason, we apply the wavelet-based filtration with the goal of time-frequency localization of the CO_2 signal trend [29–31].

This study is divided into the following sections: The Introduction provides the motivation and current state of the art on the topic of wavelet-based filtration procedure for denoising the predicted CO_2 waveforms in SH within IoT. The next section describes building automation and data collection with KNX technology, preprocessing of the collected data, predictive modeling (using IBM SPSS Modeler 18), additive noise cancelation with wavelet transformation, description of experiments, and evaluation of the obtained results. Finally, the results are discussed with comparisons to existing solutions.

2. Materials and Methods

The practical implementation of prediction and filtration of CO_2 waveforms are divided into the following parts (Figure 1):

1. Building automation and data collection (using KNX technology);
2. Preprocessing of the collected data;
3. Predictive modeling (using IBM SPSS Modeler 18);
4. Filtration (wavelet transformation);
5. Evaluation of the obtained results.

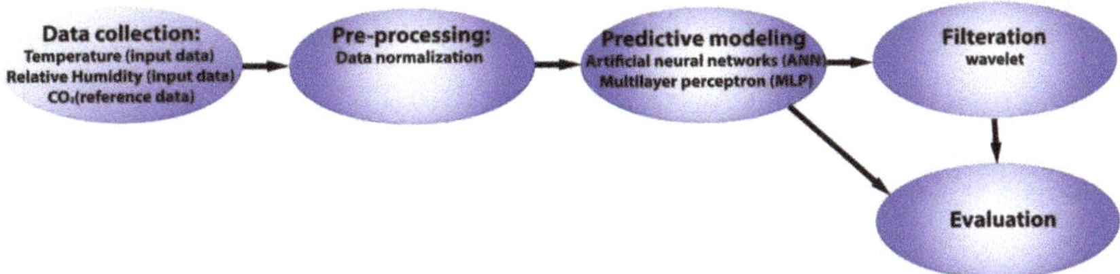

Figure 1. Block diagram of the analysis.

2.1. *Building Automation and Data Collection Using KNX Technology*

KNX is a worldwide standard (EN 50090, ISO/IEC 14543) for building automation. The creators and owners of KNX technology are the KNX Association. Product certification based on the KNX standard guarantees the compatibility of products of different companies (Siemens, ABB, Schneider Electric, WAGO, and others), which represents a high level of flexibility. KNX technology is a decentralized system, i.e., the KNX bus system that does not need a PC or a central control unit to operate. All of the information and data are stored in microprocessors of the individual KNX modules (KNX bus participants) which communicate with each other on the same level, the so-called multi-master communication. Commissioning is done using the Engineering Tool Software (ETS) software. KNX can provide a variety of applications for lighting control, sun protection, heating, cooling, ventilation, energy management, convenience control, etc. The KNX application is used to control the operational-technical functions of office buildings, shopping centers, medical facilities, institutions, banks, industrial locations, etc. This control system does not only bring the comfort of operation but above all, it is an efficient tool for efficient control of operational technical functions.

To simulate SH operation, KNX test panels (containing KNX modules) were placed in the laboratory EB312 at the new FEI (Faculty of Electrical Engineering and Computer Science) building at the VSB Technical University of Ostrava. This location often holds educational classes, or it is visited by staff and researchers. Using the modules displayed in Figure 2, it was possible to simulate the control of operational functions in SH. The measurements of CO_2 accumulation, indoor temperature, and humidity were performed using the MTN6005-0001 module. The measuring range of this device is listed below:

- CO_2 sensor, 300 to 9999 ppm;
- Temperature sensor, 0 °C to +40 °C;
- Air humidity sensor, 20% to 100%.

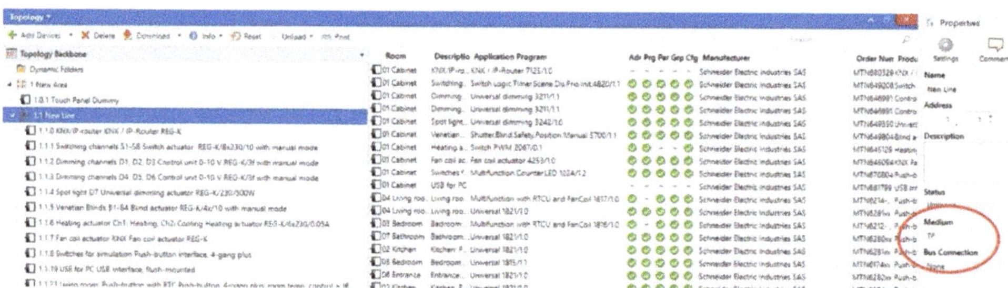

Figure 2. Created KNX topology and selected medium TP KNX in the software (SW) tool ETS 5.

In KNX topology, there are the following four available communication media for the actual transmission of data telegrams between individual KNX modules: Twisted Pair (TP) (also known as KNX bus), powerline (PL), radio frequency (RF), and ethernet (IP). Each communication medium can be used in combination with one or more configuration modes. Economical operation and comfort in the control of operational-technical functions are the main priorities of implementations within family houses. Therefore, TP was selected as the backbone structure of this implementation (Figure 2).

The operation of the individual components of KNX technology is ensured by the means of group addresses (Figure 3).

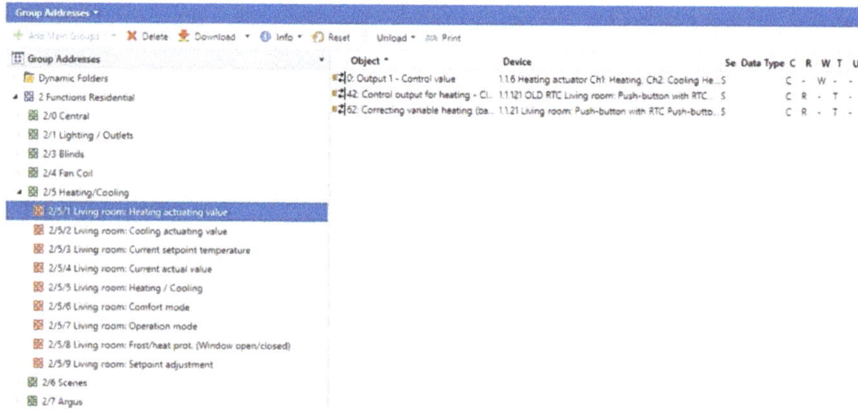

Figure 3. Group address structure created to simulate the management of operational and technical functions in the Engineering Tool Software (ETS) tool.

The connection of KNX technology and IBM cloud technology is ensured in this work by our developed software [32], which enables the communication between IBM Watson IoT platform and KNX smart installation (Figure 4). Message queuing telemetry transport (MQTT) protocol is used as a communication protocol.

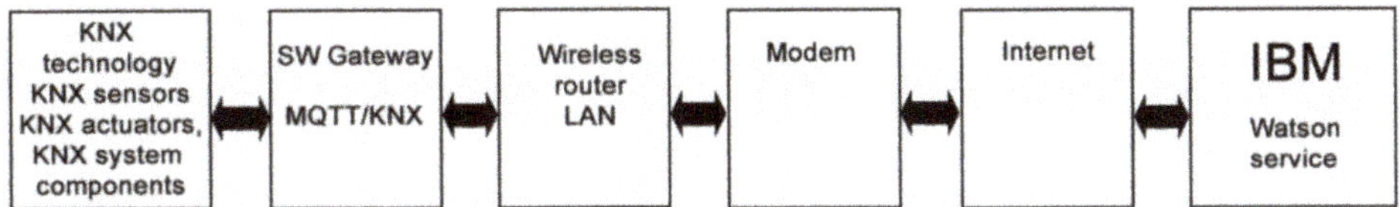

Figure 4. Block diagram of created SW connection between KNX technology on the building of new FEI VSB-TU Ostrava and IBM IoT Watson platform [29].

2.2. *Preprocessing of the Collected Data*

Data normalization (using feature scaling) was selected as the preprocessing stage. Feature scaling (min–max normalization) is rescaling the range of features to the scale of zero to one. Mainly, the feature scaling is applied because the gradient descent converges much faster with feature scaling than without it. The general formula for a min–max is given as [29]:

$$\text{Normalized value} = \frac{\text{current value} - \text{minimum value}}{\text{maximum value} - \text{minimum value}} \tag{1}$$

2.3. *Predictive Modeling (Using IBM SPSS)*

Predictive models are based on variables (predictors) that are most likely to influence the outcome (prediction) [33]. This article employs a predictive model that is categorized as machine learning with a supervised learning strategy. Machine learning can be described as the process of computers making intelligent decisions by learning and recognizing patterns based on the sample data. In a supervised learning strategy, the machine establishes a pattern between the problem and the answer by learning from a set of solved (labeled) examples. Once the pattern is established the machine is able to solve similar problems [34]. ANNs are one of the most popular modeling methods used in predictive applications (such as [35–42]), due to their power flexibility and ease of use. In general, ANNs obtain their knowledge from the learning process and then use interneuron connection strengths (known as synaptic weights) to store the obtained knowledge [43,44]. One of the most commonly used classes of ANNs is a multilayer perceptron (MLP). MLP is a feedforward neural network. In addition, input and output layers of the MLP can contain multiple hidden layers (at least one) and each can contain multiple neurons. The MLP utilizes backpropagation for training [45–47]. Due to its multiple layers and nonlinear activation, MLP can distinguish data that is not linearly separable [48].

The MLP ANN was implemented in the IBM SPSS Modeler 18 software tool. Figure 5 displays the developed data stream. Initially, the input data was imported to the data stream (using Excel node). The filter and type were utilized to select relevant input data, assign correct variable types (continuous, categorical, etc.), and predefining inputs and the outputs. The data stream continues with a partitioning node with a predefined ratio of 40% training, 30% testing, and 30% validation. Commonly, K-fold, V-fold, N-fold, and partitioning methods are used to evaluate the performance of the developed models in the IBM SPSS modeler. K-fold, V-fold, and N-fold are splitting methods that divide the dataset into as many parts as there are possible values for a split field. Splitting results in every input vector are used for training and validation (by building multiple models). Unlike splitting, partitioning is used to evaluate the performance of a single model. It randomly divides the input dataset into three parts of training, testing, and validation. It provides a good indication of model performance by using one sample to generate the model and a separate sample to test and evaluate it. In general, partitioning is an optimal validation method for building a single model with the large datasets. Using validation partition, the built models can perform predictions using only predictors.

In the next stage, the partitioned data are fed into the ANN modeling node, and the IBM SPSS modeler algorithm guide [49] mathematically describes its MLP model as followings:

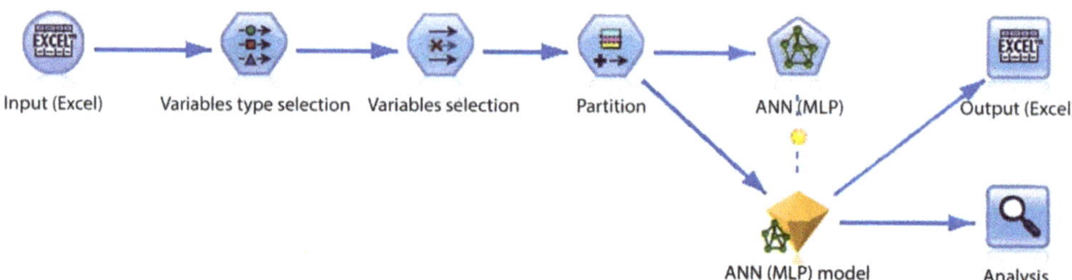

Figure 5. Developed data stream in IBM SPSS modeler.

Input layer $j_0 = p$ units, $a_{0:j}, \ldots, a_{0:j0}$, with $a_{0:j} = xj$, where j is the number of neurons in the layer and X is the input.

***i*th hidden layer** j_i units, $a_{i:1}, \ldots, a_{i:j_i}$, with $a_{1:k} = \gamma_i(C_{i:k})$ and $C_{i:k} = \sum_{j=0}^{j_{i-1}} \omega_{I:j1}$, $\mathrm{k}^{a_{i-1:j}}$, where $a_{i-1:0} = 1$, γ_i is the activation function for the layer I, and $\omega_{I:j1}$ is weight leading from layer i−1. At this layer the model uses hyperbolic tangent as an activation functions given by $\gamma_{(C)} = \tan\mathrm{h}(\mathrm{c})\frac{e^c - e^{-c}}{e^c + e^{-c}}$.

Output layer $j_I = R$ units, $a_{I:1}, \ldots, a_{I:J_I}$, with $a_{\mathrm{I}:k} = \gamma_I(C_{I:k})$ and $C_{I:k} = \sum_{J=0}^{J_1} \omega_{I:j}, \mathrm{k}^{a_{i-1:j}}$, where $a_{i-1:0} = 1$. For continuous prediction signal at this layer, the model uses identity ($\gamma_{(C)} = \mathrm{c}$) as an activation function.

Training or estimation of the weights is divided into the following three stages:

- Initialization of the weights (using alternated simulated annealing and training procedure);
- Computing the derivative of the error function with respect to the weights (via the error backpropagation algorithm);
- Updating the estimated weights (via gradient descent method).

The resulting model (displayed as nugget gem) can export its predictions to Excel files (using excel node) or analyze them using built-in functions such as plots and analysis nodes.

2.4. Wavelet Filtration

In the wavelet transformation, we operate with variable complex window functions, which are assigned into wavelet families. These groups of the wavelet functions differ from each other by the frequency features and morphological structure for the extraction of the specific features from the noisy signals. In this context, a selection of a suitable wavelet function is essential for the proper trend detection. Furthermore, it is important to mention that the wavelet transformation allows for the CO_2 signal decomposition into individual decomposition levels, keeping certain trends and detail information, while the rest is irreversibly suppressed [50,51]. On the basis of this procedure, we can build the wavelet-based filter bank, allowing for the CO_2 signal decomposition in multiple levels. In this context, it is crucial to select an appropriate decomposition level to detect the CO_2 signal trend and simultaneously suppress the signal details, representing the image noise, or improper prediction of the ANNs. Another aspect of wavelet-based filtering is the settings of filtration. Wavelet-based filtration is based on the fact that some approximation and detail coefficients from the wavelet transformation represent the signal noise, other than the signal trend. Those which contain the signal noise are suppressed by applying the thresholding procedure, where we need to select a suitable threshold and thresholding. These aspects of the wavelet transformation, including the type of the wavelet, level of decomposition, and thresholding rules, are input parameters on the basis of which we build the filtration procedure for the wavelet-based smoothing of CO_2 signals. In this paper, we present a comparative analysis of different settings of the wavelet analysis to achieve the most suitable procedure for the CO_2 signal smoothing with the aim of improving the accuracy of ANN prediction. All the settings are objectively verified based on selected evaluation parameters against the reference CO_2 signals [52–54].

Concept of Wavelet Transformation

As we stated before, the CO_2 signal standard contains time-varying frequency content, which makes this signal nonstationary. Therefore, we can use the time-frequency localization via using the window function, called wavelets. In this context, it is important to note that this window function is related to varying resolution in the time and frequency domain. According to the principle of uncertainty, the longer the window is, the better the frequency resolution we achieve and vice versa.

This predetermines the fact that it is impossible to achieve a perfect simultaneous localization of the low- and high-frequency components by using the window function with a constant length. In this context, on the one hand, the discrete wavelet transformation (DWT) enables the following benefits: The dynamic window function enables optimization of the time-frequency localization of different scales, enables the CO_2 signal decomposition in different levels with different level of the details and trend suppression, and enables use of the complex window function in the comparison with elementary window functions which are used in the short-time Fourier transformation (STFT). On the other hand, we need to consider the limitations of the wavelet transformation. Mainly, it is plenty of settings, including the mother's wavelets, level of decomposition, and the thresholding rules. These parameters are different for individual applications, and there is not a versatile way to determine the best setting for a particular task [55,56]. The definition of the DWT is given as follows:

$$d_{j,n} = \langle x(t),\ \psi_{j,n}(t)\rangle = \int_{-\infty}^{\infty} x(t)\ \overline{\psi_{j,n}(t)}dt\,, \tag{2}$$

where $\psi_{j,n}(t)$ represents the mother wavelet with the following definition:

$$\psi_{j,n}(t) = a_0^{-\frac{j}{2}}\psi\left(a_0^{-j}t - nb_0\right). \tag{3}$$

In this definition, the parameter a_0 stands for the scaling and b_0 is translation. These parameters are selected so that $\psi_{j,n}(t)$ has the orthogonal bases. Using $a_0 = 2$ and $b_0 = 1$ we obtain the equation for the mother wavelet as follows:

$$\psi_{j,n}(t) = 2^{-\frac{j}{2}}\psi\left(2^{-j}t - n\right). \tag{4}$$

Using Equation (2), the orthonormal wavelet transformation is given as follows:

$$d_{j,n} = \langle x(t),\ \psi_{j,n}(t)\rangle = 2^{-\frac{j}{2}}\int_{-\infty}^{\infty} x(t)\ \overline{\psi_{j,n}\left(2^{-j}t - n\right)}\,dt. \tag{5}$$

The inverse transformation is given by:

$$x(t) = \sum_{j=1}^{J}\sum_{n=1}^{N} d_{j,n}\psi_{j,n}(t). \tag{6}$$

The discrete wavelet transform is computed in the consecutive steps by applying low-pass and high-pass filters for the definition of the approximation and detail coefficients for each level of decomposition. This decomposition scheme represents a binary tree in the form of the Mallat algorithm with the goal of the multiresolution analysis as a filter bank. At each decomposition level, these half-band filters pass the signal with a half frequency band. This decimation by two down-sampling halves the time resolution and the signal is represented by half of the original samples. By using this approach, we achieve arbitrarily optimal time resolution in high frequencies and the frequency resolution in low frequencies. The process of decomposition is repeated until the maximal level of the decomposition is reached. This level is depended on the signal length. The inverse reconstruction of the original signal is done via sequences of the approximation and detail coefficients begins at the last decomposition level [57,58]. The signal inverse reconstruction undergoes all the levels of the decomposition.

2.5. Evaluation Methods

Accuracy The accuracy of the built models was obtained using the following expression:

$$\text{Accuracy} = \frac{1}{n}\sum_{m=1}^{M}\left(1 - \frac{\left|y_i^{(m)} - \hat{y}_i^{(m)}\right|}{max_m\left(y_i^{(m)}\right) - min_m\left(y_i^{(m)}\right)}\right). \tag{7}$$

Mean Square Error (MSE) This measures the average of the error squares between two signals. It is given by the following mathematical expression [59]:

$$MSE = \frac{1}{n}\sum_{i=1}^{n}(y_i - \hat{y}_i)^2. \tag{8}$$

Linear Correlation (LC) This corresponds to a degree of dependence (correlation) between two variables. It is given by the mathematical description [60]:

$$\text{Linear Correlation} = \frac{n\sum_{i=1}^{n} y_i\hat{y}_i - \sum_{i=1}^{n} y_i \sum_{i=1}^{n} \hat{y}_i}{\sqrt{n\sum_{i=1}^{n} y_i^2 - (\sum_{i=1}^{n} y_i)^2}\sqrt{n\sum_{i=1}^{n} \hat{y}_i^2 - (\sum_{i=1}^{n} \hat{y}_i)^2}}. \tag{9}$$

3. Results

3.1. Data Collection

The measurements were performed in the laboratory EB312 on the premises of the new FEI building of the VSB Technical University of Ostrava. The data collection started on May 2 at 10:08:06 and ended on May 10 at 11:52:45 (a weeklong data interval). Using the developed software, the data collection rate can vary between one to ten samples per minute. Resulting in a total of 55,241 samples. This location often holds educational classes, or it is visited by staff and researchers. However, during the days 4th (Saturday), 5th (Sunday), and 8th (public holiday "Victory in Europe Day") of May the measurement room (laboratory EB312 in new FEI building of VSB Technical University of Ostrava) remained unoccupied.

3.2. Predictions and Evaluations

Table 1 shows the obtained result from evaluating the validation partition with respect to the reference signal. The accuracy, MSE, and LC coefficient were used for objective evaluation of the developed models. The lowest accuracy was obtained by Model Number 7 (accuracy, 91.6%; LC, 0.956; and MSE, 2.525×10^{-3}) and Model Number 3 resulted in the highest prediction accuracy (accuracy, 96.7%; LC, 0.983; and MSE, 9.78×10^{-4}). It can be observed (Table 1) that most of the obtained models result in similar accuracies. Therefore, a number of neurons do not significantly impact prediction accuracy. The complete and detailed analysis of these prediction results can be found in [32].

Table 1. Prediction results using an interval of May 2 at 10:08:06 and ended on the May 10 (based on validation partition).

Model Number	Number of Neurons		Accuracy (%)	LC	MSE
	Layer 1	Layer 2			
1	10	10	95.2	0.976	2.368×10^{-3}
2	25	50	95.4	0.976	1.383×10^{-3}
3	50	25	96.7	0.983	9.78×10^{-3}
4	100	50	95.7	0.978	1.589×10^{-3}
5	50	100	96.0	0.979	1.205×10^{-3}
6	100	20	95.9	0.979	1.693×10^{-3}
7	20	80	91.6	0.956	2.525×10^{-3}
8	60	40	95.2	0.975	1.597×10^{-3}
9	10	200	94.0	0.969	2.067×10^{-3}
10	100	500	90.0	0.947	3.22×10^{-3}
11	500	20	95.9	0.979	1.77×10^{-3}
12	20	500	93.1	0.964	2.73×10^{-3}

3.3. Filtration and Evaluation

3.3.1. Wavelet Settings for CO_2 Signal Prediction

In this section, we describe the wavelet transformation settings for the CO_2 signals smoothing. The predicted signals mostly contain additive signal noise, exhibiting steep fluctuations that have a nature to deteriorate the CO_2 signal trend. Therefore, we aimed to eliminate the signal details which do not have the origin of the ambient CO_2 concentration, but are the product of the ANN, depending on the number of the neurons in the ANN.

In our analysis, we use the one-dimensional (1D) model of the wavelet transformation, which is a one-dimensional function, serving for the noise suppression. Wavelet transformation transforms the original CO_2 signal samples on the sequence of the wavelet coefficients. The wavelet-based filtration is consequently based on the thresholding of the wavelet coefficients. An essential part of the analysis is selecting the suitable settings of the wavelet filtration with the goal of optimal extraction of the CO_2 signal trend part while suppressing other details. Before applying the wavelet filtration, we tested and adjusted wavelet filter parameters. On the basis of the testing, we used the following settings for the signal smoothing (Table 2).

Table 2. Wavelet filtration settings for CO_2 signal smoothing.

Threshold	Wavelet Filter Thresholding	Thresholding Rescaling
rigrsure	soft thresholding	for no rescaling

In this work, we use filtering based on adaptive threshold selection using the principle of Stein's unilateral risk estimate (SURE), which is called rigrsure. A threshold is for the soft threshold estimator. Starting with an estimate of risk for a particular threshold value, t, the algorithm minimizes the risks to yield a threshold value. For the soft thresholding, values for both positive and negative coefficients are "shrinked" towards zero.

Another crucial part of the analysis deals with optimized settings of the mother's wavelets which appear as the most suitable for the analysis since individual families of the wavelets differ among each other by their morphological features, allowing for the extraction of specific signal features. In this context, it is supposed that unappropriated wavelet selections would lead to a bad CO_2 signal approximation, and therefore an unsuitable prediction. On the basis of the experimental testing, we selected three, the most significant wavelets, a well approximating CO_2 signal trend (Table 3).

Table 3. Wavelet functions for analysis of CO_2 prediction.

Daubechies	'Db6's
Coiflets	'coif1'e
Symlets	'sym1'

On the basis of the experimental testing, we found the wavelet scaling to be one of the most significant parameters which significantly influences the resulting prediction. Within the wavelet smoothing, the CO_2 signal is decomposed into a finite number of levels which gradually suppress the details of the CO_2 signal. This task controls the ANN prediction. After experimental testing, we decided to use the following levels of the decomposition: n = 3, 6, and 10. These levels are, consequently, used for the building of the prediction model.

Since we use different settings of the wavelet transformation and different topology of the ANN, we need to objectively evaluate the efficiency and robustness of each setting to report the configuration for the CO_2 signal prediction. For the purpose of this objective evaluation, we use MSE and correlation coefficient for objective testing. The MSE represents an error function calculated between the native predicted signal from ANN $Y(i)$ and smooth signal $\hat{Y}(i)$ from the wavelet-based filtration.

The next evaluation parameter is the correlation coefficient which measures a level of the linear dependency between native predicted signal and the wavelet smooth signal. In difference with MSE, the correlation coefficient is a normed parameter in the range: $[0;1]$, where zero stands for no correlation, whereas one stand for the full correlation.

3.3.2. Optimization of Wavelet Settings and Testing

In this section, we present testing and optimization of the wavelet settings from the view of wavelet filtration settings and levels of the decomposition, as well as for the different time ranges of the native CO_2 signals and selection of the ANN. In our analysis, we bring a comparative analysis of the testing trend-level (n) of the wavelet-based CO_2 concentration predicted waveform decomposition where n = 3, 6, and 10. Each such level performs the signal decomposition into approximation and detail coefficients according to the Mallat decomposition scheme. In each decomposition level, a part of the CO_2 signal energy is stored in approximation coefficients, representing the CO_2 signal trend and the rest of the energy is kept in the detail coefficients. In this way, part of the signal energy is suppressed in the detail coefficients. Regarding the decomposition level, we use the fact that the higher the decomposition level, the more energy is removed from the CO_2 trend and the resulting signal morphology is more distorted. We experimentally set the minimal energy which should be stored in the signal trend as 75% (n = 10), minimal CO_2 signal change 5% (n = 6) from the original CO_2 signal, and their average (n = 6). These decomposition settings are used for the comparative analysis of the best wavelet settings for improvement of the CO_2 prediction accuracy.

As the input signals, we use one-day and week signals. Each of these CO_2 signals was predicted by using one of the twelve different settings of the ANN network, differing in the number of neurons within their hidden layers. This testing also points out suitable settings of the ANN.

The following Figures 6–8 report CO_2 concentrations acquired and predicted within different time periods. Signals are processed by using all the levels of the decomposition to evaluate the effect of the decomposition on the wavelet-based smoothing. Figures 6–8 report application of the mother's wavelet Db6 (Daubechies). All of the signals are compared with the reference signal.

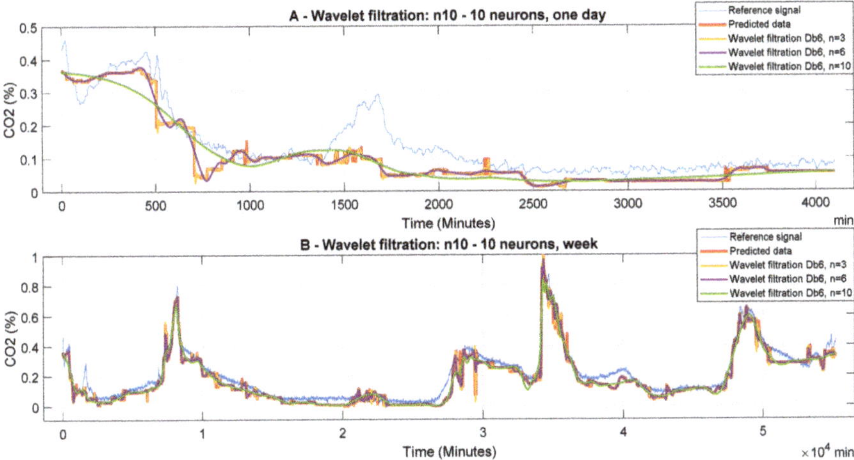

Figure 6. CO_2 concentration for different time period using model 1 (n10 ot 10 neurons). Wavelet-based filtration is performed for Db6 with level of decomposition, n = 3, 6, and 10.

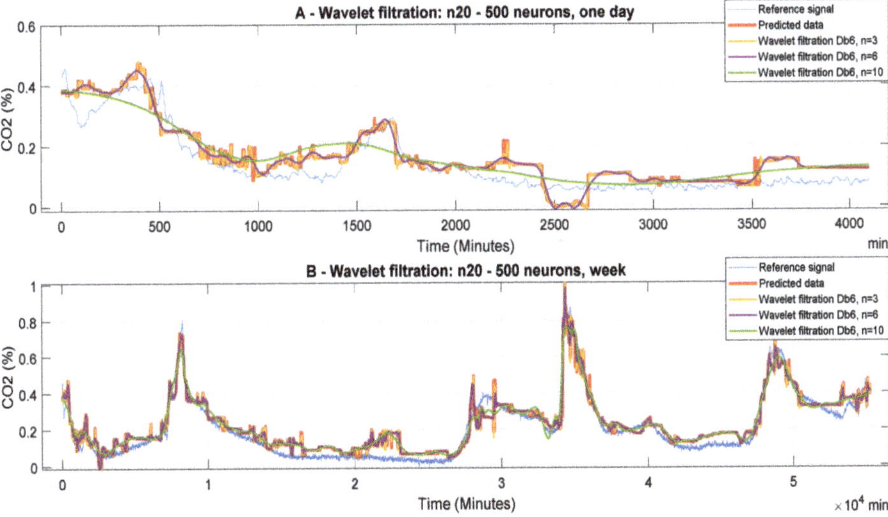

Figure 7. CO_2 concentration for different time periods using Model 12 (n20 to 500 neurons). Wavelet-based filtration is performed for Db6 with level of decomposition, n = 3, 6, and 10.

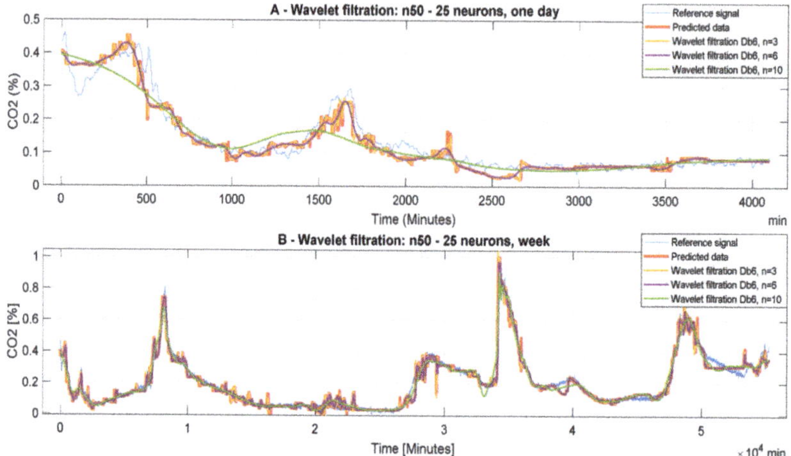

Figure 8. CO_2 concentration for different time periods using Model 3 (n50 to 25 neurons). Wavelet-based filtration is performed for Db6 with level of decomposition, n = 3, 6, and 10. Concentration for different time periods using ANN model 12.

Figure 6 represents the filtered signals which were processed by the ANN, containing 10 neurons in the first hidden layer and 10 neurons in the second layer; it compares the predicted and the reference signals. In the first case (Figure 6A), it is one-day data. It is apparent that n = 10 is not suitable for the wavelet settings due to significant distortion of the filtered signal as comparing with the reference signal. Thus, there is high suppression of the CO_2 information. The graph shows that the signal predicted by the ANN and the consequent filtered signals in all cases of wavelet filtration show a considerable deviation as compared with the reference signal. In the second case (Figure 6B), we report a week prediction. In this case, we observe, for all variants of wavelet filtration, a relatively accurate prediction of the CO_2 concentration as compared with the reference signal, taking into account the purpose of the predictions. A more accurate assessment of the quality of prediction and subsequent filtration is performed using MSE analysis and correlation analysis.

Figure 7 represents filtered signals, predicted with the ANN with 20 neurons in the first hidden layer and 500 neurons in the second layer (model 12). In this particular case, we can observe a more accurate prediction against the reference signal. In the first case (Figure 7A), which belongs to analyzed daily signals, it is possible to observe the generated parasitic data for all possibilities of wavelet filtration. Again, the predicted data and the wavelet filtration data are inaccurate with respect to the reference signal. Their next evaluation will be given by objective evaluation. We can also note large distortions and loss of important information when n = 10. In the second case (Figure 7B), which is related to the predicted weekly data, we receive the most accurate prediction regarding the reference. By comparing the prediction signal in Figures 9 and 10, it becomes apparent that the settings of the ANN have a significant impact on the prediction accuracy

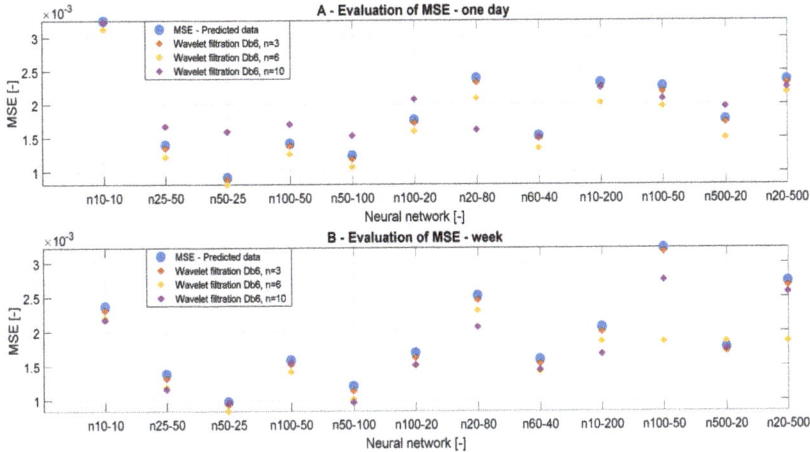

Figure 9. MSE values, a comparison of reference signal with the wavelet-based filtration signals.

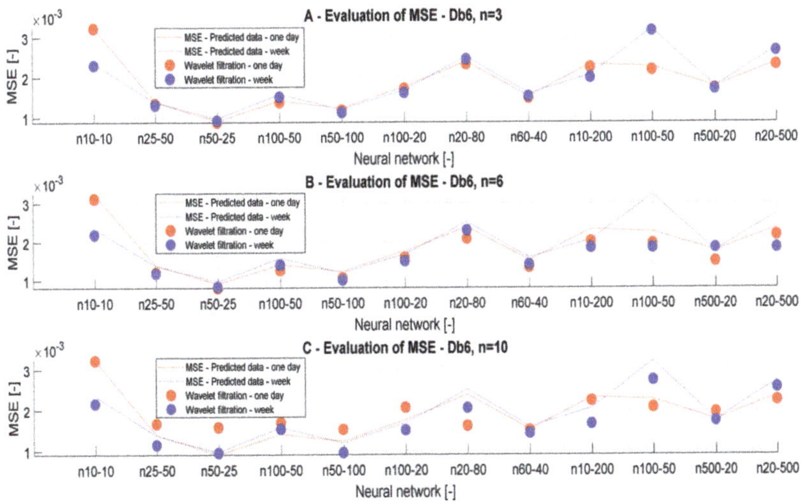

Figure 10. MSE values, a comparison of the reference signal with different wavelet settings.

Figure 8 refers to the ANN with 50 neurons in the first hidden layer and 25 neurons in the second layer (Model 3). Here, in each (Figure 8A,B) setting the decomposition level n = 10 is the worst, but in (Figure 8A) important information is suppressed in the analyzed data and information are lost in (Figure 8B), this signal distortion is not very large and for the purpose of predicting the CO_2 concentration this decomposition level is satisfactory. For the data where the decomposition levels n = 3 and 6 has been used, it is not possible to determine with precision which of the decomposition levels are more acceptable, MSE and correlation analysis should decide again. Within the testing of the wavelet-based settings, it was observed that the neuron's settings of ANN have a significant impact on the accuracy of the CO_2 prediction. In this context, we can note that the ANN Model 3 has the best accuracy of the wavelet-based filtration. Testing has shown that the use of weekly data is the most appropriate for the intended purposes of CO_2 concentration prediction. The next important finding was that the level of decomposition n = 10 is suitable for the weekly CO_2 prediction data, but unsuitable for daily data due to excessive suppression of the signal details, and thus bad overall prediction. These findings are only preliminary results. Objective findings should be, consequently, done by using analysis based on the MSE and correlation coefficient.

3.3.3. MSE Analysis

In this section, we present the results of the MSE analysis. Figures 9 and 10 represent MSE results for the filtration with Db6, n = 3, n = 6, and n = 10 for individual predicted CO_2 signals. The MSE values should converge to zero in each case. We present the MSE values in the dependence of the time range of the CO_2 measurements, but also on the ANN settings and the wavelet settings. Tables 4 and 5 summarize the MSE values. MSE should primarily verify optimal settings of the decomposition level. In each case of MSE, we compared the results with the reference signal.

Table 4. Mean square error (MSE) values for wavelet-based filtration of CO_2 signals (one day).

Model Number	Number of Neurons		Day			
	Layer 1	Layer 2	ref	n = 3	n = 6	n = 10
1	10	10	3.241×10^{-3}	3.231×10^{-3}	3.122×10^{-3}	3.229×10^{-3}
2	25	50	1.399×10^{-3}	1.346×10^{-3}	1.216×10^{-3}	1.671×10^{-3}
3	50	25	9.132×10^{-4}	8.826×10^{-4}	8.087×10^{-4}	1.593×10^{-3}
4	100	50	1.421×10^{-3}	1.382×10^{-3}	1.264×10^{-3}	1.707×10^{-3}
5	50	100	1.240×10^{-3}	1.183×10^{-3}	1.069×10^{-3}	1.535×10^{-3}
6	100	20	1.767×10^{-3}	1.722×10^{-3}	1.597×10^{-3}	2.075×10^{-3}
7	20	80	2.393×10^{-3}	2.326×10^{-3}	2.089×10^{-3}	1.618×10^{-3}
8	60	40	1.531×10^{-3}	1.487×10^{-3}	1.344×10^{-3}	1.516×10^{-3}
9	10	200	2.328×10^{-3}	2.254×10^{-3}	2.021×10^{-3}	2.249×10^{-3}
10	100	50	2.274×10^{-3}	2.188×10^{-3}	1.968×10^{-3}	2.082×10^{-3}
11	500	20	1.773×10^{-3}	1.726×10^{-3}	1.497×10^{-3}	1.960×10^{-3}
12	20	500	2.367×10^{-3}	2.315×10^{-3}	2.177×10^{-3}	2.253×10^{-3}

Table 5. MSE values for wavelet-based filtration of CO_2 signals (week).

Model Number	Number of Neurons		Week			
	Layer 1	Layer 2	ref	n = 3	n = 6	n = 10
1	10	10	2.368×10^{-3}	2.307×10^{-3}	2.195×10^{-3}	2.172×10^{-3}
2	25	50	1.383×10^{-3}	1.310×10^{-3}	1.182×10^{-3}	1.156×10^{-3}
3	50	25	9.780×10^{-4}	9.295×10^{-4}	8.382×10^{-4}	9.618×10^{-4}
4	100	50	1.589×10^{-3}	1.524×10^{-3}	1.415×10^{-3}	1.545×10^{-3}
5	50	100	1.205×10^{-3}	1.129×10^{-3}	1.013×10^{-3}	9.649×10^{-4}
6	100	20	1.693×10^{-3}	1.619×10^{-3}	1.504×10^{-3}	1.516×10^{-3}
7	20	80	2.525×10^{-3}	2.457×10^{-3}	2.306×10^{-3}	2.067×10^{-3}
8	60	40	1.597×10^{-3}	1.530×10^{-3}	1.417×10^{-3}	1.444×10^{-3}
9	10	200	2.067×10^{-3}	1.996×10^{-3}	1.857×10^{-3}	1.675×10^{-3}
10	100	50	3.220×10^{-3}	3.165×10^{-3}	1.857×10^{-3}	2.753×10^{-3}
11	500	20	1.770×10^{-3}	1.712×10^{-3}	1.857×10^{-3}	1.746×10^{-3}
12	20	500	2.730×10^{-3}	2.667×10^{-3}	1.857×10^{-3}	2.569×10^{-3}

Tables 4 and 5 presents the MSE values in the dependence on the setting of ANN. In Table 4, we present the results of the one-day data. According to the analysis, we note that the level of the decomposition n = 6 is most accurate. In the case of n = 10 (Table 4), we achieved the greatest values of MSE. Thus, we can state that this level of decomposition is unsuitable. In Table 5, we present MSE values for weekly data. In this case, we achieved the best MSE values for n = 10. MSE values for n = 3 (Table 5) are most different from the reference signal. The ANN models 10 and 12 showed the best MSE values for n = 6 (Table 5). By comparing values of MSE analysis for the ANN Model 3 of a one-day signal (Table 4), we get a clear conclusion about the influence of the decomposition level. The principle of the MSE analysis works on the principle than lower values indicate better results, and therefore the decomposition level n = 6 is the most suitable based on the real results. From the MSE analysis, we can conclude that weekly data are not affected by the decomposition level used and it is appropriate to use the decomposition level around n = 6 to 10 for these data, whereas for daily signals it is suitable to use the decomposition level n = 3 to 6.

Figure 10 presents the same results, but they are classified by type of wavelet. On the basis of the MSE analysis, we found that the best alternative for CO_2 settings is using weekly predicted data, using the ANN Model 3 and settings of the decomposition level n = 6 to 10, because these values represent a compromise regarding the decomposition level and settings of the ANN.

3.3.4. Correlation Analysis

To confirm the results of the MSE analysis, we also used the correlation analysis, investigating the linear dependency between the reference signals and the filtered signals. Tables 6 and 7 summarize the correlation results for different wavelets, time periods, and the number of neurons. Figures 11 and 12 bring the graphical trend representation of the correlation coefficient for different wavelet settings.

Table 6. Correlation analysis for the reference signals and wavelet-based filtration (day).

Model Number	Number of Neurons		Day			
	Layer 1	Layer 2	ref	n = 3	n = 6	n = 10
1	10	10	0.913	0.914	0.919	0.913
2	25	50	0.943	0.945	0.951	0.927
3	50	25	0.960	0.961	0.964	0.925
4	100	50	0.945	0.946	0.952	0.931
5	50	100	0.952	0.955	0.959	0.927
6	100	20	0.947	0.949	0.955	0.925
7	20	80	0.909	0.911	0.919	0.927
8	60	40	0.931	0.932	0.939	0.929
9	10	200	0.909	0.913	0.924	0.916
10	100	50	0.924	0.927	0.935	0.916
11	500	20	0.938	0.940	0.951	0.930
12	20	500	0.923	0.925	0.931	0.922

Table 7. Correlation analysis for the reference signals and wavelet-based filtration (week).

Model Number	Number of Neurons		Week			
	Layer 1	Layer 2	ref	n = 3	n = 6	n = 10
1	10	10	0.976	0.977	0.979	0.980
2	25	50	0.977	0.978	0.980	0.981
3	50	25	0.983	0.984	0.986	0.984
4	100	50	0.978	0.979	0.982	0.980
5	50	100	0.980	0.981	0.983	0.983
6	100	20	0.979	0.981	0.983	0.983
7	20	80	0.957	0.958	0.960	0.964
8	60	40	0.976	0.977	0.979	0.980
9	10	200	0.969	0.970	0.973	0.979
10	100	50	0.947	0.949	0.952	0.960
11	500	20	0.980	0.981	0.983	0.981
12	20	500	0.965	0.966	0.969	0.971

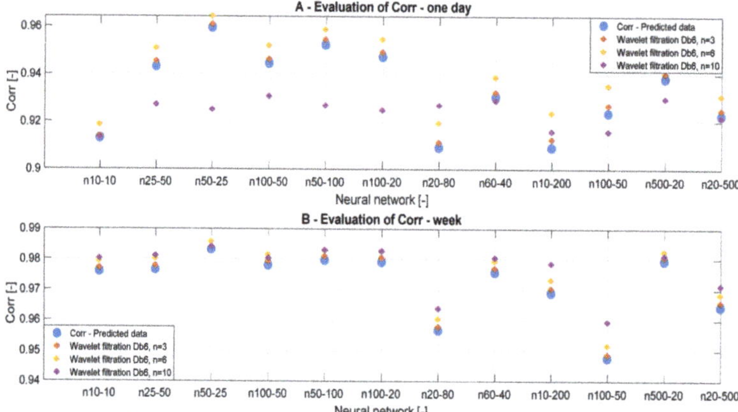

Figure 11. Correlation analysis of the reference signal and individual wavelet settings for two time periods: (**A**) one day and (**B**) week.

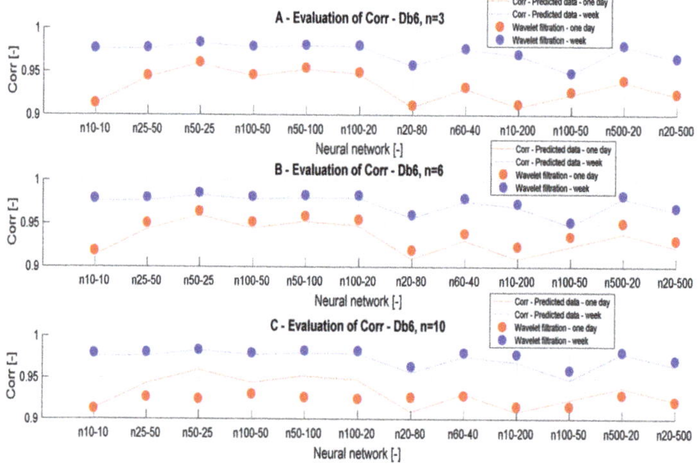

Figure 12. Correlation results, a comparison of the reference signal with wavelet-based filtration results.

On the basis of the analysis from Table 4 and Figure 11, we can state that the decomposition level has a greater impact on the analysis of daily data. The results of the correlation analysis are the same as the result of the MSE analysis and it can be stated that the conclusions were correct. On the basis of the correlation analysis, we can confirm the conclusion that the decomposition level n = 10 causes a significant signal distortion in daily data. This phenomenon is especially noticeable in Figure 11A. In contrast, in the analysis of weekly data in Figure 11B, the decomposition level n = 10 appears to be the best. This could be caused by the type of the signal, particularly its length. Figure 12 presents the correlation analysis results which are classified by the filtration method used.

3.3.5. Analysis of Wavelet Selection for CO_2 Prediction

In this section, we evaluate the influence of the selected wavelet on the quality of the predicted signal. As we stated before, we analyze the Db6 wavelet from the family Daubechies, coif1 from the family Coiflets, and sym1 from the family Symlets, when the level of the decomposition and filter settings were the same for all of the cases. This testing should bring an answer to the question of the influence of characteristic wavelet features on the predicted CO_2 signal. We only work with the n = 3 since the individual wavelet types must be compared.

Figure 13 represents the information about the output filtered signal with the use of different wavelets. In this case, it is relatively apparent that the selection of different wavelets does not have a significant impact on the filtered results for all of the time periods. In addition, it is obvious that the ANN Model 1 does not give satisfactory results for CO_2 prediction due to an insufficient similarity of the reference signals with the predicted results. This phenomenon is mainly observable in the case of daily data. This fact causes signal loss and distortion.

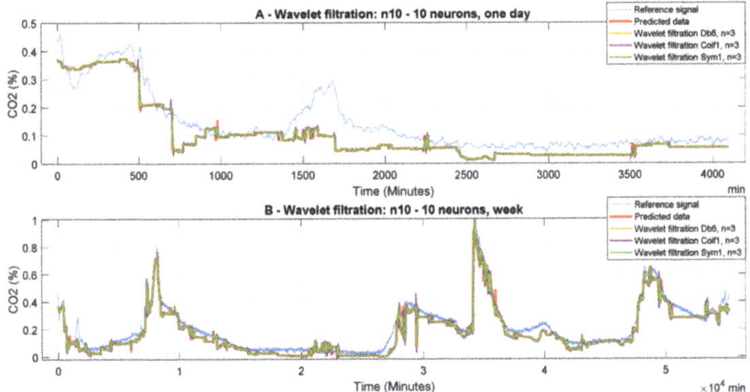

Figure 13. Dependence of CO_2 prediction on the time period for the ANN Model 1.

Figures 14 and 15 also present the information that individual wavelets do not bring significantly comparable information. These graphical representations confirm previous conclusions about the settings of ANN regarding the preciseness of the CO_2 prediction.

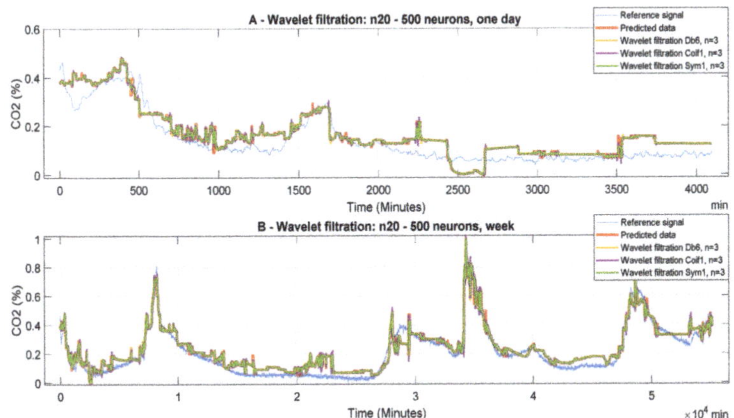

Figure 14. The dependency of CO_2 concentration on the time period for the neural network model 12 (n20 to 500).

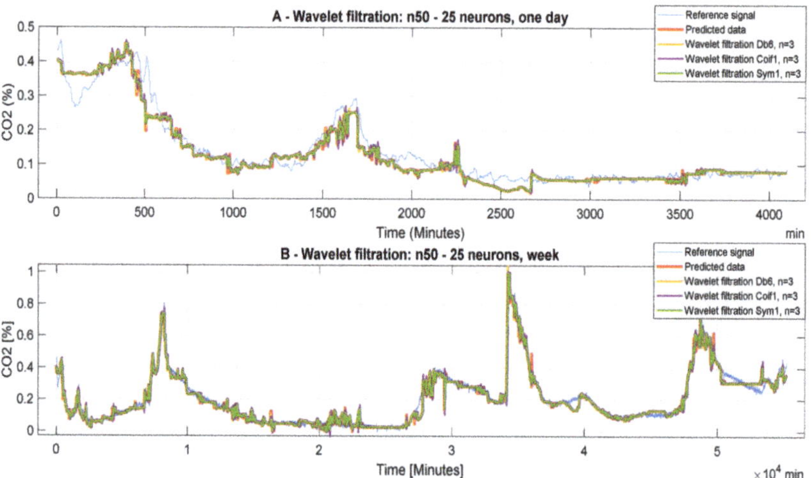

Figure 15. The dependency of CO_2 concentration on the time period for the neural network with 400 neurons.

3.3.6. MSE Analysis for Different Wavelets

In Tables 8 and 9, we present the MSE analysis for different wavelets (Db6, Coif1, and Sym), for various time intervals and the ANN settings. The graphical distribution of these results is presented in Figures 16 and 17. According to these results, we obtain similar results for these wavelets. These wavelets were selected based on the testing among other wavelets. These wavelets achieved the best results as in the MSE analysis, and also in the correlation analysis.

Table 8. MSE values for the reference signals and wavelet-based filtered signals.

Model Number	Number of Neurons		Day			
	Layer 1	Layer 2	ref	n = 3	n = 6	n = 10
1	10	10	3.249×10^{-3}	3.231×10^{-3}	3.227×10^{-3}	3.224×10^{-3}
2	25	50	1.399×10^{-3}	1.346×10^{-3}	1.349×10^{-3}	1.345×10^{-3}
3	50	25	9.132×10^{-4}	8.826×10^{-4}	8.803×10^{-4}	8.776×10^{-4}
4	100	50	1.421×10^{-3}	1.382×10^{-3}	1.383×10^{-3}	1.376×10^{-3}
5	50	100	1.240×10^{-3}	1.183×10^{-3}	1.185×10^{-3}	1.179×10^{-3}
6	100	20	1.767×10^{-3}	1.722×10^{-3}	1.724×10^{-3}	1.716×10^{-3}
7	20	80	2.393×10^{-3}	2.326×10^{-3}	2.322×10^{-3}	2.313×10^{-3}
8	60	40	1.530×10^{-3}	1.487×10^{-3}	1.484×10^{-3}	1.479×10^{-3}
9	10	200	2.328×10^{-3}	2.254×10^{-3}	2.259×10^{-3}	2.251×10^{-3}
10	100	50	2.274×10^{-3}	2.188×10^{-3}	2.192×10^{-3}	2.186×10^{-3}
11	500	20	1.773×10^{-3}	1.726×10^{-3}	1.716×10^{-3}	1.708×10^{-3}
12	20	500	2.367×10^{-3}	2.315×10^{-3}	2.317×10^{-3}	2.313×10^{-3}

Table 9. MSE values for the reference signals and wavelet-based filtered signals.

Model Number	Number of Neurons		Week			
	Layer 1	Layer 2	ref	n = 3	n = 6	n = 10
1	10	10	2.368×10^{-3}	2.307×10^{-3}	2.306×10^{-3}	2.300×10^{-3}
2	25	50	1.383×10^{-3}	1.310×10^{-3}	1.315×10^{-3}	1.307×10^{-3}
3	50	25	9.780×10^{-4}	9.295×10^{-4}	9.308×10^{-4}	9.256×10^{-4}
4	100	50	1.590×10^{-3}	1.524×10^{-3}	1.527×10^{-3}	1.521×10^{-3}
5	50	100	1.205×10^{-3}	1.129×10^{-3}	1.133×10^{-3}	1.125×10^{-3}
6	100	20	1.693×10^{-3}	1.619×10^{-3}	1.625×10^{-3}	1.615×10^{-3}
7	20	80	2.525×10^{-3}	2.457×10^{-3}	2.458×10^{-3}	2.448×10^{-3}
8	60	40	1.597×10^{-3}	1.530×10^{-3}	1.531×10^{-3}	1.522×10^{-3}
9	10	200	2.067×10^{-3}	1.996×10^{-3}	1.999×10^{-3}	1.991×10^{-3}
10	100	50	3.220×10^{-3}	3.165×10^{-3}	1.999×10^{-3}	3.152×10^{-3}
11	500	20	1.770×10^{-3}	1.712×10^{-3}	1.999×10^{-3}	1.708×10^{-3}
12	20	500	2.730×10^{-3}	2.667×10^{-3}	1.999×10^{-3}	2.664×10^{-3}

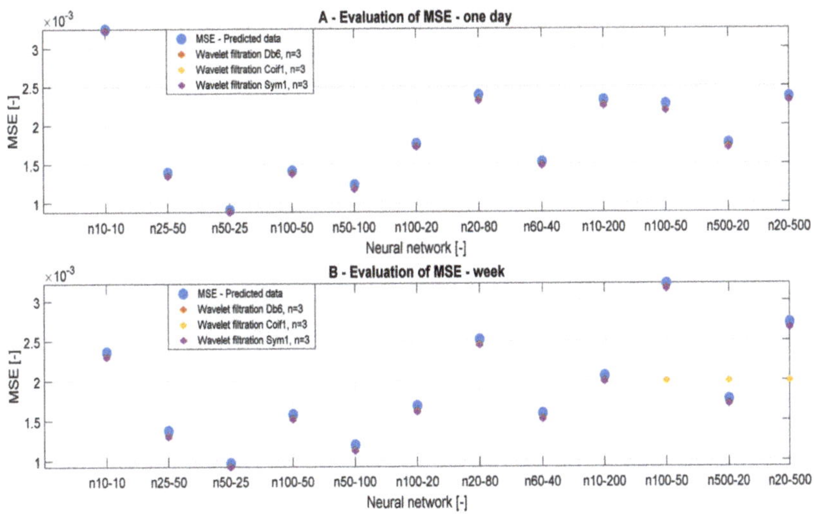

Figure 16. MSE values, a comparison of the reference signals with the wavelet-based filtration.

When we use the definition of MSE, it is obvious that MSE differences are insignificant, and it is not possible to objectively classify which wavelet is the most suitable for the analysis. Similar results are also achieved in the case of the weekly predictions.

Figure 17 presents the trend evaluation of MSE values for different wavelets with the level of decomposition n = 3 depending on the ANN configuration.

For the one-day prediction, we achieved that MSE results are comparable for all the wavelets. As well as for longer predictions, we obtain the results which do not exhibit significant differences. Figure 17 also presents MSE analysis results. In this case, we compare trend evaluation for each wavelet on the ANN settings and the prediction periods. These results show a monotonic tendency of the prediction accuracy depending on the number of neurons and the prediction time. In most cases, it applies the stated conclusion, only for ANN models 4, 11, and 12, we see specific values for wavelet filtration using wavelet coif1.

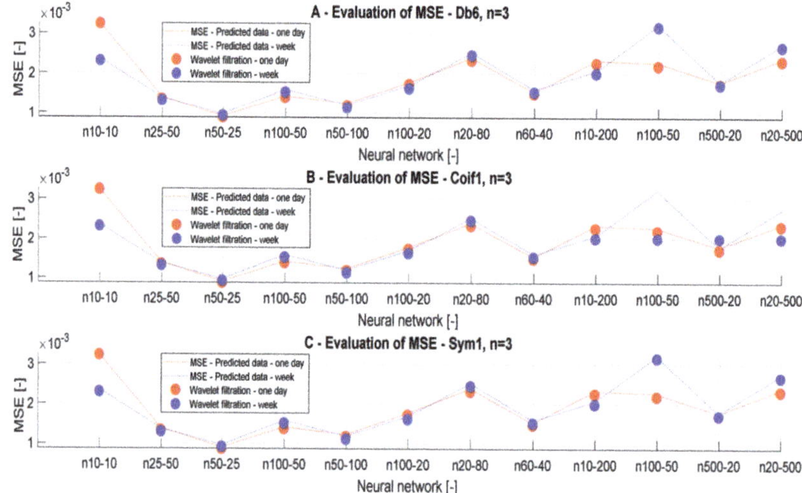

Figure 17. Trend characteristics for MSE values, a comparison of the reference signals with the wavelet-based filtration for different neural network settings and time prediction.

3.3.7. Correlation Analysis for Different Wavelets

In this section, we present the results of the correlation analysis for different wavelets. This analysis should confirm the results of the MSE analysis. Figures 18 and 19 show (Tables 10 and 11) the trend correlation characteristics for different wavelets with the level of the decomposition n = 3, depending on the ANN settings and the time interval of prediction. In Figure 18B, we have a comparison of the correlation analysis for weekly data. As we present the correlation results, it is obvious that differences are negligible. These results prove that the selection of the mother's wavelet for the wavelet-based prediction does not have a substantial impact on the prediction accuracy.

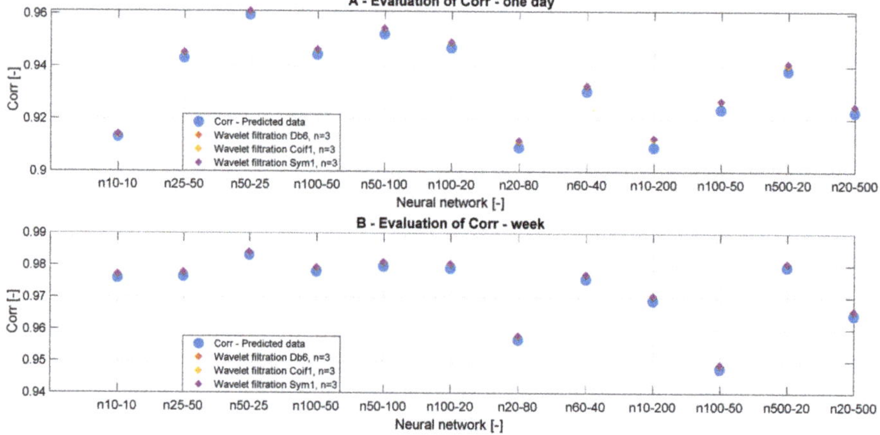

Figure 18. Analysis of correlation values, a comparison for different wavelets depending on various time periods and the neural network settings.

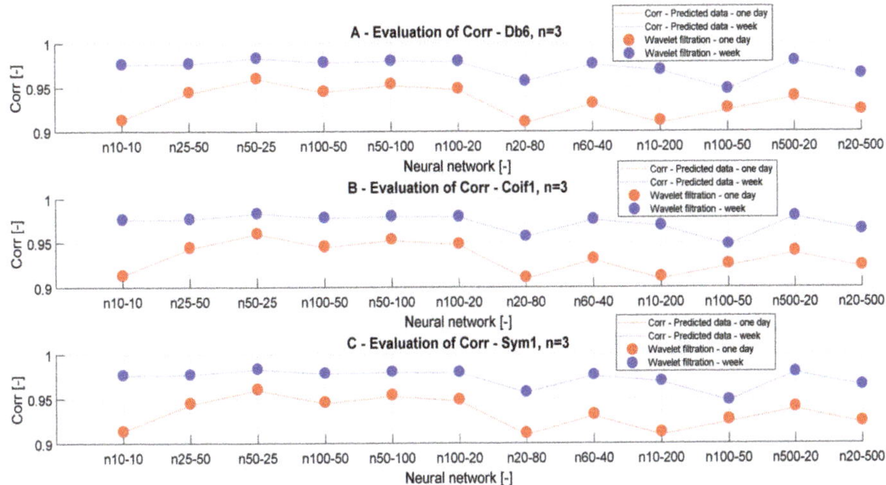

Figure 19. Analysis of correlation values, a comparison of individual wavelet settings for each time period of prediction and the number of neurons.

Table 10. Correlation analysis for the reference signals and wavelet-based filtration.

Model Number	Number of Neurons		Day			
	Layer 1	Layer 2	ref	n= 3	n= 6	n= 10
1	10	10	0.913	0.914	0.914	0.914
2	25	50	0.943	0.945	0.945	0.945
3	50	25	0.960	0.961	0.961	0.961
4	100	50	0.945	0.946	0.946	0.947
5	50	100	0.952	0.955	0.954	0.955
6	100	20	0.947	0.949	0.949	0.949
7	20	80	0.909	0.911	0.912	0.912
8	60	40	0.931	0.932	0.933	0.933
9	10	200	0.909	0.913	0.912	0.913
10	100	50	0.924	0.927	0.927	0.927
11	500	20	0.938	0.940	0.941	0.941
12	20	500	0.923	0.925	0.925	0.925

Table 11. Correlation analysis for the reference signals and wavelet-based filtration.

Model Number	Number of Neurons		Week			
	Layer 1	Layer 2	ref	n= 3	n= 6	n= 10
1	10	10	0.976	0.977	0.977	0.977
2	25	50	0.977	0.978	0.978	0.978
3	50	25	0.983	0.984	0.984	0.984
4	100	50	0.978	0.980	0.979	0.979
5	50	100	0.980	0.981	0.981	0.981
6	100	20	0.979	0.981	0.980	0.981
7	20	80	0.957	0.958	0.958	0.958
8	60	40	0.976	0.977	0.977	0.977
9	10	200	0.969	0.970	0.970	0.971
10	100	50	0.948	0.949	0.949	0.949
11	500	20	0.980	0.981	0.980	0.981
12	20	500	0.965	0.966	0.966	0.966

4. Discussion

In this paper, we report the comparative analysis of various wavelet settings with the goal of improving the CO_2 signal prediction. The predicted results from the ANN standard contain glitches and artifacts which more or less deteriorate the quality of the CO_2 signal accuracy. Therefore, we analyze the hybrid system that consists of the ANN prediction, with the consequent filtration procedure, based on the wavelet transformation.

In the wavelet analysis, we are mainly focused on the wavelet decomposition level (n = 3, 6, and 10) and different mother's wavelets, including Daubechies (db6's), Coiflets (coif1'e), and Symlet (sym1). We test these various wavelet settings for ten modifications of the ANN architecture with the goal of evaluating the best wavelet settings and the most suitable ANN settings for the CO_2 signal prediction.

The experimental testing is done for the one-day and one-week data of the CO_2 concentration. All the evaluations are performed objectively, where we compare the predicted results from the ANN and wavelet filtration against the reference CO_2 signals. Such a procedure objectively evaluates the quality and robustness of each wavelet setting for the CO_2 signal prediction optimization.

In this objective analysis, we use the following two metrics: (1) MSE, which is a type of the error function, expressing a difference between the reference signal and prediction and (2) LC, expressing the linear dependence level between the reference signal and prediction. On the basis of the MSE analysis, we conclude n = 6 as the most accurate for the CO_2 prediction as compared with other decompositions. Regarding the time period of the CO_2 signal, we found the week prediction to be the most accurate with n = 6 and 10. Consequently, we verify the MSE results by the correlation analysis. On the basis of the correlation analysis, we confirm the conclusion that the decomposition level n = 10 causes a significant signal distortion in the daily data. Furthermore, n = 6 appears as the best compromise of wavelet settings for the CO_2 prediction in the case of one-day data. Contrarily, in the case of the week-data, we noticed n = 10 as the best compromise.

On the one hand, the experimental settings and testing show the tendency and potential of the wavelet-based filtration for the optimization of CO_2 prediction accuracy. On the other hand, there are still open research questions regarding wavelet applications. Mainly, the potential of this system could be optimized by incorporating the methods of artificial intelligence with the goal of the autonomous selection of the most appropriate settings for the wavelet-based filtration.

5. Conclusions

This article proposed the implementation of a new method to determine the occupancy of monitored areas in IB by predicting the course of CO_2 concentration from the measured indoor temperature and indoor relative humidity. The article introduced the procedure of programming KNX modules using the ETS 5 SW tool to simulate the control of operational and technical functions in SH. Additionally, KNX-IoT connectivity was implemented to store and, subsequently, process the measured data using ANN (MLP).

A crucial part of the proposed method was to increase the accuracy of CO_2 predictions by wavelet transformation by suppressing the additive noise from the predicted signal. In this article, we present the comparative analysis of different settings of the wavelet transformation for the CO_2 signal prediction. We mainly pay attention to the effect of the level decomposition and type of mother's wavelet on the prediction accuracy. We experimentally set these settings and, consequently, evaluated settings impact on the prediction accuracy. In selected experiments, the accuracy of the prediction was better than 98%. On the basis of the above results and documented experiments, it can be stated that the main goal of the authors, "finding the optimal setting of individual parameters of the wavelet transform in the additive noise suppression application with an emphasis on increasing the accuracy of predictions of CO_2 concentration from measured values of indoor temperature and indoor relative humidity" were unambiguously met.

Future trends in our analysis could be aimed at building the optimization scheme based on artificial intelligence employing either genetic algorithms or methods of evolutionary computing with the goal of optimal selection of wavelet settings. Such optimization could bring new possibilities of CO_2 modeling, and autonomous classification of the wavelet settings for particular CO_2 signals.

Future work could deal with IB automation in the context of optimized management of operational-technical functions to reduce operating costs, increase control and comfort (for example

HVAC control, light control, and blinds control) with different technological systems (for example KNX RF technology, Bacnet technology, Lonworks technology, and Loxone technology) and platforms (for example IoT, SH, SHC, and SC) based on occupancy recognition of humans in IB spaces and indoor monitoring of human positioning. Our objectives are optimization and practical implementation of a novel design method to monitor human daily living activities in the SH, indirect methods for human presence monitoring in the IB, and lifelong learning of occupant behavior in SH systems within the ethical and privacy-preservation approach to SH.

Author Contributions: J.V., O.M.G, J.K., and K.F., methodology; J.V., O.M.G., J.K., and K.F., software; J.V., O.M.G., J.K., and K.F., validation; J.V., O.M.G., J.K., and K.F., formal analysis; J.V., O.M.G., J.K., and K.F., investigation; J.V., O.M.G., J.K., and K.F., resources; J.V., data creation; J.V., O.M.G., J.K., and K.F., writing—original draft preparation; J.V., O.M.G., J.K., and K.F., writing—review and editing; J.V., O.M.G., J.K., and K.F., visualization; J.V., O.M.G., J.K., and K.F., supervision; J.V., project administration; J.V., funding acquisition J.V. and J.K. All authors have read and agreed to the published version of the manuscript.

Acknowledgments: The work and the contributions were supported by the project SV450994 Biomedical Engineering Systems XV'. This work was supported by the Student Grant System of VSB Technical University of Ostrava, grant number SP2019/118. This work was supported by the European Regional Development Fund in the Research Centre of Advanced Mechatronic Systems project, project number CZ.02.1.01/0.0/0.0/16_019/0000867 within the Operational Programme Research, Development and Education.

Conflicts of Interest: The funders had no role in the design of the study; in the collection, analyses, or interpretation of data; in the writing of the manuscript, or in the decision to publish the results.

Abbreviations

The following abbreviations are used in this manuscript:

ANN	Artificial neural network
MLP	Multilayer perceptron
CC	Cloud computing
IoT	Internet of Things
SH	Smart home
KNX	Konnex (standard EN 50090, ISO/IEC 14543)
MQTT	Message queuing telemetry transport
ETS	Engineering tool software
SW	Software
CO_2	Carbon dioxide
CoAP	Constrained application protocol
XMPP	Extensible messaging and presence protocol
IBM SPSS	Statistical package for the social sciences from the company IBM
LMA	Levenberg–Marquardt algorithm
BRM	Bayesian regulation method
ppm	Parts per million
LMS	Least mean square
AF	Adaptive filtration
SCADA	Supervisory Control and Data Acquisition
SCG	Scaled conjugate gradient
DWT	Discrete wavelet transformation
STFT	Short-time Fourier transformation
TP	Twisted pair (also known as KNX bus)
PL	Powerline
RF	Radio frequency
IP	Industrial protocol, ethernet
MSE	Mean square error

LC	Linear correlation
EB312	Laboratory room at the VSB TU Ostrava
FEI	Faculty of Electrical Engineering and Computer Science
VSB	TU Ostrava - VSB - Technical University of Ostrava
db6's	Daubechies
coif1'e	Coiflets
sym1	Symlet
HVAC	Heating, ventilation, and air conditioning

References

1. Directive (EU) 2018/844 of the European Parliament and of the Council of 30 May 2018. Available online: https://eur-lex.europa.eu/legal-content/EN/TXT/PDF/?uri=CELEX:32018L0844&from=CS (accessed on 5 December 2019).
2. Vanus, J.; Machac, J.; Martinek, R.; Bilik, P.; Zidek, J.; Nedoma, J.; Fajkus, M. The design of an indirect method for the human presence monitoring in the intelligent building. In *Human-Centric Computing and Information Sciences*; Springer: Berlin, Germany, 2018; ISSN 2192-1962. [CrossRef]
3. Vanus, J.; Belesova, J.; Martinek, R.; Nedoma, J.; Fajkus, M.; Bilik, P.; Zidek, J. Monitoring of the daily living activities in smart home care. In *Human-Centric Computing and Information Sciences*; Springer: Berlin, Germany, 2017; ISSN 2192-1962. [CrossRef]
4. Bastos, D.; Shackleton, M.; El-Moussa, F. Internet of Things: A Survey of Technologies and Security Risks in Smart Home and City Environments. In Proceedings of the Living in the Internet of Things: Cybersecurity of the IoT—2018, London, UK, 28–29 March 2018; pp. 1–7.
5. Asensio, J.; Criado, J.; Padilla, N.; Iribarne, L. Emulating home automation installations through component-based web technology. In *Future Generation Computer Systems*; ScienceDirect: Amsterdam, The Netherlands, 2017; ISSN 0167-739X. [CrossRef]
6. Petnik, J.; Vanus, J. Design of Smart Home Implementation within IoT with Natural Language Interface. *IFAC-PapersOnLine* **2018**, *51*, 174–179. [CrossRef]
7. Dwivedi, A.D.; Srivastava, G.; Dhar, S.; Singh, R. A decentralized privacy-preserving healthcare blockchain for iot. *Sensors* **2019**, *19*, 326. [CrossRef] [PubMed]
8. Polap, D. Analysis of skin marks through the use of intelligent things. *IEEE Access* **2019**, *7*, 149355–149363. [CrossRef]
9. Winnicka, A.; Kęsik, K.; Połap, D.; Woźniak, M.; Marszałek, Z. A Multi-Agent Gamification System for Managing Smart Homes. *Sensors* **2019**, *19*, 1249. [CrossRef] [PubMed]
10. Vanus, J.; Koziorek, J. Employment of Dynamic Time Warping method for monitoring states in Smart Home. In Proceedings of the 4th International Workshop on Computer Science and Engineering-Winter, WCSE 2014, Hong Kong, 26–28 December 2014.
11. Vanus, J. The use of the adaptive noise cancellation for voice communication with the control system. *Int. J. Comput. Sci. Appl.* **2011**, *8*, 54–70.
12. Vanus, J.; Stýskala, V. Application of optimal settings of the LMS adaptive filter for speech signal processing. In *Proceedings of the International Multiconference on Computer Science and Information Technology, Wisla, Poland, 18–20 October 2010*; Volume 5, pp. 767–774.
13. Vaňuš, J.; Stýskala, V. Application of variations of the LMS adaptive filter for voice communications with control system. *Teh. Vjesn.* **2011**, *18*, 555–580.
14. Vanus, J.; Belesova, J.; Martinek, R.; Bilik, P.; Zidek, J.; Koval, L. Development of Software Tool for Operational and Technical Functions Control in the Smart Home with KNX technology. *IFAC-PapersOnLine* **2016**, *49*, 431–436. [CrossRef]
15. Vanus, J.; Fajkus, M.; Martinek, R.; Zabka, S.; Stolarik, M. Smart home room's occupancy monitoring using fiber bragg grating sensor. In *Optical Sensors 2019, Proceedings of the SPIE Optics + Optoelectronics, Prague, Czech Republic, 11 April 2019*; Spie-Int Soc Optical Engineering: Bellingham, WA, USA, 2019.
16. Vanus, J.; Martinek, R.; Nedoma, J.; Fajkus, M.; Cvejn, D.; Valicek, P.; Novak, T. Utilization of the lms algorithm to filter the predicted course by means of neural networks for monitoring the occupancy of rooms in an intelligent administrative building. *IFAC Papersonline* **2018**, *51*, 378–383. [CrossRef]

17. Vanus, J.; Martinek, R.; Bilik, P.; Zídek, J.; Dohnalek, P.; Gajdos, P. New method for accurate prediction of CO2 in the Smart Home. In Proceedings of the 2016 IEEE International Instrumentation and Measurement Technology Conference Proceedings, Taipei, Taiwan, 23–26 May 2016; pp. 1–5.
18. Vanus, J.; Krestanova, A.; Kubicek, J.; Gorjani, O.; Penhaker, M.; Oczka, D. Using Wavelet Transformation for Prediction CO2 in Smart Home Care Within IoT for Monitor Activities of Daily Living. In Proceedings of the 11th International Conference on Computational Collective Intelligence, Hendaye, France, 4–6 September 2019.
19. Vanus, J.; Kubicek, J.; Gorjani, O.M.; Koziorek, J. Using the IBM SPSS SW tool with wavelet transformation for CO2 prediction within IoT in smart home care. *Sensors* **2019**, *19*, 1407. [CrossRef]
20. Chen, W.; Bai, M.; Song, H. Seismic noise attenuation based on waveform classification. *J. Appl. Geophys.* **2019**, *167*, 118–127. [CrossRef]
21. Deo, R.N.; Cull, J.P. Denoising time-domain induced polarisation data using wavelet techniques. *Explor. Geophys.* **2016**, *47*, 108–114. [CrossRef]
22. Dolabdjian, C.; Fadili, J.; Leyva, E.H. Classical low-pass filter and real-time wavelet-based denoising technique implemented on a DSP: A comparison study. *Eur. Phys. J. Appl. Phys.* **2002**, *20*, 135–140. [CrossRef]
23. Kang, S.; Zhang, H.; Kang, Y. Application of signal processing and neural network for transient waveform recognition in power system. In Proceedings of the 2010 Chinese Control and Decision Conference, Xuzhou, China, 26–28 May 2010; pp. 2481–2484.
24. Rahman, M.A.; Rashid, M.A.; Ahmad, M. Selecting the optimal conditions of Savitzky–Golay filter for fNIRS signal. *Biocybern. Biomed. Eng.* **2019**, *39*, 624–637. [CrossRef]
25. Bridwell, D.A.; Cavanagh, J.F.; Collins, A.G.E.; Nunez, M.D.; Srinivasan, R.; Stober, S.; Calhoun, V.D. Moving Beyond ERP Components: A Selective Review of Approaches to Integrate EEG and Behavior. *Front. Hum. Neurosci.* **2018**, *12*, 106. [CrossRef]
26. Cohen, M.X. Comparison of linear spatial filters for identifying oscillatory activity in multichannel data. *J. Neurosci. Methods* **2017**, *278*, 1–12. [CrossRef]
27. Parameshwaran, D.; Subramaniyam, N.P.; Thiagarajan, T.C. Waveform complexity: A new metric for EEG analysis. *J. Neurosci. Methods* **2019**, *325*, 108313. [CrossRef]
28. Varshavskiy, I.E.; Krasnova, A.I.; Polivanov, V.V. Efficiency Estimation of the Noise Digital Filtering Algorithms. In Proceedings of the 2019 IEEE Conference of Russian Young Researchers in Electrical and Electronic Engineering (EIConRus), Saint Petersburg/Moscow, Russia, 28–31 January 2019.
29. Su, Z.; Ye, L. An intelligent signal processing and pattern recognition technique for defect identification using an active sensor network. *Smart Mater. Struct.* **2004**, *13*, 957–969. [CrossRef]
30. Vivier, J.; Mehablia, A. New Artificial Network Approach for Membrane Filtration Simulation. *Chem. Biochem. Eng. Q.* **2012**, *26*, 241–248.
31. Wei, A.-L.W.; Zeng, G.M.; Huang, G.H.; Liang, J.; Li, X.D. Modeling of a permeate flux of cross-flow membrane filtration of colloidal suspensions: A wavelet network approach. *Int. J. Environ. Sci. Technol.* **2009**, *6*, 395–406. [CrossRef]
32. Vanus, J.; Gorjani, O.M.; Bilik, P. Novel Proposal for Prediction of CO2 Course and Occupancy Recognition in Intelligent Buildings within IoT. *Energies* **2019**, *12*, 4541. [CrossRef]
33. Rouse, M. Predictive Modeling. Available online: https://searchenterpriseai.techtarget.com/definition/predictive-modeling (accessed on 1 November 2019).
34. Han, J.; Pei, J.; Kamber, M. *Data Mining: Concepts and Techniques*; Elsevier: Amsterdam, The Netherlands, 2011.
35. Moosavi, S.R.; Wood, D.A.; Ahmadi, M.A.; Choubineh, A. ANN-Based Prediction of Laboratory-Scale Performance of CO2-Foam Flooding for Improving Oil Recovery. *Nat. Resour. Res.* **2019**, *28*, 1619–1637. [CrossRef]
36. Zarei, T.; Behyad, R. Predicting the water production of a solar seawater greenhouse desalination unit using multi-layer perceptron model. *Sol. Energy* **2019**, *177*, 595–603. [CrossRef]
37. Yilmaz, I.; Kaynar, O. Multiple regression, ANN (RBF, MLP) and ANFIS models for prediction of swell potential of clayey soils. *Expert Syst. Appl.* **2011**, *38*, 5958–5966. [CrossRef]
38. Heidari, E.; Sobati, M.A.; Movahedirad, S. Accurate prediction of nanofluid viscosity using a multilayer perceptron artificial neural network (MLP-ANN). *Chemom. Intell. Lab. Syst.* **2016**, *155*, 73–85. [CrossRef]
39. Abdi-Khanghah, M.; Bemani, A.; Naserzadeh, Z.; Zhang, Z. Prediction of solubility of N-alkanes in supercritical CO2 using RBF-ANN and MLP-ANN. *J. CO2 Util.* **2018**, *25*, 108–119. [CrossRef]

40. Ahmed, S.A.; Dey, S.; Sarma, K.K. Image texture classification using Artificial Neural Network (ANN). In Proceedings of the 2011 2nd National Conference on Emerging Trends and Applications in Computer Science, Shillong, India, 4–5 March 2011; pp. 1–4.
41. Zarei, F.; Baghban, A. Phase behavior modelling of asphaltene precipitation utilizing MLP-ANN approach. *Pet. Sci. Technol.* **2017**, *35*, 2009–2015. [CrossRef]
42. Behrang, M.; Assareh, E.; Ghanbarzadeh, A.; Noghrehabadi, A. The potential of different artificial neural network (ANN) techniques in daily global solar radiation modeling based on meteorological data. *Sol. Energy* **2010**, *84*, 1468–1480. [CrossRef]
43. Haykin, S. *Neural Networks: A Comprehensive Foundation*, 2nd ed.; Macmillan College Publishing: New York, NY, USA, 1998.
44. Ripley, B.D. *Pattern Recognition and Neural Networks*; Cambridge University Press (CUP): Cambridge, UK, 1996.
45. Hastie, T.; Tibshirani, R.; Friedman, J. *The Elements of Statistical Learning: Data Mining, Inference, and Prediction*; Springer: New York, NY, USA, 2009.
46. Rosenblatt, F. *Principles of Neurodynamics. Perceptrons and the Theory of Brain Mechanisms*; Defense Technical Information Center (DTIC): Baer Fort, VA, USA, 1961.
47. Rumelhart, D.E.; Hinton, G.E.; Williams, R.J. *Learning Internal Representations by Error Propagation*; The MIT Press: Cambridge, MA, USA, 1985.
48. Cybenko, G. Approximation by superpositions of a sigmoidal function. *Math. Control Signals Syst.* **1989**, *2*, 303–314. [CrossRef]
49. IBM SPSS Modeler 18 Algorithms Guide. Available online: Ftp://public.dhe.ibm.com/software/analytics/spss/documentation/modeler/18.0/en/AlgorithmsGuide.pdf (accessed on 1 November 2019).
50. Islam, M.R.; Rahim, M.A.; Islam, M.R.; Shin, J. Genetic algorithm based optimal feature selection extracted by time-frequency analysis for enhanced sleep disorder diagnosis using eeg signal. In Proceedings of the SAI Intelligent Systems Conference 2019, London, UK, 5–6 September 2019.
51. Strömbergsson, D.; Marklund, P.; Berglund, K.; Saari, J.; Thomson, A. Mother wavelet selection in the discrete wavelet transform for condition monitoring of wind turbine drivetrain bearings. *Wind Energy* **2019**, *22*, 1581–1592. [CrossRef]
52. Chen, H.; Fan, Y. Identification and wavelet estimation of weighted ATE under discontinuous and kink incentive assignment mechanisms. *J. Econom.* **2019**, *212*, 476–502. [CrossRef]
53. Cohen, M.X. A better way to define and describe Morlet wavelets for time-frequency analysis. *NeuroImage* **2019**, *199*, 81–86. [CrossRef] [PubMed]
54. Feli, M. Abdali-Mohammadi, F. 12 lead electrocardiography signals compression by a new genetic programming based mathematical modeling algorithm. *Biomed. Signal Process. Control* **2019**, *54*, 101596. [CrossRef]
55. Liu, J.; Zhang, C.; Zhu, Y.; Ristaniemi, T.; Parviainen, T.; Cong, F. Automated detection and localization system of myocardial infarction in single-beat ECG using Dual-Q TQWT and wavelet packet tensor decomposition. *Comput. Methods Programs Biomed.* **2020**, *184*, 105120. [CrossRef] [PubMed]
56. Feng, Z.; Yu, X.; Zhang, D.; Liang, M. Generalized adaptive mode decomposition for nonstationary signal analysis of rotating machinery: Principle and applications. *Mech. Syst. Signal Process.* **2020**, *136*, 106530. [CrossRef]
57. Babichev, S.; Sharko, O.; Sharko, A.; Mikhalyov, O. Soft Filtering of Acoustic Emission Signals Based on the Complex Use of Huang Transform and Wavelet Analysis. *Adv. Intell. Syst. Comput.* **2020**, *1020*, 3–19.
58. Hasan, F.S. Chaotic signals denoising using empirical mode decomposition inspired by multivariate denoising. *Int. J. Electr. Comput. Eng.* **2020**, *10*, 1352–1358.
59. Lehmann, E.L.; Casella, G. *Theory of Point Estimation*; Springer Science & Business Media: Berlin, Germany, 2006.
60. Ijiri, Y. The linear aggregation coefficient as the dual of the linear correlation coefficient. *Econom. J. Econom. Soc.* **1968**, *36*, 252–259. [CrossRef]

Review

Hardware for Recognition of Human Activities: A Review of Smart Home and AAL Related Technologies

Andres Sanchez-Comas [1,*], Kåre Synnes [2,*] and Josef Hallberg [2]

[1] Department of Productivity and Innovation, Universidad de la Costa, Barranquilla 080 002, Colombia
[2] Department of Computer Science, Electrical and Space Engineering, Luleå Tekniska Universitet, 971 87 Luleå, Sweden; josef.hallberg@ltu.se
* Correspondence: asanchez@cuc.edu.co (A.S.-C.); kare.synnes@ltu.se (K.S.); Tel.: +57-317-495-2457 (A.S.-C.)

Received: 29 June 2020; Accepted: 20 July 2020; Published: 29 July 2020

Abstract: Activity recognition (AR) from an applied perspective of ambient assisted living (AAL) and smart homes (SH) has become a subject of great interest. Promising a better quality of life, AR applied in contexts such as health, security, and energy consumption can lead to solutions capable of reaching even the people most in need. This study was strongly motivated because levels of development, deployment, and technology of AR solutions transferred to society and industry are based on software development, but also depend on the hardware devices used. The current paper identifies contributions to hardware uses for activity recognition through a scientific literature review in the Web of Science (WoS) database. This work found four dominant groups of technologies used for AR in SH and AAL—smartphones, wearables, video, and electronic components—and two emerging technologies: Wi-Fi and assistive robots. Many of these technologies overlap across many research works. Through bibliometric networks analysis, the present review identified some gaps and new potential combinations of technologies for advances in this emerging worldwide field and their uses. The review also relates the use of these six technologies in health conditions, health care, emotion recognition, occupancy, mobility, posture recognition, localization, fall detection, and generic activity recognition applications. The above can serve as a road map that allows readers to execute approachable projects and deploy applications in different socioeconomic contexts, and the possibility to establish networks with the community involved in this topic. This analysis shows that the research field in activity recognition accepts that specific goals cannot be achieved using one single hardware technology, but can be using joint solutions, this paper shows how such technology works in this regard.

Keywords: smart home; AAL; ambient assisted living; activity recognition; hardware; review

---

1. Introduction

Smart home (SH) technology moved in the last decade beyond a research field into a commercial enterprise. In the beginning, SH technology was applied strongly in security and surveillance, energy-saving, and entertainment, among others. Nowadays, the landscape has expanded with technologies such as the Internet of Things (IoT), artificial intelligence (AI), and computing techniques, helping to focus research and development (R&D) on working in fields such as improving the standard of living and autonomy for elder or disabled people, among others [1], this raise questions such as what can houses do for inhabitants' needs, and how. A smart home can improve inhabitants' lives when it is capable of sensing, anticipating, and responding to their daily activities, assisting them in a socially appropriate and timely way [2]. A basic smart home system is composed of an Internet connection, a smart home gateway, and devices connected as multiple nodes in the system [3], with nodes as sensors and actuators with wired or wireless communication [4]. This amount of data generation requires data processing techniques, allowing research areas such as ubiquitous and mobile computing to emerge as vital components of surveillance, security, and ambient assisted living, requiring research on human activity recognition. Some research fields have emerged as well, such as wearable sensor-based activity monitoring as a result of sensors deployed over the human body, and dense sensor-based activity monitoring from sensor network technologies, smart sensors, or smart appliances, among others [5].

The concept of "activity" itself, as what can be performed by a person, is the core for constructing applications or concepts like ambient assisted living (AAL) [6]. The complexity of the activity recognition problem increases with the complexity of the activity. Researchers are focusing on complex activity recognition, for example, using a computer, which involves other activities such as typing, using a mouse, sitting [7], etc., as well as those activities with longer duration composed of multiple actions and sequences of simple activities [8]. Thus arose the need to develop solutions around smart home concepts using hardware and software capable of capturing residents' behavior and understanding their activities, informing them of risk situations, or taking action for their satisfaction [9]. Event recognition and emotion recognition are also part of this technology concept [10]. The smart home is considered as a technology that can help reduce the cost of living and care for the elderly and disabled population, and improve their quality of life. This concept is also applicable to solutions like energy saving, security management, and risk detection, such as fire, e.g., using such technologies as video monitors, alarms, planners and calendars, reminders, sensors, or actuators, among others [9]. All of the above complements the vision of Mark Weiser [11], allowing those research fields as pervasive or ubiquitous as computing bear vanguard systems such as AAL [12], which are context-aware, personalized to individual needs, adaptive to changing individual needs, ubiquitous in our everyday environment, and remain transparent in individual daily life [13]. The importance of developing these systems lies in their capacity to empower people's activities through digital environments capable of sensing, adapting, and responding to human needs. In addition, these systems identify actions, habits, gestures, emotions, and establish a pervasive and unobtrusive human–machine communication interaction [13].

Both smart home and AAL needs for activity recognition developments are based on hardware and software capabilities. It is worth noting that activity recognition depends on data gathered from sensor systems, but the core is the data processing system based on software development. Therefore, trying new approaches, models, and algorithms with new data captured each time could be expensive. That is why activity recognition datasets freely accessible for R&D, gathered from research at specialized facilities by research institutes, helped to generate an explosion of knowledge in computer science around artificial intelligence problems, methods, tools, and techniques. A recent review of datasets for activity recognition is presented by [14]. CASAS (Center for Advanced Studies in Adaptive Systems) and UCI Human Activity Recognition dataset, among others, are the most popular for activity recognition system development, used by [15,16], respectively. Despite such advances in research, it is good to study all perspectives of the research context in AR to boost technological advances further in smart home and AAL. This review contributes by complementing the knowledge pool of software solutions with a broad overview of the hardware technology used for activity recognition applied in the field of smart homes and ambient assisted living. This work, covering the hardware technology used for AR, is not exhaustive but does give an extensive overview of recent technology in smart home and AAL. However, this paper focuses only on published studies in which researchers tested software development on hardware technology they used themselves, as this review seeks mainly to provide a road map for hardware solutions of activity recognition for smart home and AAL.

As activity recognition has been a growing research area in the last decade, while exploring the scientific literature retrieved from the search, we found several reviews related to hardware for smart home and AAL. Health is an interesting sector for these fields. Kötteritzsch [17] analyzed ambient assisted living systems in urban areas, focusing on assistive technologies for older adults, this work identified three categories to help classify AAL systems and pointed out challenges and future trends.

Kötteritzsch [17] also found six hardware technologies proposed for use in AAL for older adults: wireless sensor network (WSN), camera, global positioning system (GPS), radiofrequency, and laser. Ni [18] presented a survey of elders independent living, characterizing the main activities considered in smart home scenarios, sensors, and data processing methods to facilitate service development. He offered some guidelines to help select sensors and processing techniques, grouping them into five categories for smart home environments for independent elders: environmental, wearable, inertial, and vital signs sensors. Acampora [13] discussed the emergence of AAL techniques in the health domain, examining body area networks (BANs) and dense/mesh sensor networks as infrastructure and sensor technologies in ambient sensor architecture. He summarized the hardware required for developing ambient intelligence systems based on special boards with Bluetooth and Zigbee for communication among the sensors, and sensors like accelerometer/gyroscope, blood glucose, blood pressure, electrocardiogram (ECG), electroencephalogram (EEG), electromyography (EMG), pulse oximetry, and humidity and temperature sensors. Expanding the study landscape, Kumari [6] presented a survey on increasing trends of wearables and multimodal interfaces for human activity recognition, discussing basic requirements, architectures, the current market situation, and developments using wearable sensors and bio-potential signals. Bejarano [1] reviewed the literature from 2010 to 2015 on technical and functional aspects of identifying common aspects such as architecture, network topology, scientometric information, and components of a smart home system, and described the uses, among other aspects. Peeton [19] investigated what kind of technologies exist to monitor activities at home for elderly people living independently, identifying five main groups of monitoring technologies: passive infrared (PIR) motion sensors, body-worn sensors, video monitors, pressure sensors, and sound sensors.

Although several review papers have been published over the years, considering the wealth of literature about applications, architectures, component functionality, and analysis comparing performance among the studies published in different sector applications, there are no broad studies related to the hardware used in activity recognition for smart home and AAL. This study is exploratory and has limitations; this paper does not study accuracy and performance, as they depend on more variety of data processing techniques, which are not the focus of the study. Nor was the level of acceptance, as many of these works were at a low development level, and many were laboratory tests. Even so, we do not limit the scope; we study and characterize the different uses or applications in which the hardware technology was used. We believe that this study provides an insightful overview of the hardware being used in AR, refreshes the knowledge in this area, and provides a different organization of the technology for smart home and AAL. This work is not a data summarization since bibliometrics networks allowed us to identify gaps in the new relationships between technologies, informing researchers and developers on current practices of how the available hardware is being used to develop useful applications on activity recognition for smart home and AAL. Besides, knowledge about what has not yet been tried can be retrieved, prompting valuable insights for novel development approaches and future research, promoting new combinations of ideas or uses of hardware technology through innovative strategies like the Medici effect [20], and contributing to possibly disruptive innovations. This research work can help future researchers to identify new systems based on the hardware being created in the AR field, map those developments, and strengthen their research. We also aimed to identify new research questions as input for new AR hardware development and highlight possible approaches that could potentially impact needs in the near future for smart home and AAL applications.

In this review, it is likely that, due to the broad and interdisciplinary nature of this applied technology and its research area, some relevant articles have been disregarded because they are not clearly identifiable in the titles or abstracts, or due to our inclusion and exclusion criteria or the choice of our key terms to build the query strings, also, due to the journal's database selected. The review method is described in detail in Section 2. We present in Section 3, a brief scientometric and relational analysis of the research works chosen for the review, as well as the AR technology used in smart homes and AAL. The discussion in Section 4 points out interesting gaps in hardware technology combinations, and new potential studies around hardware technology for activity recognition are proposed. The review concludes in Section 5.

2. Review Method

This work, conducted as a systematic literature review, was executed in four stages, following PRISMA [21] guidelines, and the review approach for structuring the information was gathered from [22]. We applied software for visualizing bibliometric networks [23] in the first stage for the construction of query strings; the second stage focused on gathering potential results in the Web of Science (WoS) database; the third focused on excluding and including results based on criteria. Finally, the fourth stage consisted of characterizing the selected literature. The search was initially guided by wide concepts, but firmly focused on four technological areas of interest: smart home, smart environment, activity recognition, and ambient assisted living. The review did not consider gray literature.

Although WoS has many characteristics in common with Scopus in terms of indexed journals based on quality, they also differ, according to [24], in coverage and accuracy. We considered even though Scopus covers more journals than WoS, according to [25], Scopus tends to neglect indexing more papers, causing the loss of possible relevant works for our study. WoS has a stronger tendency to index journals in the science and technology field [26], as well as better accuracy in journal classification [24]. The above, in conjunction with the review method and the inclusion and exclusion criteria, helped to reduce the efforts of exploring quality scientific information, as the review seeks to capture a broad panorama of AR hardware technology with recent experimentation.

2.1. Query String Construction

Seeking to minimize the risk of overlooking relevant papers due to the choice of our key terms and to cover as many contributions as possible, a bibliometric networks analysis conducted in VOSviewer software [27] allowed us to get the best relevant terms used around the four areas of interest using titles, abstracts, and key terms. We retrieved from the WoS database the 100 most cited articles, and terms from all articles by the three most relevant authors as indicated by Google Scholar profiles from each area of interest: smart home (SH) and smart environment (SE), activity recognition (AR), and ambient assisted living (AAL). We generated different networks in VOSviewer to see the most mentioned words related to the more relevant terms, and to identify those that were semantically related and used once or a few times. This analysis helped select those terms that were synonymous with the areas of interest, common terms, strongly related terms, and synonyms, as shown in Figure 1.

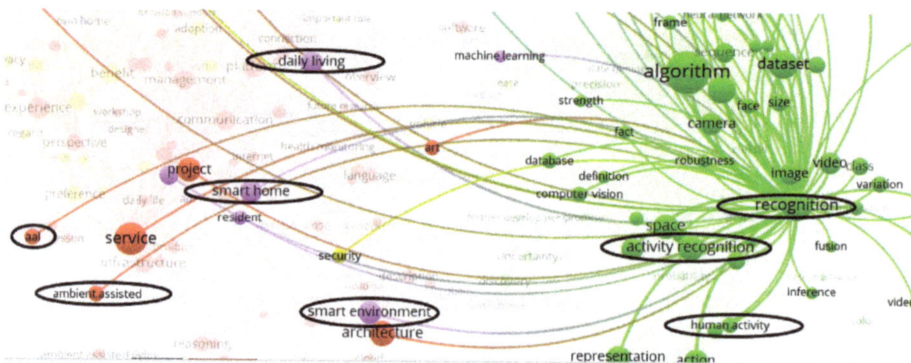

Figure 1. Some terms selected in network visualization of the bibliometric analysis generated in VOSviewer software.

Per area of interest (SH, SE, AR, AAL), we grouped and counted the selected terms to check duplication across the analysis, and chose common terms from concepts formed by one or more words. Finally, from the four terms (smart home, smart environment, activity recognition, ambient assisted living), we built three primary query strings (Table 1). Seeking to minimize the number of results per query and simplify the search, we supported the relationship of terms in the bigger bibliometric network visualization shown in Figure 2. As can be seen, AAL and smart home/environment are in the same cluster (red), and activity recognition is in a different cluster (green). Then, we combined the three primary query strings into two final query strings (FQ):

- FQ1: (AAL query) × (AR query)
- FQ2: (SH query) × (AR query)

Table 1. Transformation of common terms detected in the bibliometric networks analysis.

Interest Area	Common Term from VOSviewer	Duplication Frequency	Chosen Terms	Primary Query Strings
Ambient assisted living (AAL)	ALL	11	AAL Ambient assisted Assistance Assistive	**AAL query:** AAL OR "ambient assisted" OR assistance OR assistive
	Ambient assisted	11		
	Assisted	4		
	Ambient	4		
	Ambient assisted living	3		
	Assisted technology	2		
	ALL platform	1		
	ALL service	1		
	ALL system	1		
Smart home (SH)	Smart home	9	Smart home Environment Device House	**SH query:** Smart AND (home OR environment OR house OR device) OR intelligence
	Smart home technology	6		
	Smart home system	5		
	Smart home device	3		
	Smart house	1		
	Smart device	1		
Smart environment (SE)	Smart environment	6	Smart, environment, intelligence, home	
	Home environment	3		
	Intelligent environment	2		
	Smart environment	1		
	Intelligence	1		
Activity recognition (AR)	Activity	18	Activity Recognition "Human activity" "Human action" "Event detection" Action	**AR query:** Activity OR recognition OR "human activity" OR action OR "human action" OR "event detection"
	Recognition	14		
	Human activity	7		
	Human activity recognition	4		
	Activity recognition system	3		
	Action recognition	2		
	Human action recognition	2		
	Recognition system	1		
	Human action	1		

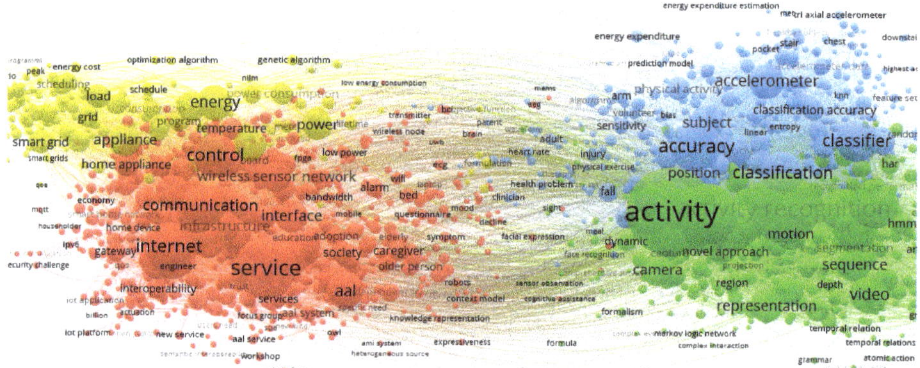

Figure 2. Biggest bibliometric network visualization of mixing papers retrieved from the World of Science (WoS) around the terms smart home, smart environment, activity recognition, and ambient assisted living.

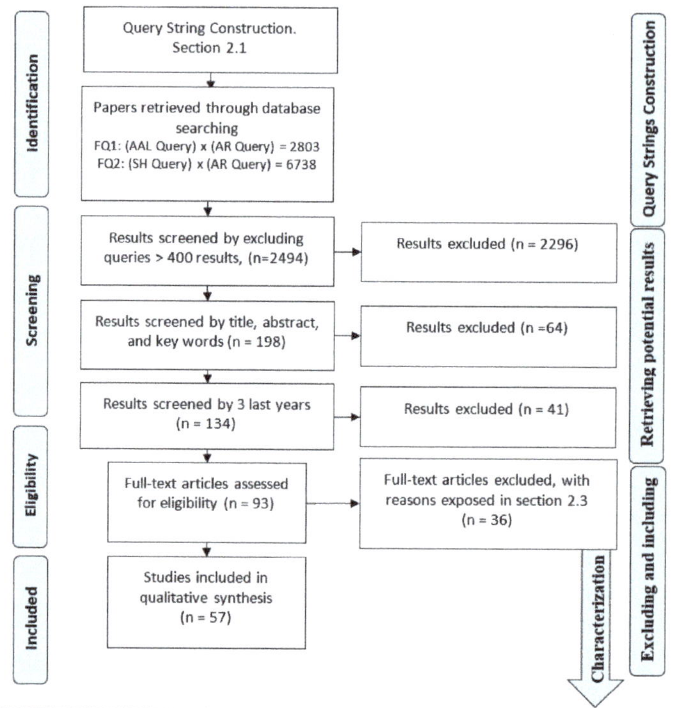

Figure 3. Diagram of the review method conducted based on PRISMA and operative structure in [28].

2.2. Gathering Potential Results

Testing FQ1 and FQ2 in WoS showed that the results were too big (Figure 3), so we decided to build 32 more reasonable small queries, from which we excluded queries with more than 400 results, considering them not reasonable to look at. For those with fewer than 400, based on the classification criteria used in [19], we checked the title and abstract shown in the results listed by the database as relevant or at least possibly relevant. For this, we used the match criterion "if it was about a technique or the use of technology and if the database was self-generated, but not acquired from a public one," gathering 196 potential papers (Figure 3). As this amount was not suitable, we selected 2016, 2017, and 2018 as the last three years of the technology concept, obtaining 131 articles.

2.3. Including and EXCLUDING Results

In order to reduce even more the number of papers to be characterized, the aim at this stage was to get a final list, so for those papers still marked as dubious, we checked the whole paper to see whether it matched or not, looking for exclusions, using the following criteria:

- Proposal schemes and approaches, simulated scenarios or datasets, use of open or popular or well-known datasets, without proved experiment.
- Proposals of methodologies, approaches, frameworks or related that do not mention explicit testbeds, prototypes, or experimentation with hardware.
- Home automation applications, brain or gait activity recognition, health variables, or proposals to improve systems or cognitive activity.

As the focus of this work was to get the latest hardware technologies used in activity recognition research around smart homes and ALL, we considered the following criteria:

- The paper used hardware to acquire information for AR in the same research work.
- Datasets in the research work generated in the same experiment were used.
- Commercial technology, self-built devices, or developed prototypes were used.
- Tested approaches with self-built datasets using virtual or physical sensors on smartphones, smartwatches, smart bands, etc.
- There was a focus on testing and using hardware, acquired, or self-developed as part of the research.

As example, papers like "3D Printed 'Earable' Smart Devices for Real-Time Detection of Core Body Temperature" [29] were not included, because the main objective was only temperature detection, and not recognition of human activities.

2.4. Characterization of the Selected Literature

This final stage consisted of more profoundly analyzing the information and filling the technical characterization tables, which consisted of mainly gathering information about the hardware systems for activity recognition, their uses, the population or commercial target, the types of technologies, hardware references or models, and scientometric behaviors as guidance to establish research networking. We selected 56 papers to be part of this review. A complete view of the whole review process is shown in Figure 3.

3. Results

The main goal of this work is to gather information and provide knowledge about the hardware technologies used in activity recognition research for smart home and AAL as well as a road map for project development for companies or entrepreneurs who may want to get into this field. This section provides a significant overview of how hardware technology is being used. Activity recognition in smart home and ALL development of hardware is recent; the first documents gathered on the WoS database showed that publications in the field do not have even a decade, as shown in Figure 4. Due to the timing of journal publication, it is possible that hardware technology for activity recognition in smart homes and AAL started to be used more since 2010. There is no doubt that R&D in activity recognition for smart home and ALL is a trend that has increased year to year.

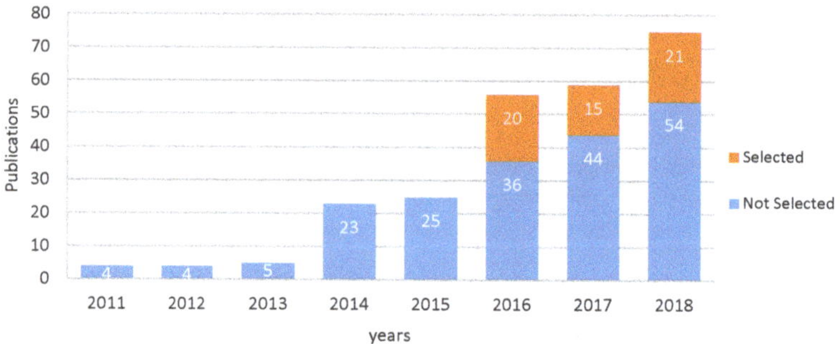

Figure 4. The trend in numbers of publications from papers initially selected for all years included in the database.

From the selected papers, the WoS analysis tool shows that only 2.5% of countries published reports on deploying hardware in activity recognition for smart home and ambient assisted living. Of those, 71% of the authors were concentrated in England, China, USA, and Spain, and 32% of authors were in Australia, Germany, India, Japan, North Ireland, and Saudi Arabia. Only 23% of countries reported one author with one publication (Figure 5). Latin American and Africa did not appear in the analysis, which does not mean that these regions are not working in this field, but may be due to the focused database (WoS) used for the review. For example, in a study published by Latin American researchers [30], they use the Emotiv Insight wearable for emotion recognition to study emotional levels during task execution, applying a different data mining approach.

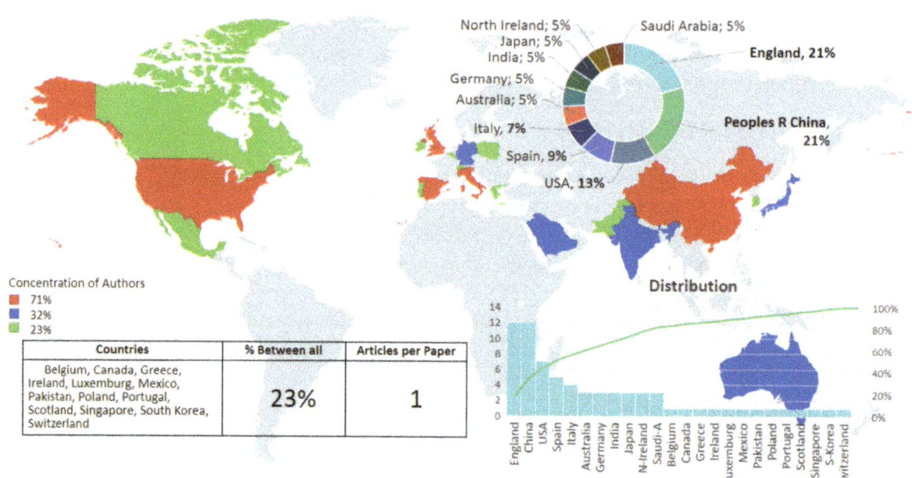

Figure 5. Worldwide map with an overview of the concentration and distribution of selected works.

There is no marked difference between the lowest and highest numbers of publications in journals. Despite that, we have to highlight that the Sensors Journal has the most publications, and IEEE, MDPI, IEIC, ACM, and Springer have a strong presence as publishers in this field of research. All journals with publications reported in this study are shown in Figure 6.

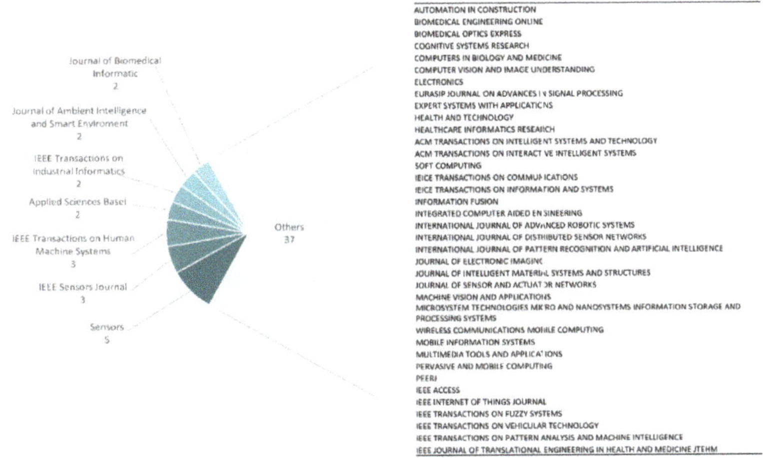

Figure 6. Overview of the journal distribution of selected papers.

The indexed categorization of WoS research areas has a marked fact (Figure 7), with engineering and computer science as the main areas of published works, followed by telecommunication and instrumentation categories. This is consistent with the type of hardware and software technology used to achieve the goals of activity recognition in smart home and AAL, as these are at the heart of the technology. Figure 7 also allows appreciating other research areas from which these hardware developments in AR for smart home and AAL are also carried out, such as physic, chemistry, biochemical, biology, medical, among others.

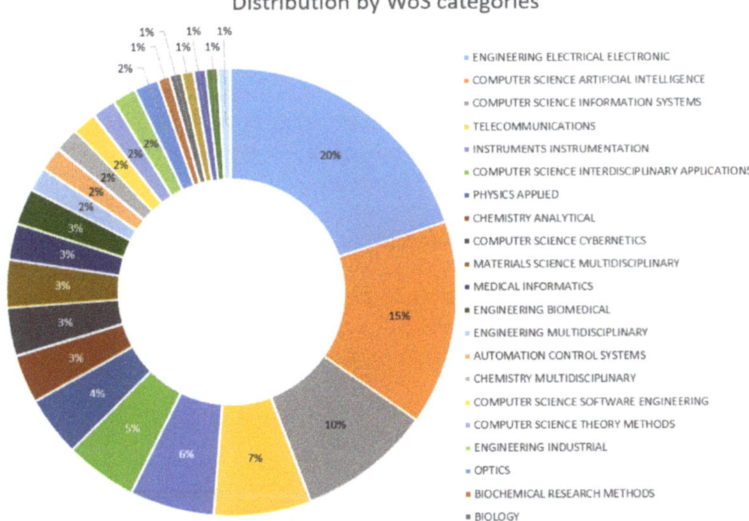

Figure 7. WoS research area distribution of the selected works.

Smart home technology became a focus of the product market beyond a research topic [9]. This study found six groups of technologies; the four biggest are video, smartphone, wearable, and electronic components, and the other two are prominent in development: Wi-Fi and assistive robotics. Figure 8 shows the distribution of these technologies, and whether they are self-developed hardware or commercial end-user hardware without modification already available on the market as a final product. It shows the most used technologies in the research works reviewed as well.

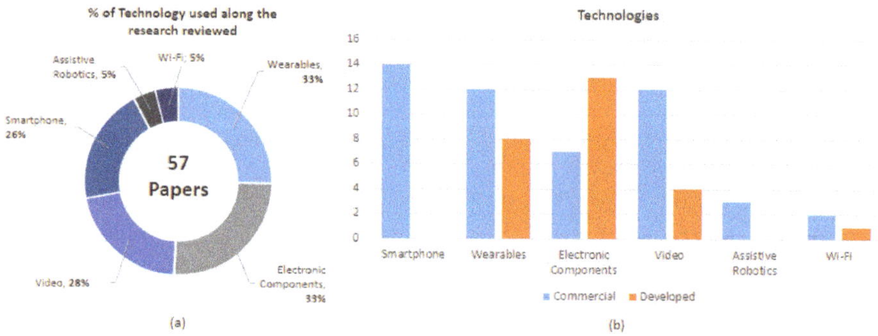

Figure 8. Analysis of hardware technology distribution. (**a**) Percentage of uses along the works reviewed. (**b**) Based on self-developed or commercial device-based solutions.

Developing and prototyping hardware is an attractive alternative in activity recognition research for smart home and AAL, to build systems from scratch using kits, boards, or chipsets as Arduino, Seeeduino, Raspberry, low-power microcontroller (MCUs), and sensors which later require data acquisition units to process the data. Almost 50% of the studies use this type of hardware solution. On the other hand, 60% also use components based on "plug and play" devices and systems with low levels of configuration just for connecting and gathering data before process it, like wall-mounted PIR sensors [31], microphones [32], infrared cameras [33], active tags [34], and radio-frequency identification (RFID) systems [35]. We found some interesting developments around video solutions, not using regular video cameras as would be expected, but specialized video hardware. Many applications that use wearables are based on commercial smartwatches, but others are based on self-developed smart bands or commercial wearables sensor devices like Shimmer. Smartphone applications are used on commercial devices run on Android, iOS, and Windows Mobile. We put smartphones in a different category from wearables; even though we can hold them in our pockets, handbags, and hands, smartphones are not be worn on the body, as wristwatches, rings, glasses, and necklaces are, following the categorization of wearables defined in [36] as accessories, clothing, and on-skin. Despite close use of smartphones and wearables such as smartwatches in daily life nowadays, this review found that not all applications of wearables are based on integration with smartphones; many studies analyzed the use of electronic components as a built-in solution for creating one's own wearables. Even so, these groups are just a broad categorization to facilitate an analysis of how this technology is being used together. It is worth highlighting that almost all studies had solutions using different technologies, so those are categorized into more than one group, as shown in Figure 9, showing a general view of the studies integrating different types of technology.

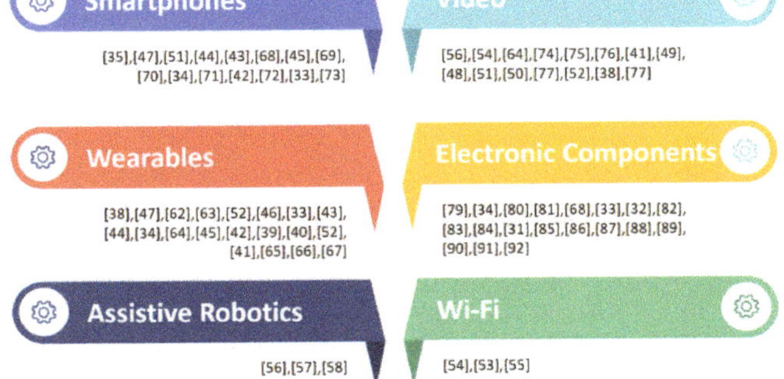

Figure 9. Identified categories of hardware technology used for activity recognition in smart home and ambient assisted living and its contributions.

3.1. Wearables

New products like smart bands and smartwatches from big tech companies like Samsung, Apple, and Microsoft put on the map the concept of wearable technology. Wearable sensors are devices composed of tiny sensors that can be worn in an article of clothing or more unobtrusively, such as embedded rings, shirts, or watches, which gather body and context information to process or transmit it [6]. Wearable wireless sensor technology attracted social and corporate interest in areas such as enhancing independent living for disabled people, support for physical training and monitoring work, but even more in health care applications such as posture and movement recognition, real life-vision, rehabilitation systems, and respiratory and stress biofeedback assessment, among others [6]. The above may be due to emerging IoT technology and smart devices, sensors, and data processing hardware becoming commodities; on the other hand, the rising cost of healthcare systems induces wearable health tech research and new developments. Some wearable health devices are health regulatory authorized and successfully deployed, such as Nymy™, Samsung Gear™, and Apple Watch, not used for specialized or critical health issues but just to get biomedical signal data for daily life analysis [37]. We note commercial efforts in developing bendable smartphones, which can fall in the wearables zone. However, these are far from being used on the wrist due to the folded and flexible touchscreen display prototype level, besides that, none was found in this study.

A significant percentage of the papers based their experiments on self-developed technology or development tools for a wearable solution. Only 50% of the selected studies used commercial devices; others preferred to use modules, sensor boards, and related items. Accelerometers are a common factor among almost all of the studies, followed by gyroscopes. The rapid and low-cost accessibility, such as the flexibility of technology to build customized wearable combinations, allowed measuring variables in other parts of the body, such as heart rate in the chest [38]. On the other hand, interesting commercial wearable sensor bands like the Shimmer device are mentioned in more than one study [39–41].

The combination of wearables and smartphone technology is not apparent; only 37% of the studies used this combination, and just with specific smartwatch devices. Many wearables like smartwatches need to work with a smartphone, extending the functionality of the smartphone beyond data transmission, receiving and confirming text and audio messages, and taking and making calls. However, these smartwatches can work on their own for other purposes without being paired with a smartphone [7].

Mixing smartwatches with video capture and processing technology seems to be a field of work for various researchers. For the rest, it seems to be sufficient to use wearable technology alone to assess activity recognition for smart home and AAL, maybe to try simplicity in technological solutions. Commercial devices from big companies, such as Samsung Galaxy Gear Live [42], Microsoft Band 2 [43], and Intel Basis Peak [44], are mentioned in several studies, as well as other commercial alternatives like Empatica E3 [33], Fitbit [44], HiCling [34], Pebble [45], and Google Glass [33,46] (see Table 2).

3.2. Smartphones

Android seems to be a favorite platform to support activity recognition systems for smart home and AAL, not to say this is more effective than others, but this OS appears in most of the studies, except in [35,47], which used a smartphone but did not say which one, and [35], which used iOS. We did not identify any use of Windows Phone or any other mobile operative system. We did not identify a preferred model of Android phones. Besides, the use of wearable technology jumps out, and the elderly are the main benefiting population. Of the smartphone sensors, accelerometers are the most used, followed by GPS. Beyond generic AR applications for smart home and AAL, there is a focus on smartphones working in localization, occupancy, fall detection, posture recognition, and for the elderly population, disabled people, and health care (see Table 3).

3.3. Video

Activity recognition for smart home and AAL developed in video-based technology is popular. From the selected studies, 60% used RGB-D sensors, which are based mostly on the Kinect platform from Microsoft; only [48] uses an RGB camera from FLIR Systems. The authors of [49] combine RGB-D cameras with Vicon Systems cameras, and the authors of [48] use thermal cameras. Thermal cameras are used alone in [50] and with smartphones in [51]. There did not seem to be any interest in using video cameras combined with other technologies, more than with wearables [52] and infrared cameras [38] (see Table 4).

3.4. Electronic Components

Electronic components such as sensor boards, microcontrollers, board processors, electronic modules, communication devices, development toolkits, chipsets, and related devices, are mainly used to build from scratch or complement any function that a commercial device cannot provide. Electronic components appear in almost 30% of the selected research and they are one of the four main technologies used to build activity recognition for smart homes and AAL. Table 5 offers a complete overview of the types of hardware and some references, and models researchers worked with. Just a few works based on electronic components use other kinds of technology identified in this paper, such as [34], which uses active tags with smartphones and wearables, and [33], which uses a Raspberry board and an infrared camera taken from a Pupil Labs eye tracker and adapted for Google Glass. Electronic components are used for special activity recognition functions such as fall detection, localization, mobility, occupancy, posture recognition, and health, targeted to the elderly population.

3.5. Wi-Fi

The scientific community is concerned about nonintrusive activity recognition solutions. In this regard, this study presents an interesting way to apply AR for smart home and AAL: by using radio waves (Table 6). The above seems to be a promising solution by using a widely deployed technology, Wi-Fi routers. The authors of [53] captured information generated during radio wave propagation in indoor environments using wireless signals through a smart radio system that turns radio waves generated by Wi-Fi signals in an intelligent environment able to capture changes in multipath radio profiles, detecting motion and monitoring indoor events, even through walls in real time.

The authors of [54] present a human activity sense system for indoor environments called HuAc, based on a combination of Kinect and Wi-Fi. The system can detect even in conditions of occlusion, weak light, and activities with different perspectives such as forward kick, side kick, bending, walking, answering a phone, squatting, drinking water, and gestures like horizontal arm wave. In addition, this system also detects other activities such as two-handed waving, high throwing, tossing paper, drawing a tick mark, drawing an x, clapping hands, and high arm-waving.

The authors of [55] also use Wi-Fi links for evaluating passive occupancy inference problems. They set up signal processing methods and tools with electronic components to adapt this in a commercial Wi-Fi router. Based on the analysis of channel state information (CSI) collected from multiple-input-multiple-output (MIMO) using orthogonal frequency division multiplexing (OFDM) radio interfaces in off-the-shelf Wi-Fi networks, the system is capable of detecting localization of two independent human bodies moving arbitrarily through the working area of the system.

3.6. Assistive Robotics

High technological level assistive robotics is used for developing applications on activity recognition for smart home and AAL, based on commercial robots and mainly focused on applications for health care and the elderly population. All studies use interactive robots manufactured in Germany, Japan, and the United States, as shown in Figure 10. Only the PR2 robot is being used in the same country [56], while Care-O-bot3 is used on collaboration between Portugal and Spain [57], and Pepper is used in the UK [58] (see Table 7).

The uses of PR2 [56] combine the robot with video capture through an RGB-D adapted to the robot's head; with this camera, the robot can sense people's movement. RGB-D sensors recognize people's movements and anticipate future activity as a reactive response, called activity prediction. This is aimed at making smarter robots that can assist humans in making tasks more efficient or take on tasks that humans are unable to perform. Care-O-bot 3 is used in [57], in which AR is used to teach the robot to perform assisting tasks and behave in response to some tasks. The robot can identify some human activities thanks to the use of a fully sensorized system and ceiling-mounted cameras deployed in a house. The study mainly seeks to develop a robot personalization platform for end-users, as a robot system to teaching and learning for care workers and related helpers, and as a trusted companion for older adults as well. The above is a perfect example of how activity recognition systems can be matched with other technologies to achieve better living conditions.

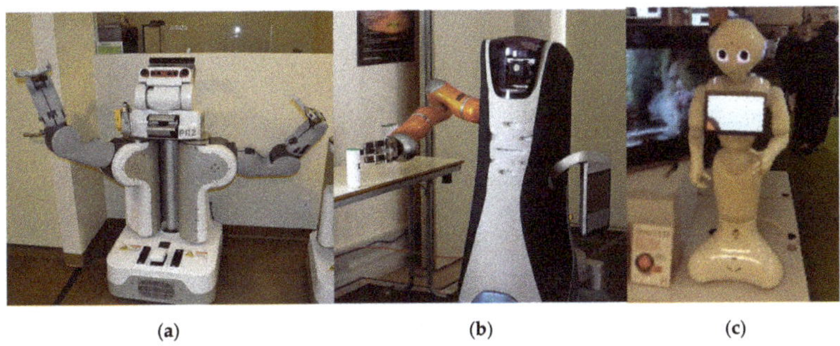

(a) (b) (c)

Figure 10. Assistive robots identified in activity recognition research: (**a**) PR2 robot [59]; (**b**) Pepper robot [60]; (**c**) Care-O-bot3 [61].

PHAROS is a platform developed which uses the Pepper robot [58] to assist caregivers in teaching and evaluating the movements of adults in their daily physical activities. The PHAROS system identifies the elder person and, based on his physical condition, recommends a series of personalized and planned exercises. In a scheduled way, the robot is capable of capturing the attention of older adults, showing on the screen and describing by audio the exercises he should perform. Pepper's camera provides the video input to recognize the activity and extract the skeletal data by Openpose software, which helps to label the activity being performed, and sends it to a module that registers the health status, and based on that, gives recommended exercises.

Table 2. Characterization of wearable technology used in selected papers.

Wearable Technology Used				Context of the Proposal			AR Solution		
Model	Type	Sensor	Body Part	Combination	Applications	Target	Commercial	Developed	Ref.
Customized	Wearable sensor band	Accelerometer + heart rate sensor	Chest + limb	Video	Generic AR applications	All		X	[38]
Customized	Wearable sensor band	Accelerometer + Gyroscope + Magnetometer	Arms	Smartphone	Generic AR applications + Localization	Elderly		X	[47]
Customized	Wearable sensor band	Accelerometer + Gyroscope	Arm	-	Generic AR applications	Health		X	[62]
Customized	Wearable sensor band	Accelerometer	Hand	-	Emotion recognition	Health		X	[63]
Customized	Skin sensor	Electro-dermal activity (EDA)	Skin	Video	Emotion recognition	Health		X	[52]
Google Glass Explorer	SmartGlass	Video capture	Head	-	Localization	Elderly	X		[46]
Google Glass-based + Head tracking device + Empatica E3 sensor armband	SmartGlass + smart band	IMU + Audio + Video	Head + Arm	Electronic components + Smartphone + Video	Generic AR applications	Elderly	X	X	[33]
Microsoft Band 2	Smartwatch	Accelerometer	Arms	Smartphone	Generic AR applications	All	X		[43]
Fitbit + Intel Basis Peak	Smartwatch	Heart rate monitoring + Skin temperature monitoring	Hand	Smartphone	Posture recognition	All	X		[44]
HiCling	Smartwatch	Optical sensor + Accelerometer + Captive skin touch sensor	Arms	Electronic Components + Smartphone	Fall detection	All	X		[34]
NS	Smartwatch	Accelerometer + Gyroscope	Arms	Video	Generic AR applications	All	X		[64]
Pebble SmartWatch	Smartwatch	3-axis integer accelerometer	Arms	Smartphone	Fall detection	Elderly	X		[45]
Samsung Galaxy Gear Live	Smartwatch	Accelerometer + Heart rate sensor	Arms	Smartphone	Mobility	All	X		[42]
Shimmer	Wearable sensor band	Accelerometer	Wrist	-	Generic AR applications	Elderly	X		[39]
Shimmer	Wearable sensor band	Accelerometer + Gyroscope	Abs	-	Fall detection	Elderly	X		[40,52]
Shimmer	Wearable sensor band	Accelerometer + Gyroscope	Wrist	Video	Generic AR applications	Elderly	X		[41]
WiSE	Wearable sensor band	Accelerometer	Arms	-	Generic AR applications	Sport		X	[65]
WiSE	Wearable sensor band	Electrodes + Accelerometer	Arms	-	Generic AR applications	Sport		X	[66]
Microsoft Sens Cam	Wearable camera	Video capture	Chest	-	Generic AR applications	All	X		[67]

Table 3. Characterization of smartphone technology used in selected papers.

Smartphone Uses		Context of the Proposal			AR Solution		
Model	**Sensor Applied**	**Combination**	**Applications**	**Target**	**Commercial**	**Developed**	
Smartphone	Data transmission	-	Generic AR applications	Elderly	X		[35]
Smartphone	Data transmission + App	Wearable	Generic AR applications + Localization	Elderly		X	[47]
Android	App	Video	Health conditions	All	X		[51]
Android	Accelerometer	Wearable	Posture recognition	All	X		[44]
Android	Mic	Wearable	Generic AR applications	All	X		[43]
Android	Data transmission + App	Electronic Components	Occupancy	All		X	[68]
Android	Data transmission + App	Wearable	Generic AR applications	Elderly	X		[45]
Google NEXUS 4	Accelerometer	-	Generic AR applications	Health	X		[69]
Google NEXUS 5	Accelerometer + Mic + Magnetometer	-	Occupancy	All	X		[70]
HTC802w	Accelerometer + GPS	Electronic Components + Wearable	Fall detection	All	X		[34]
IPod Touch	Accelerometer	-	Mobility	Disabled	X		[71]
LG Nexus 5	Accelerometer + Mic + GPS + Wi-Fi	Wearable	Mobility	All	X		[42]
Samsung ATIV	Accelerometer Gyroscope	-	Posture recognition	All	X		[72]
Samsung Galaxy S4	Accelerometer + Mic	Electronic Components + Wearable + Video	Generic AR applications	Elderly	X	X	[33]
Xolo era 2x and Samsung GT57562	Accelerometer	-	Generic AR applications	All	X		[73]

Table 4. Characterization of video technology used in selected papers.

Video Technology		Context of the Proposal			AR Solution		
Type	Model	Combination	Applications	Target	Commercial	Developed	Ref.
RGB-D Sensor	Kinect	Assistive robotics	Care	All	X		[56]
	Kinect	Wi-Fi	Generic AR applications	All	X		[54]
	Kinect	Wearable	Generic AR applications	All	X		[64]
	Kinect	-	Posture recognition	All	X		[74]
	Kinect	-	Care	Disabled + Elderly	X		[75]
	Kinect	-	Generic AR applications	All	X		[76]
	Kinect	Wearable	Generic AR applications	Elderly	X		[41]
RGB-D Sensor + Vicon System camera	Kinect + Vicon System camera	-	Fall detection	Elderly	X		[49]
RGB-D sensor + Thermal camera	Thermal camera PI450 Grasshopper RGB GS3-U3-28S5C-C FLIR	-	Fall detection	Elderly		X	[48]
Thermal camera	FLIR One for Android	Smartphone	Health conditions	All	X		[51]
	FLIR E60 thermal infrared camera	-	Care	Elderly	X		[50]
Video camera	-	-	Fall detection	Elderly	X		[77]
	-	Wearable	Emotion recognition	Health		X	[52]
Video camera + Infrared camera	-	Wearable	Generic AR applications	All		X	[38]
Optical sensor	Agilent ADNS-3060 Optical mouse sensors	-	Care	Elderly		X	[78]

Table 5. Characterization of electronic component technology used in selected papers.

Electronic Components Used		Context of the Proposal			AR Solution		
Technologies	**Reference/Model**	**Combination**	**Applications**	**Target**	**Commercial**	**Developed**	**Ref.**
TAG RFID + RFID antennas + RFID reader	Smartrack FROG 3D RFID + RFID reader antennas + Impinj Speedway R-420 RFID reader	-	Fall detection	Elderly	X		[79]
Active tags	-	Smartphone + Wearable	Fall detection	All	X		[34]
Grid-EYE + Ultrasonic sensor + Arduino	Grid-EYE (AMG8853, Panasonic Inc.) hotspot detection + Ultrasonic HC-SR04 + Arduino Mega	-	Fall detection	Elderly		X	[80]
Gird-EYE + Rotational platform + Time of flight (ToF) ranging sensor + Arduino	Gird-EYE AMG 8853 Panasonic VL53L0X + Arduino Nano	-	Localization + Occupancy	All		X	[81]
HC-SR04 + PIR module + BLE module	-	-	Occupancy	All	X		[68]
Infrared camera Raspberry	Pupil Labs eye tracker Raspberry Pi 2	Smartphone + Wearable	Generic AR applications	Elderly	X	X	[33]
Microphone	-	-	Generic AR applications	All	X		[32]
Zigbee transceiver ultra-low-power microcontroller	CC2520 + MSP430F5438 chipsets.	-	Localization	All		X	[82]
Capacitive sensing	OpenCapSense sensing toolkit	-	Posture recognition	Health		X	[83]
XBee Pro + Series Pro 2B antennas + Laser diode	Part 2 XBee Pro + Series Pro 2B antennas + NR	-	Fall detection	Elderly		X	[84]
PIR sensors	-	-	Fall detection	Elderly	X		[31]
PIR sensor + Motion sensor + Data sharing device	NR sensor + PogoPlug		Generic AR applications	All	X		[85]
S-band antenna + Omnidirectional	-	-	Generic AR applications	Health	X		[86]
Seeeduino + Temperature and humidity sensor + Light sensor + Ranging sensor + Microphone	Seeeduino Arch-Pro + HTU21D + Avago ADPS-9960 + GP2Y0A60SZ + Breakout board INMP401	-	Generic AR applications	All		X	[87]
Sensor node consisting of nine PIR sensors arranged in a grid shape + CC2530 Zigbee module	CC2530 used to sample PIR signals and communicate with the sink node	-	Localization	Elderly		X	[88]
Strain gauge sensor + IMU sensor	SGT-1A/1000-TY13 StrainGauges + LSM9DS1 9axis IMU	-	Health conditions	Elderly		X	[89]
Measurement setup: low-noise amplifier (LNA), data-acquisition unit (DAQ) + Switching SP64T + Downconverter unit + Path antennas		-	Localization	Elderly		X	[90]
Portable brain-activity measuring equipment NIRS-EEG probes and NIRS-EEG unit + Thermometer + Laser range finder + Kinect + Pyroelectric sensor + Wireless LAN system + Sensor arrangement cameras + Microphones + Infrared devices		-	Mobility	Disabled		X	[91]
Tunable RF transceivers NI USRP-2920 + MIMO cable Wireless energy transmitter + PCB antennas		-	Generic AR applications	Health		X	[92]

Table 6. Wi-Fi devices used as the main component of activity recognition.

Wi-Fi uses		Context of the Proposal		AR Solution		
Technology	**Reference**	**Combination**	**Applications**	**Commercial**	**Developed**	**Ref.**
Wireless router	Commercial TP link	Video	Generic AR applications	X		[54]
Commercial Wi-Fi device	NS	-	Generic AR applications	X		[53]
Wi-Fi + Chipset	NS	Electronic components	Localization		X	[55]

Table 7. Characterization of assistive robotics used in selected papers.

Assistive Robotics Technology for AR					
Technology	**Combination**	**Goal**	**Target**	**AR Solution**	**Ref.**
PR2 robot	Video + RGB-D sensor	Care	All	Commercial	[56]
Care-O-bot3	-	Care	Elder	Commercial	[57]
Pepper robot	-	Care	Elder	Commercial	[58]

4. Analysis and Discussion

In the previous section, we described six main types of hardware technology used for activity recognition applied to the smart home and AAL research field. The majority of the reviewed works reported several goals of AR, with fall detection as the main one, followed by localization. Other AR applications were posture, mobility, occupancy, and emotion recognition. Many works did not report a specific goal, only a system capable of reaching it, or at least the authors of this review did not detect them, goals tagged as generic AR applications for smart home and AAL. Figure 11 shows an overview of how these goals are aimed at specific populations such as older adults through fall detection, localization, and care, and the disabled population through mobility, care, and health conditions. Surprisingly, emotion recognition seems to affect healthcare more than social or entertainment applications. Recognition of activities, events, and gestures is used to assess caregiving through behavioral patterns for health diagnostics. Generic AR applications refer to studies that did not mention a specific application or practical use.

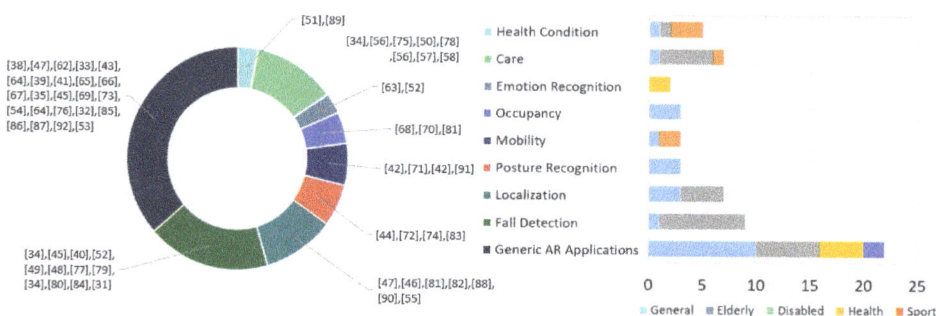

Figure 11. Distribution of activity recognition application in smart home and AAL and its relationship with target populations.

Results show specific relationships between types of technology and application focus of activity recognition for smart home and AAL. Figure 12 shows this relation through a relation network in which the size of the node means the frequency of technology use or application focus, and the thickness of the lines shows a greater or lesser relationship between both groups. Some reviewed works show applications such as occupancy based on technologies like electronic components and smartphones. In [68], the Android phone is used for data transmission through an app, with ultrasonic and passive infrared sensors, achieving height detection as a unique bio-feature, and efficient differentiation of multiple residents in a home environment. Other research also used electronic components and smartphones for medical treatment of health conditions, monitoring vital signs like respiratory rate. For example, [51] combined those technologies with video technology to achieve accurate respiratory rate tracking using an app phone for visualization and processing thermal images from a thermal camera (Flir One for Android).

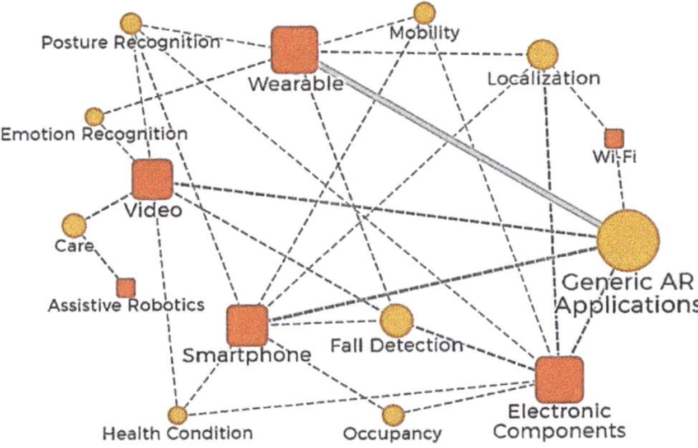

Figure 12. Relationships between technology (square) and research focus (circle) for activity recognition.

For care applications, researchers combined video and assistive robot technology, using activity recognition as input for activity prediction to help the robot perform actions in response to human activity; a similar goal was achieved in [56], combining a PR2 robot with RGB-D sensor Kinect technology. Using only video technology can also help in elderly care; video helped estimate locations and perform behavioral analysis under low-resolution constraints as an alternative to PIR sensors or high-resolution cameras. For example, [78] used an Agilent ADNS-3060 optical sensor (30 × 30 pixels) installed in a service apartment for senior citizens, projecting pattern identification for recovery periods through caregiver monitoring.

Through video technology combined with wearables, some researchers project the use of emotion recognition applications such as monitoring and regulation emotions for patients in smart health environments, this is achieved by [52] using an electro-dermal activity (EDA) sensor with a low-power camera and Bluetooth data transmission. Fairly accurate recognition of emotions such as happy, neutral, and angry was achieved using only wearables, as done in [63], using the built-in accelerometer of a smart band. It is possible to achieve posture recognition using video, wearables, smartphones, and electronic components. An application like this could prevent decubitus ulcers through electronic components such as capacitive sensing, as the research work of [83], which a wired grid in a bedsheet with an OpenCapSense sensing unit, to help detect prolonged posture, allowing caregivers to be aware of this situation. Posture recognition using smartphones and wearables at the same time allows the mitigation of fake alarms in activity recognition. In [44], physiological sensors of smart bands like Fitbit and Intel Basis Peak are used to detect vital signs alarms; before the system sends an alarm, the user gives feedback about the situation through a screen and speech recognition mobile app, improving the accuracy of the activity recognition system and starting real-time communication with caregivers. Even for ambiguous posture detection, video technology is used for recognizing activities such as calling, drinking water, using a remote control, and pouring water.

Wearable, smartphone, and electronic component technologies also help to build solutions for activity recognition on mobile applications for smart home and AAL. In [42], a group of sensors such as accelerometer and heart rate sensors from a smartwatch, as well as a mic, accelerometer, GPS, and Wi-Fi traces from a smartphone was used to generate mobility pattern information from activities like walking, running, driving a vehicle, riding a bicycle, and inactive or sedentary activities.

Localization applications also use wearables and smartphones, achieving location-agnostic activity recognition. In [47], used customized sensor bands (accelerometer, rotation, and magnetic) placed on the arm and foot, using a smartphone as data transmission into the place, addressing home monitoring and privacy concerns for fine-grained lifelogging and cognitive prosthesis. Privacy concerns in localization use Wi-Fi commodity with some electronic components for passive occupancy inference [55], achieving detection and localization of two independent targets moving around the transmitter/receiver locations.

However, we point out from this relationship network analysis some interesting potential technological developments:

- Video technology can help in mobility and localization by using wearables as a way of alerting.
- Due to the prominent Wi-Fi results, research should extend to occupancy detection, fall detection, and posture for care.
- Assistive robots with wearables, smartphones, and electronic components can be used for vital sign monitoring and alerts for remote care.
- Wearables can be used for occupancy applications and care of health conditions.

Some technologies that are less articulated with other technologies to develop solutions toward activity recognition for smart home and AAL, can be identified through a more in-depth relationship network analysis, as well as other points of interest stand out around the sensors or specific devices used for each technology identified in the present work (video, electronic components, wearables, smartphones, Wi-Fi, and assistive robots). The big panorama of deploying hardware technology for activity recognition for smart home and AAL shown in Figure 13, shows nodes with different colors representing the types of technologies, sensors, and devices. In this deeper relationship network, the size of each node represents the frequency of hardware use among the works reviewed, and the thickness of the lines between nodes represents how much these technologies are used in collaboration. This network relation uses a "has a" node hierarchy, like this: "Technology" has a "particular type of technology", which has "sensors" and "other devices". These last two are more detailed hardware info than the first two, which brings a better panorama about the hardware is being used in AR.

Video, electronic components, smartphones, and wearables show a trend of hardware used for AR in SH and AAL, these are the most frequently used among the technological solutions deployed; the relationship network (Figure 13) shows how these interact strongly through each type of technology. RGB-D sensors, video, and audio capture devices, infrared cameras, controller devices, optical sensors, wearable sensor bands, and smartwatches show amounts of collaborative solutions. Many papers included detailed information about sensors or devices used, highlighting strong collaborative solutions using apps for processing data, ultrasonic sensors, infrared and PIR modules, proximity sensors, temperature sensors, IMU, magnetometers, EEG, and heart rate monitoring. Other less strong, but still collaborative, are technologies like apps for data transmission, Bluetooth, Grid-EYE and time of flight, laser range finders, microphones, humidity sensors, and light sensors.

There is a potential roadmap for developing new solutions using technologies that are not currently being used very collaboratively with others, which researchers should study in future work, such as wearable cameras, strain gauges, skin temperature sensors, EDA sensors, smart glasses, GPS, electromyography (EMG) sensors, and Zigbee. Other technologies are far from joint solution deployment: assistive robots, Wi-Fi for passive detection, and capacitive sensors. Notice the novel technologies applied in activity recognition such as radiofrequency systems over S-band antennas, RF transceivers, antennas data acquisition systems, and RFID. The above may be due to the highly specialized knowledge needed to use and adapt these technologies for specific uses, more than data transmission. This last analysis shows that specific research goals in activity recognition cannot be achieved using one single hardware technology, but can through joint solutions. We consider essential try to integrate these technologies with others that are commonly used to expand the goal achievement of applications such as fall detection, localization, posture and occupancy recognition, care and health condition monitoring, and other potential applications.

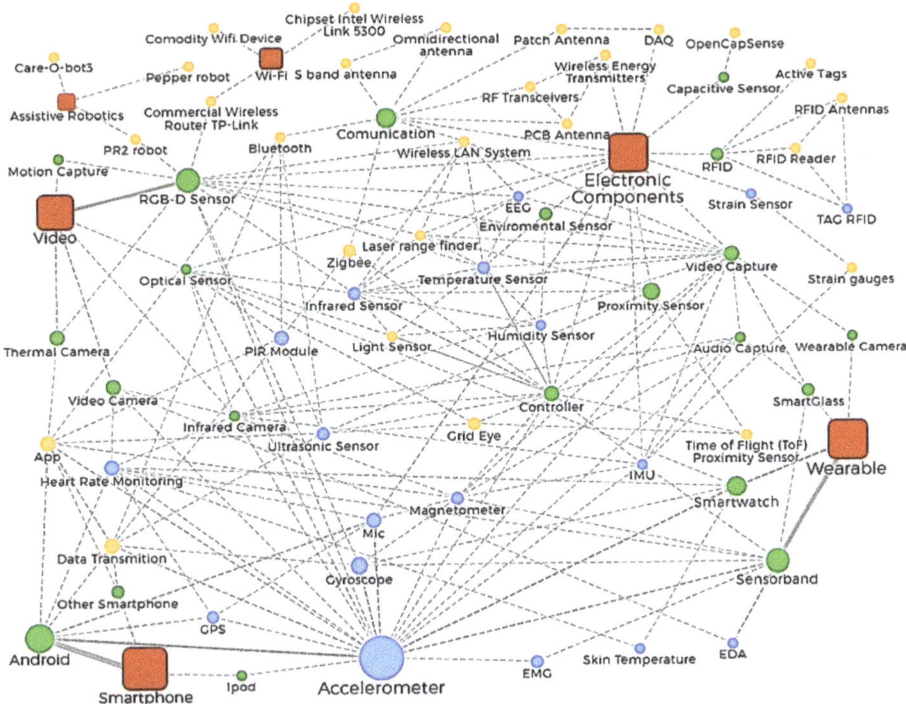

Figure 13. Relationship network analysis for hardware solutions deployed in activity recognition for smart home and AAL. Technology (orange square), a particular type of technology (green circle), itemized sensors (pink circle), and other specific devices (blue circle).

Through this work, we identify how several hardware technologies are deployed for activity recognition around smart homes and AAL. We can now evaluate and determine which ones to develop and start to experiment from a secure starting point to address some societal issues, and to further close the knowledge gap in this field. This is the case of the Smart Home CUC laboratory starting in Colombia, for which this study will serve as raw information to plan the infrastructure, technology acquisition, and networking, and cross some research approaches (localization, mobility, etc.) with populations (elders, athletes, disabled, etc.) and local needs.

As this literature review was not planned to be deep but instead wide in coverage, it highlights some questions to be addressed in future works in order to give a broad and clear panorama of advances in technologies in this field, such as the following:

- How are large-scale house projects for activity recognition planned?
- Through technological surveillance, how can we extend our understanding of promising advances such as smart floors, smart beds, and smart walls?
- Which types of tested hardware technology are giving better results?
- How can researchers design testbeds? It is crucial to have an overview of how to design this type of experiment and increase the credibility for approval by scientific networks of new paper proposals.

- What is the cost-benefit relationship in achieving effectiveness in each focus of activity recognition?
- Which commercial technology gives the best effective results in activity recognition so that it can be taken to market?

All of these could open the door to new studies around activity recognition, helping reduce the time to market solutions.

5. Conclusions

This paper provides a detailed review of hardware technology used in activity recognition research related to smart homes and ambient assistive living (AAL) applications published in the last three years and indexed on the WoS database. The reviewed papers showed four main groups of hardware technology: smartphones, wearables, electronic components, and video. Half of the research approaches focus on fall detection, care, posture recognition, mobility, occupancy, emotion recognition, and health conditions. In contrast, the other half are not developed for any specific function, just for exploring and exploiting the available technology. RGB-D sensors and thermal and video cameras are the main video hardware to capture information. Android is the mobile operating system most used, usually with wearables and video technology. Two other technologies identified as emerging fields of study for applications in activity recognition in smart home and AAL are Wi-Fi and assistive robots. The first one has potential as a non-intrusive and invisible technology. Assistive robots are used to assist and guide human activity for health, and activity recognition is being implemented as a function of this type of robot.

From a relationship network analysis between types of technology and applications for activity recognition in smart homes and AAL, the review points out some interesting new potential developments combining some technologies. One of these is the use of video technology to help mobility and localization with wearables as a way of alerting. Another is to extend research to occupancy detection, fall detection, and posture for care due to the prominent Wi-Fi results. Another new solution is to use assistive robots with wearables, smartphones, and electronic components for vital sign monitoring and alerts for remote care, also the use of wearables for occupancy and care of health conditions.

Through a more in-depth relationship analysis of hardware uses in terms of sensors or specific devices used in each technology identified, the review also detected some lack of articulation of developing solutions toward activity recognition: wearable cameras, strain gauges, skin temperature sensors, EDA sensors, smart glasses, GPS, EMG sensors, and Zigbee. Others far from joint solution deployment are assistive robots and Wi-Fi for passive detection with technologies such as capacitive sensors, S-band antennas, RF transceivers, antenna data acquisition systems, and RFID. Assistive robots and Wi-Fi can be combined with others commonly used to expand the spectrum of applications for activity recognition in smart homes and AAL, with devices such as RGB-D sensors, video, and audio capture devices, infrared cameras, controller devices, optical sensors, wearable sensor bands, smartwatches, Android phones, apps for processing data, ultrasonic sensors, infrared and PIR modules, proximity sensors, temperature sensors, IMU, magnetometers, EEG, and heart rate monitors.

Further research could also expand and update the notion about hardware uses for activity recognition, for instance in other sources like Scopus, Google Scholar, or patent databases, as part of technological surveillance for monitoring these advances, and to study the effectiveness of these developments and find novel combinations and promising hardware that can help accelerate innovations in activity recognition field for smart home and ambient assisted living.

Funding: This research has been supported under the REMIND project Marie Sklodowska-Curie EU Framework for Research and Innovation Horizon 2020, under grant agreement No. 734355 Project REMIND.

Conflicts of Interest: The authors declare no conflict of interest.

References

1. Bejarano, A. Towards the Evolution of Smart Home Environments: A Survey. *Int. J. Autom. Smart Technol.* **2016**, *6*, 105–136.
2. Mantoro, T.; Ayu, M.A.; Elnour, E.E. Web-enabled smart home using wireless node infrastructure. In Proceedings of the MoMM '11: 9th International Conference on Advances in Mobile Computing and Multimedia, Ho Chi Minh City, Vietnam, 5–7 December 2011; pp. 72–79. [CrossRef]
3. Qu, T.; Bin, S.; Huang, G.Q.; Yang, H.D. Two-stage product platform development for mass customization. *Int. J. Prod. Res.* **2011**, *49*, 2197–2219. [CrossRef]
4. Kim, C.G.; Kim, K.J. Implementation of a cost-effective home lighting control system on embedded Linux with OpenWrt. *Pers. Ubiquitous Comput.* **2014**, *18*, 535–542. [CrossRef]
5. Chen, L.; Hoey, J.; Nugent, C.D.; Cook, D.J.; Yu, Z. Sensor-based activity recognition. *IEEE Trans. Syst. Man Cybern. Part C Appl. Rev.* **2012**, *42*, 790–808. [CrossRef]
6. Kumari, P.; Mathew, L.; Syal, P. Increasing trend of wearables and multimodal interface for human activity monitoring: A review. *Biosens. Bioelectron.* **2017**, *90*, 298–307. [CrossRef] [PubMed]
7. Younes, R.; Jones, M.; Martin, T.L. Classifier for activities with variations. *Sensors* **2018**, *18*, 3529. [CrossRef] [PubMed]
8. Liangying, P.; Chen, L.; Wu, X.; Guo, H.; Chen, G. Hierarchical Complex Activity Representation and Recognition Using Topic Model and Classifier Level Fusion. *IEEE Trans. Biomed. Eng.* **2017**, *64*, 1369–1379.
9. Amiribesheli, M.; Benmansour, A.; Bouchachia, A. A review of smart homes in healthcare. *J. Ambient Intell. Humaniz. Comput.* **2015**, *6*, 495–517. [CrossRef]
10. Bang, J.; Hur, T.; Kim, D.; Huynh-The, T.; Lee, J.; Han, Y.; Banos, O.; Kim, J.I.; Lee, S. Adaptive data boosting technique for robust personalized speech emotion in emotionally-imbalanced small-sample environments. *Sensors (Switzerland)* **2018**, *18*, 3744. [CrossRef]
11. Weiser, M. The computer for the 21st century. *Sci. Am.* **1991**, *265*, 94–105. [CrossRef]
12. Malkani, Y.A.; Memon, W.A.; Dhomeja, L.D. A Low-cost Activity Recognition System for Smart Homes. In Proceedings of the 2018 IEEE 5th International Conference on Engineering Technologies and Applied Sciences (ICETAS), Bangkok, Thailand, 22–23 November 2019; pp. 1–7.
13. Acampora, G. A Survey on Ambient Intelligence in Health Care. *Proc. IEEE* **2012**, *40*, 1301–1315.
14. De-La-Hoz-Franco, E.; Ariza-Colpas, P.; Quero, J.M.; Espinilla, M. Sensor-based datasets for human activity recognition—A systematic review of literature. *IEEE Access* **2018**, *6*, 59192–59210. [CrossRef]
15. Espinilla, M.; Medina, J.; Calzada, A.; Liu, J.; Martínez, L.; Nugent, C. Optimizing the configuration of an heterogeneous architecture of sensors for activity recognition, using the extended belief rule-based inference methodology. *Microprocess. Microsyst.* **2017**, *52*, 381–390. [CrossRef]
16. Mehedi, M.; Uddin, Z.; Mohamed, A.; Almogren, A. A robust human activity recognition system using smartphone sensors and deep learning. *Future Gener. Comput. Syst.* **2018**, *81*, 307–313.
17. Kötteritzsch, A.; Weyers, B. Assistive Technologies for Older Adults in Urban Areas: A Literature Review. *Cognit. Comput.* **2016**, *8*, 299–317. [CrossRef]
18. Ni, Q.; Hernando, A.B.G.; de la Cruz, I.P. The Elderly's Independent Living in Smart Homes: A Characterization of Activities and Sensing Infrastructure Survey to Facilitate Services Development. *Sensors* **2015**, *15*, 11312–11362. [CrossRef]
19. Peetoom, K.K.B.; Lexis, M.A.S.; Joore, M.; Dirksen, C.D.; De Witte, L.P. Literature review on monitoring technologies and their outcomes in independently living elderly people. *Disabil. Rehabil. Assist. Technol.* **2015**, *10*, 271–294. [CrossRef]
20. Johansson, F. *Medici Effect: What Elephants and Epidemics can Teach Us about Innovation*; Harvard Business School Press: Boston, MA, USA, 2020. ISBN 978163362947.
21. Moher, D. Preferred Reporting Items for Systematic Reviews and Meta-Analyses: The PRISMA Statement. *Ann. Intern. Med.* **2013**, *151*, 264. [CrossRef]
22. Sanchez, A.; Neira, D.; Cabello, J. Frameworks applied in Quality Management—A Systematic Review. *Rev. Espac.* **2016**, *37*, 17.
23. Van Eck, N.J.; Waltman, L. Visualizing Bibliometric Networks. In *Measuring Scholarly Impact*; Springer: Cham, Switzerland, 2014. ISBN 9783319103778.

24. Wang, Q.; Waltman, L. Large-scale analysis of the accuracy of the journal classification systems of Web of Science and Scopus. *J. Informetr.* **2016**, *10*, 347–364. [CrossRef]
25. Franceschini, F.; Maisano, D.; Mastrogiacomo, L. Customer requirement prioritization on QFD: A new proposal based on the generalized Yager's algorithm. *Res. Eng. Des.* **2015**, *26*, 171–187. [CrossRef]
26. AlRyalat, S.A.S.; Malkawi, L.W.; Momani, S.M. Comparing Bibliometric Analysis Using PubMed, Scopus, and Web of Science Databases. *J. Vis. Exp.* **2019**, *152*, e58494. [CrossRef] [PubMed]
27. Perianes-Rodriguez, A.; Waltman, L.; van Eck, N.J. Constructing bibliometric networks: A comparison between full and fractional counting. *J. Informetr.* **2016**, *10*, 1178–1195. [CrossRef]
28. Sanchez-Comas, A.; Neira, D.; Cabello, J.J. Marcos aplicados a la Gestión de Calidad—Una Revisión Sistemática de la Literatura. *Espacios* **2016**, *37*, 17.
29. Ota, H.; Chao, M.; Gao, Y.; Wu, E.; Tai, L.C.; Chen, K.; Matsuoka, Y.; Iwai, K.; Fahad, H.M.; Gao, W.; et al. 3D printed "earable" smart devices for real-time detection of core body temperature. *ACS Sens.* **2017**, *2*, 990–997. [CrossRef] [PubMed]
30. Mendoza-Palechor, F.; Menezes, M.L.; Sant'Anna, A.; Ortiz-Barrios, M.; Samara, A.; Galway, L. Affective recognition from EEG signals: An integrated data-mining approach. *J. Ambient Intell. Humaniz. Comput.* **2019**, *10*, 3955–3974. [CrossRef]
31. Bilbao, A.; Almeida, A.; López-de-ipiña, D. Promotion of active ageing combining sensor and social network data. *J. Biomed. Inform.* **2016**, *64*, 108–115. [CrossRef]
32. Lee, J.S.; Choi, S.; Kwon, O. Identifying multiuser activity with overlapping acoustic data for mobile decision making in smart home environments. *Expert Syst. Appl.* **2017**, *81*, 299–308. [CrossRef]
33. Damian, I.; Dietz, M.; Steinert, A.; André, E.; Haesner, M.; Schork, D. Automatic Detection of Visual Search for the Elderly using Eye and Head Tracking Data. *KI Künstl. Intell.* **2017**, *31*, 339–348.
34. Zhang, S.; Mccullagh, P. Situation Awareness Inferred From Posture Transition and Location. *IEEE Trans. Hum. Mach. Syst.* **2017**, *47*, 814–821. [CrossRef]
35. Rafferty, J.; Nugent, C.D.; Liu, J. From Activity Recognition to Intention Recognition for Assisted Living Within Smart Homes. *IEEE Trans. Hum. Mach. Syst.* **2017**, *47*, 368–379. [CrossRef]
36. Amft, O.; Laerhoven, K. Wearable Section Applications Title Computing Here. *IEEE Pervasive Comput.* **2017**, *19*, 80–85. [CrossRef]
37. Athavale, Y.; Krishnan, S. Biosignal monitoring using wearables: Observations and opportunities. *Biomed. Signal. Process. Control.* **2017**, *38*, 22–33. [CrossRef]
38. Augustyniak, P.; Ślusarczyk, G. Graph-based representation of behavior in detection and prediction of daily living activities. *Comput. Biol. Med.* **2018**, *95*, 261–270. [CrossRef]
39. Ni, Q.; Zhang, L.; Li, L. A Heterogeneous Ensemble Approach for Activity Recognition with Integration of Change Point-Based Data Segmentation. *Appl. Sci.* **2018**, *8*, 1695. [CrossRef]
40. Ahmed, M.; Mehmood, N.; Nadeem, A.; Mehmood, A.; Rizwan, K. Fall Detection System for the Elderly Based on the Classification of Shimmer Sensor Prototype Data. *Healthc. Inform. Res.* **2017**, *23*, 147–158. [CrossRef]
41. Clapés, A.; Pardo, À.; Pujol Vila, O.; Escalera, S. Action detection fusing multiple Kinects and a WIMU: An application to in-home assistive technology for the elderly. *Mach. Vis. Appl.* **2018**, *29*, 765–788. [CrossRef]
42. Faye, S.; Bronzi, W.; Tahirou, I.; Engel, T. Characterizing user mobility using mobile sensing systems. *Int. J. Distrib. Sens. Netw.* **2017**, *13*, 1550147717726310. [CrossRef]
43. Garcia-Ceja, E.; Galván-Tejada, C.E.; Brena, R. Multi-view stacking for activity recognition with sound and accelerometer data. *Inf. Fusion* **2018**, *40*, 45–56. [CrossRef]
44. Kang, J.; Larkin, H. Application of an Emergency Alarm System for Physiological Sensors Utilizing Smart Devices. *Technologies* **2017**, *5*, 26. [CrossRef]
45. Maglogiannis, I. Fall detection and activity identification using wearable and hand-held devices. *Integr. Comput. Aided Eng.* **2016**, *23*, 161–172. [CrossRef]
46. Shewell, C.; Nugent, C.; Donnelly, M.; Wang, H.; Espinilla, M. Indoor localization through object detection within multiple environments utilizing a single wearable camera. *Health Technol.* **2017**, *7*, 51–60. [CrossRef] [PubMed]
47. Hardegger, M.; Calatroni, A.; Oster, G.T.; Hardegger, M.; Calatroni, A.; Tröster, G.; Roggen, D. S-SMART: A Unified Bayesian Framework for Simultaneous Semantic Mapping, Activity Recognition, and Tracking. *ACM Trans. Intell. Syst. Technol. Intell. Syst. Technol. Artic.* **2016**, *7*, 1–28. [CrossRef]

48. Ma, C.; Shimada, A.; Uchiyama, H.; Nagahara, H.; Taniguchi, R.I. Fall detection using optical level anonymous image sensing system. *Opt. Laser Technol.* **2019**, *110*, 44–61. [CrossRef]
49. Withanage, K.I.; Lee, I.; Brinkworth, R.; Mackintosh, S.; Thewlis, D. Fall Recovery Subactivity Recognition with RGB-D Cameras. *IEEE Trans. Ind. Inform.* **2016**, *12*, 2312–2320. [CrossRef]
50. Akula, A.; Shah, A.K.; Ghosh, R. ScienceDirect Deep learning approach for human action recognition in infrared images. *Cogn. Syst. Res.* **2018**, *50*, 146–154. [CrossRef]
51. Ho, Y.O.C.; Ulier, S.I.J.J.; Arquardt, N.I.M.; Adia, N.; Erthouze, B.I. Robust tracking of respiratory rate in high- dynamic range scenes using mobile thermal imaging. *Biomed. Opt. Express* **2017**, *8*, 1565–1588.
52. Fernández-Caballero, A.; Martínez-Rodrigo, A.; Pastor, J.M.; Castillo, J.C.; Lozano-Monasor, E.; López, M.T.; Zangróniz, R.; Latorre, J.M.; Fernández-Sotos, A. Smart environment architecture for emotion detection and regulation. *J. Biomed. Inform.* **2016**, *64*, 55–73. [CrossRef]
53. Xu, Q.; Safar, Z.; Han, Y.; Wang, B.; Liu, K.J.R. Statistical Learning Over Time-Reversal Space for Indoor Monitoring System. *IEEE Internet Things J.* **2018**, *5*, 970–983. [CrossRef]
54. Guo, L.; Wang, L.; Liu, J.; Zhou, W.; Lu, B. HuAc: Human Activity Recognition Using Crowdsourced WiFi Signals and Skeleton Data. *Wirel. Commun. Mob. Comput.* **2018**, *2018*, 6163475. [CrossRef]
55. Savazzi, S.; Rampa, V. Leveraging MIMO-OFDM radio signals for device-free occupancy inference: System design and experiments. *EURASIP J. Adv. Signal Process.* **2018**, *44*, 1–19.
56. Koppula, H.S. Anticipating Human Activities using Object Affordances for Reactive Robotic Response. *Proc. IEEE Int. Conf. Comput. Vis.* **2013**, *38*, 14–29.
57. Saunders, J.; Syrdal, D.S.; Koay, K.L.; Burke, N.; Dautenhahn, K. 'Teach Me-Show Me'-End-User Personalization of a Smart Home and Companion Robot. *IEEE Trans. Hum. Mach. Syst.* **2016**, *46*, 27–40. [CrossRef]
58. Costa, A.; Martinez-Martin, E.; Cazorla, M.; Julian, V. PHAROS—PHysical assistant RObot system. *Sensors (Switzerland)* **2018**, *18*, 2633. [CrossRef] [PubMed]
59. File: PR2 Robot with Advanced Grasping hands.JPG. Available online: https://commons.wikimedia.org/w/index.php?title=File:PR2_robot_with_advanced_grasping_hands.JPG (accessed on 17 December 2019).
60. File: Pepper—France—Les Quatres Temps—Darty—2016-11-04.jpg. Available online: https://commons.wikimedia.org/w/index.php?title=File:Pepper_-_France_-_Les_Quatres_Temps_-_Darty_-_2016-11-04.jpg (accessed on 17 December 2019).
61. File: Care-O-Bot Grasping an Object on the Table (5117071459).jpg. Available online: https://commons.wikimedia.org/w/index.php?title=File:Care-O-Bot_grasping_an_object_on_the_table_(5117071459).jpg (accessed on 17 December 2019).
62. Villeneuve, E.; Harwin, W.; Holderbaum, W.; Janko, B.; Sherratt, R.S. Special Section On Advances of Multisensory Services And Reconstruction of Angular Kinematics From Wrist-Worn Inertial Sensor Data for Smart Home Healthcare. *IEEE Access* **2017**, *5*, 2351–2363. [CrossRef]
63. Zhang, Z.; Song, Y.; Cui, L.; Liu, X. Emotion recognition based on customized smart bracelet with built-in accelerometer. *PeerJ* **2016**, *4*, e2258. [CrossRef]
64. Activity, H.; Using, R. Hierarchical Activity Recognition Using Smart Watches and RGB-Depth Cameras. *Sensors* **2016**, *16*, 1713.
65. Biagetti, G.; Crippa, P.; Falaschetti, L.; Turchetti, C. Classifier level fusion of accelerometer and sEMG signals for automatic fitness activity diarization. *Sensors (Switzerland)* **2018**, *18*, 2850. [CrossRef]
66. Orcioni, S.; Turchetti, C.; Falaschetti, L.; Crippa, P.; Biagetti, G. Human activity monitoring system based on wearable sEMG and accelerometer wireless sensor nodes. *Biomed. Eng. Online* **2018**, *17*, 132.
67. Wang, P.; Sun, L.; Yang, S.; Smeaton, A.F.; Gurrin, C. Characterizing everyday activities from visual lifelogs based on enhancing concept representation. *Comput. Vis. Image Underst.* **2016**, *148*, 181–192. [CrossRef]
68. Mokhtari, G.; Zhang, Q.; Nourbakhsh, G.; Ball, S.; Karunanithi, M.; Member, S. BLUESOUND: A New Resident Identification Sensor—Using Ultrasound Array and BLE Technology for Smart Home Platform. *IEEE Sens. J.* **2017**, *17*, 1503–1512. [CrossRef]
69. Chen, Z. Robust Human Activity Recognition Using Smartphone Sensors via CT-PCA. *IEEE Trans. Ind. Inform.* **2017**, *13*, 3070–3080. [CrossRef]
70. Khan, M.A.A.H.; Roy, N.; Hossain, H.M.S. Wearable Sensor-Based Location-Specific Occupancy Detection in Smart Environments. *Mob. Inf. Syst.* **2018**, *2018*, 4570182. [CrossRef]

71. Iwasawa, Y.; Eguchi Yairi, I.; Matsuo, Y. Combining human action sensing of wheelchair users and machine learning for autonomous accessibility data collection. *IEICE Trans. Inf. Syst.* **2016**, *E99D*, 1153–1161. [CrossRef]
72. Gupta, H.P.; Chudgar, H.S.; Mukherjee, S.; Dutta, T.; Sharma, K. A Continuous Hand Gestures Recognition Technique for Human-Machine Interaction Using Accelerometer and gyroscope sensors. *IEEE Sens. J.* **2016**, *16*, 6425–6432. [CrossRef]
73. Saha, J.; Chowdhury, C.; Biswas, S. Two phase ensemble classifier for smartphone based human activity recognition independent of hardware configuration and usage behaviour. *Microsyst. Technol.* **2018**, *24*, 2737–2752. [CrossRef]
74. Liu, Z.; Yin, J.; Li, J.; Wei, J.; Feng, Z. A new action recognition method by distinguishing ambiguous postures. *Int. J. Adv. Robot. Syst.* **2018**, *15*, 1729881417749482. [CrossRef]
75. Yao, B.; Hagras, H.; Alghazzawi, D.; Member, S.; Alhaddad, M.J. A Big Bang—Big Crunch Type-2 Fuzzy Logic System for Machine-Vision-Based Event Detection and Summarization in Real-World Ambient-Assisted Living. *IEEE Trans. Fuzzy Syst.* **2016**, *24*, 1307–1319. [CrossRef]
76. Trindade, P.; Langensiepen, C.; Lee, K.; Adama, D.A.; Lotfi, A. Human activity learning for assistive robotics using a classifier ensemble. *Soft Comput.* **2018**, *22*, 7027–7039.
77. Wang, S.; Chen, L.; Zhou, Z.; Sun, X.; Dong, J. Human fall detection in surveillance video based on PCANet. *Multimed. Tools Appl.* **2016**, *75*, 11603–11613. [CrossRef]
78. Eldib, M.; Deboeverie, F.; Philips, W.; Aghajan, H. Behavior analysis for elderly care using a network of low-resolution visual sensors. *J. Electron. Imaging* **2016**, *25*, 041003. [CrossRef]
79. Wickramasinghe, A.; Shinmoto Torres, R.L.; Ranasinghe, D.C. Recognition of falls using dense sensing in an ambient assisted living environment. *Pervasive Mob. Comput.* **2017**, *34*, 14–24. [CrossRef]
80. Chen, Z.; Wang, Y. Infrared–ultrasonic sensor fusion for support vector machine–based fall detection. *J. Intell. Mater. Syst. Struct.* **2018**, *29*, 2027–2039. [CrossRef]
81. Chen, Z.; Wang, Y.; Liu, H. Unobtrusive Sensor based Occupancy Facing Direction Detection and Tracking using Advanced Machine Learning Algorithms. *IEEE Sens. J.* **2018**, *18*, 6360–6368. [CrossRef]
82. Wang, J.; Zhang, X.; Gao, Q.; Feng, X.; Wang, H. Device-Free Simultaneous Wireless Localization and Activity Recognition With Wavelet Feature. *IEEE Trans. Veh. Technol.* **2017**, *66*, 1659–1669. [CrossRef]
83. Rus, S.; Grosse-Puppendahl, T.; Kuijper, A. Evaluating the recognition of bed postures using mutual capacitance sensing. *J. Ambient Intell. Smart Environ.* **2017**, *9*, 113–127. [CrossRef]
84. Cheng, A.L.; Georgoulas, C.; Bock, T. Automation in Construction Fall Detection and Intervention based on Wireless Sensor Network Technologies. *Autom. Constr.* **2016**, *71*, 116–136. [CrossRef]
85. Hossain, H.M.S.; Khan, M.A.A.H.; Roy, N. Active learning enabled activity recognition. *Pervasive Mob. Comput.* **2017**, *38*, 312–330. [CrossRef]
86. Aziz, S.; Id, S.; Ren, A.; Id, D.F.; Zhang, Z.; Zhao, N.; Yang, X. Internet of Things for Sensing: A Case Study in the Healthcare System. *Appl. Sci.* **2018**, *8*, 508.
87. Jiang, J.; Pozza, R.; Gunnarsdóttir, K.; Gilbert, N.; Moessner, K. Using Sensors to Study Home Activities. *J. Sens. Actuator Netw.* **2017**, *6*, 32. [CrossRef]
88. Luo, X.; Guan, Q.; Tan, H.; Gao, L.; Wang, Z.; Luo, X. Simultaneous Indoor Tracking and Activity Recognition Using Pyroelectric Infrared Sensors. *Sensors* **2017**, *17*, 1738. [CrossRef] [PubMed]
89. Gill, S.; Seth, N.; Scheme, E. A multi-sensor matched filter approach to robust segmentation of assisted gait. *Sensors (Switzerland)* **2018**, *18*, 2970. [CrossRef] [PubMed]
90. Sasakawa, D. Human Posture Identification Using a MIMO Array. *Electronics* **2018**, *7*, 37. [CrossRef]
91. Suyama, T. A network-type brain machine interface to support activities of daily living. *IEICE Trans. Commun.* **2016**, *E99B*, 1930–1937. [CrossRef]
92. Li, W.; Tan, B.O.; Piechocki, R. Passive Radar for Opportunistic Monitoring in E-Health Applications. *IEEE J. Trans. Eng. Health Med.* **2018**, *6*, 1–10. [CrossRef] [PubMed]

 *sensors*

Article

Sensor Failure Detection in Ambient Assisted Living Using Association Rule Mining

Nancy E. ElHady [1,*], Stephan Jonas [2,*], Julien Provost [1] and Veit Senner [1]

1 Department of Mechanical Engineering, Technical University of Munich, 85748 Garching, Germany; julien.provost@tum.de (J.P.); senner@tum.de (V.S.)
2 Department of Informatics, Technical University of Munich, 85748 Garching, Germany
* Correspondence: nancy.elhady@tum.de (N.E.E.); stephan.jonas@tum.de (S.J.)

Received: 20 October 2020; Accepted: 22 November 2020; Published: 26 November 2020

Abstract: Ambient Assisted Living (AAL) is becoming crucial to help governments face the consequences of the emerging ageing population. It aims to motivate independent living of older adults at their place of residence by monitoring their activities in an unobtrusive way. However, challenges are still faced to develop a practical AAL system. One of those challenges is detecting failures in non-intrusive sensors in the presence of the non-deterministic human behaviour. This paper proposes sensor failure detection and isolation system in the AAL environments equipped with event-driven, ambient binary sensors. Association Rule mining is used to extract fault-free correlations between sensors during the nominal behaviour of the resident. Pruning is then applied to obtain a non-redundant set of rules that captures the strongest correlations between sensors. The pruned rules are then monitored in real-time to update the health status of each sensor according to the satisfaction and/or unsatisfaction of rules. A sensor is flagged as faulty when its health status falls below a certain threshold. The results show that detection and isolation of sensors using the proposed method could be achieved using unlabelled datasets and without prior knowledge of the sensors' topology.

Keywords: ambient assisted living; enhanced living environments; sensor failure; fault detection; fault isolation; smart home; non-intrusive sensors; binary sensors; event-driven sensors

---

1. Introduction

The ageing population phenomenon is one of the toughest challenges of this century. In 2019, 1 in 11 people around the globe was over 65 years old. This number of aged people is expected to rise to 1 in 6 people by 2050. The old-age dependency ratio is the ratio of the people over 65 to people between 20 and 64 years old. Some regions will witness this demographic shift the most, e.g., Europe and North America, will have an old-age dependency ratio of 49 per 100 by 2050 [1]. This demographic shift will induce challenges to governments as well as individuals [2]. The increasing ratio of retired persons to workers requires increasing the capacity of the social system. Moreover, as people grow into older age the chances of having age-related impairments and diseases increase, which if not monitored closely could lead to much worse health complications. Thus, the health-care costs are expected to increase as the population ages as well as the need for more care-givers. Stress would also be imposed on informal caregivers, e.g., family members. In order to decrease the burden on governments and individuals, promoting healthy ageing and independent living is becoming a priority. Exploiting the vast development of the information and communication technologies (ICT) and the emergence of ambient intelligence (AmI) is the key to providing such independence to older adults.

As a result, there has been an increasing interest in establishing Ambient Assisted Living (AAL) environments [2]. One of the definitions proposed for Ambient Assisted Living is "the use of information and communication technologies (ICT) in a person's daily living and working environment to enable them to stay active longer, remain socially connected and live independently into old age" [3]. It is a multidisciplinary field that involves information and communication technologies, sociological sciences and medical research [4]. The AAL tools could be mainly categorised into health and activity monitoring tools, wandering prevention tools and cognitive orthotics tools [2]. The health and activity monitoring tools aim to monitor the activities of daily living (ADL) in an unobtrusive way, either to ensure the safety of the monitored person, or the completion of his activities, or to detect the deterioration in his cognitive and physical abilities. Wandering prevention tools were developed mainly to aid people suffering from dementia, while cognitive orthotics tools are used to aid people with cognitive decline. The AAL tools would cast some burden away from the family members of the older adults, decrease the need for qualified caregivers and have a positive impact on the psychological status of older adults as they would live independently at their homes for longer and more safely. To achieve the goals of the AAL systems, the following requirements need to be fulfilled; adaptability, interoperability, acceptability, usability and dependability [4].

Health or mobility related sensors are widely used for the monitoring purposes and represent the heart of the AAL environments [4]. Most of the sensors that are used for monitoring are event-driven binary sensors, for example the PIR sensor produces high output when motion is detected, otherwise it produces low output. Such sensors provide low level information, unlike the sophisticated information from cameras or microphones, and thus is more difficult to interpret and more prone to errors [5]. The failures that are encountered in such sensors are either fail-stop failures, where the sensor stops reporting values, or non-fail-stop failures, where the sensor reports values that do not reflect the occurring *events* that were supposed to be captured by it. Examples of the reported non-fail-stop failures that occur in AAL environments include sensors that get blocked by furniture, get remounted by the user in wrong locations, get stuck at a value or get spurious signals due to air drafts, sunlight rays or pets [6,7]. The traditional fault diagnosis methods for wireless sensor networks [8–10] are designed to deal with homogeneous, time-driven and continuous-valued sensors. However, such methods do not suit the nature of sensors installed in non-intrusive AAL environments, which are often heterogeneous, event-driven and binary sensors. This work aims to propose a sensor failure detection and isolation system for AAL environments equipped with event-driven, ambient binary sensors.

2. Related Work

A comprehensive literature review was presented by the authors of this article in [11], which focuses on the works concerned with detecting sensor failures, as well as tolerating its resulted faults, in AAL environments equipped with binary, event-driven sensors. The surveyed fault-tolerant systems focus mainly on location tracking [7,12,13] and activity recognition [6,14,15]. The sensor failure detection systems found in literature may be classified as model-based and correlation-based approaches [11]. The model-based techniques rely on deducing the location of the resident using the triggered sensors due to his movement or his performed activities. Then, this deduced location is compared with the location predicted either by his model of mobility, e.g., in [16,17] or by a localisation system, e.g., in [18,19]. The proposed model-based sensor failure detection approaches are not promising as they either use unrealistic models of resident motion that do not take into consideration previous locations and speed or install extra hardware that increases cost as well as the chances of errors. Fault detection and diagnosis frameworks that rely on modelling the sensors' and actuators' activation due to various user scenarios were presented in [20–22]. However, it can only detect failures in sensors that are involved in tasks that have sensor-actuator feedback.

The surveyed correlation-based techniques can be classified as methods based on exploiting sensor-appliance correlations, sensor-activity correlations and sensor-sensor correlations [11]. FailureSense [23] monitors the interval between motion sensor triggers and electrical appliances. Sensor failure is flagged during run-time when the monitored interval deviates from the previously learnt patterns from training datasets. The drawback of this method is that the assumption that the resident has to be physically beside the appliance to turn it on does not always hold. Idea system [24] first extracts the sensors that are triggered with each activity of daily living using an activity labelled dataset. In order to detect sensor failures, activity recognition is done, and whenever an activity is recognised while one of its sensors did not trigger, a rarity score is computed. Sensor failure alert is raised when the rarity score falls below a set threshold. The limitation of this approach is that it assumes that the activity has been correctly recognised in the first place. In addition, it requires labelled datasets for training. Following are the works based on the sensor-sensor correlations techniques. An approach based on temporal correlation and nonlinear time series analysis was investigated by Ye, Stevenson and Dobson; however, the experimental data was not enough to prove the effectiveness of this approach [25]. Same authors have proposed the use of density based clustering to detect outlier sensor triggers [26,27]. However, clustering occurs as a postprocess step on the collected data. SMART system uses simultaneous multiple classifiers, a classifier for each sensor failure. It detects a sensor failure by analysing the relative performance of these classifiers [6,28]. This approach lacks scalability and needs excessive training effort. DICE [29] extracts correlations and transitional probabilities among sensors and actuators offline. Failure is detected either when a sensor is missing from a predefined correlation or when a group of sensors fires despite having a zero transitional probability with the previous group of triggered sensors. The drawback of this approach is considering any group of triggered sensors as a correlation, even if it has only appeared once, thus questioning the reliability of correlations and making the approach more computationally complex especially when the number of installed sensors increases.

Our research work favoured adopting a correlation-based approach over a model-based approach, to avoid the disadvantages of relying on generic human mobility models, like in [16], that may not be accurate nor personalised to reflect the behaviour of the monitored person. In addition, adding extra hardware, as in [18,19], was avoided in order not to increase the implementation cost. Our proposed sensor failure detection and isolation system approach focused on sensor-sensor correlations rather than sensor-appliance and sensor-activity correlations. Sensor-appliance approaches [23] rely on assuming that there will be correlations between the activation of the electrical appliance and the triggering of the motion sensors in the areas leading to it, which is becoming less common in smart homes as most appliances can be switched on remotely. Meanwhile, failure detection using sensor-activity correlations [24] requires obtaining labelled data of performed activities to correlate the activities to the sensors during the training phase and relies on the accuracy of the activity recognition system at run-time to detect sensor failures. Our method does not need labelled datasets of sensor failures nor performed activities. It is based on extracting the nominal correlations between the installed sensors with no prior knowledge on the topology using unlabelled datasets. The association rule mining [30] technique is used to extract correlations. Unlike the approach presented in [29] that considers any proceeding triggers between sensors as a correlation, association rule mining extracts strong correlations that meet minimum relative support and confidence, which would ensure more reliability for failure detection. Association rule mining is characterised by its simplicity and good interpretability of results. There are works that have based their fault detection system on association rule mining; however, they were used to detect faults in time-series, continuous-valued data, e.g., [31,32]. Association rule mining has also been used for fault diagnosis using datasets that are already labelled with various system faults to associate which sensor signal values are responsible for corresponding system faults, e.g., [33]. In this paper, we propose a failure detection and isolation system for binary, event-driven sensors that is based on association rule mining. Association rule mining is refined to better suit our application. Postpruning is applied to get the most interesting correlations that the sensor failure detection and isolation system can rely on. The extracted correlations appear as a set of IF-THEN rules that indicate the sensors that trigger within a few seconds from each other. At run-time the set of rules are monitored and then the health status of each sensor is updated according to the satisfaction/unsatisfaction of the correlations. A sensor is flagged as faulty when its health status falls below a predefined threshold. Guidelines for the selection of the values of the parameters of association rule mining algorithm and the health status threshold are presented in Section 4.3.2. Failure detection and isolation take place at run-time; this is contrary to the approach in [26,27] that detects failure in precollected data. The approach presented in this paper is scalable; therefore, it overcomes this shortcoming found in the SMART system [6,28] which needs a large training effort to train a classifier for each sensor failure.

3. Sensor Failure Detection and Isolation System

Our sensor failure detection and isolation system consists of two stages: an offline stage and an online stage. During the offline stage, the fault-free sensor correlations are extracted from previously collected sensor dataset at the resident's home during his nominal behaviour. Meanwhile online, the fulfilment of correlations are checked as sensor *events* are triggered by the resident and accordingly failure of sensors is determined. An overview of the proposed system is shown in Figure 1.

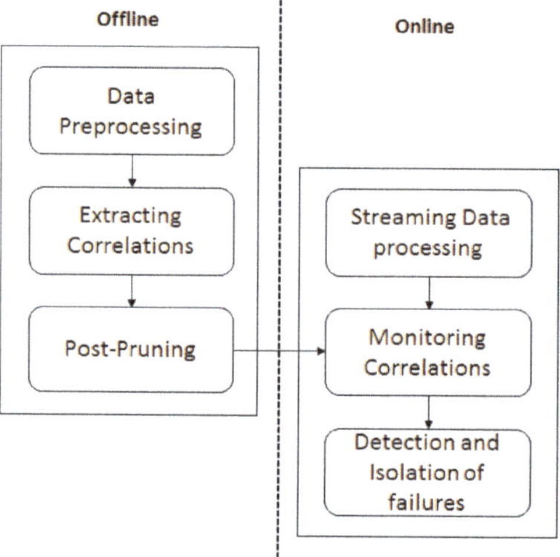

Figure 1. An overview of the proposed system.

3.1. Sensor Correlations Extraction

First, preprocessing of training data is done, followed by rules extraction using association rule mining. Afterwards, the extracted rules are further pruned to obtain the most interesting sensor correlations.

3.1.1. Data Preprocessing

The log obtained from AAL environments equipped with non-intrusive sensors consists of a series of *events*. Each *event* has a time stamp, sensor ID and the corresponding sensor *event* trigger. An example of a sensor *event* is 13 January 2011 10:28:14.65 M030 ON, which implies that sensor M030 has been positively triggered at the given time stamp. In order to extract correlations using association rule mining, the transformation of the time-stamped sensor *event* triggers dataset into a set of transactions takes place over a couple of steps. The first step consists of creating a multivariate time-series, where the value of each sensor is logged at every time stamp of the dataset in a separate sensor signal variable. Formally, let $s_{i,t} \in \{0,1\}$ be the value of the i-th sensor at timestamp $t \in T$. The set T is the set of timestamps of the log. For n sensors, concatenation produces the multivariate time-series S.

$$S = \{(s_{1,t}, s_{2,t}, ..., s_{n,t})\}_{t \in T} \tag{1}$$

Next, removal of all-zero rows is done. Formally, it corresponds to removing all-zero row vectors from the time-series S.

$$V := S \setminus \{(0_{1,t}, 0_{2,t}, ..., 0_{n,t})\}_{t \in T} \tag{2}$$

Figure 2a shows an example for a multivariate time-series created from an AAL log. At each row, a sliding window is used to group the sensors that have a signal value of 1 within the size w seconds of the sliding window via logical ORing. The output of the window will be a single transaction that has the time stamp of the start of the window. Formally, the value of the i-th sensor in the transaction computes to:

$$d_{i,t} = sgn\left(\sum_{j \in [t, t+w]} v_{i,j}\right) \tag{3}$$

The sliding window is run over the multivariate time-series data to output a transactional database as illustrated in Figure 2, where each transaction presents the sensors that appear to be ON within w seconds from each other. The obtained sensors transactional database will be used in the upcoming correlations extraction step.

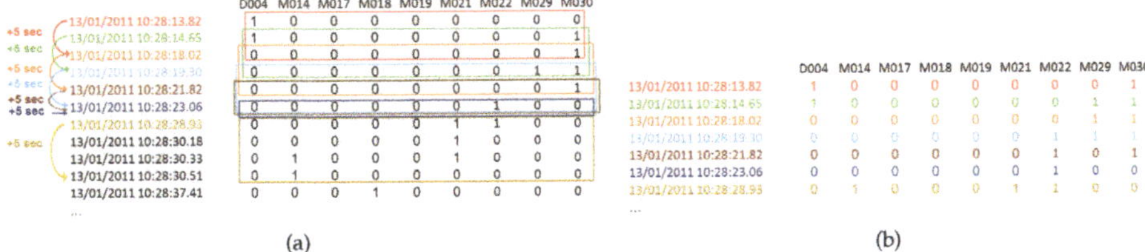

Figure 2. (**a**) Sliding window of size $w = 5$ s, is run over the multivariate time-series data. (**b**) Transactional database.

3.1.2. Extracting Correlations

Correlations between fault-free sensors are extracted using the association rule mining technique. It is a data mining technique that was introduced by Agrawal et al. [30] and is commonly used on large transactional databases to find correlations between its items. Its most famous application is the market basket analysis, where the transactions of a supermarket are analysed to find which items are usually bought together by customers. Similarly, we aim to detect which sensors are most likely to be simultaneously active implying strong correlations.

A formal representation of the association rule mining problem is as follows. Let $I = \{I_1, I_2, .., I_m\}$ be a set of binary features denoted as items. Let the dataset T consist of a set of transactions $T = \{T_1, T_2, .., T_n\}$, where each transaction is a binary vector of items, e.g., if transaction T_1 contains only two items I_1 and I_3, then T_1 will have $T_1[1] = 1$, $T_1[3] = 1$ and the rest of T_1 vector are zeros. An association rule has the form of $X \rightarrow Y$, where the antecedent $X \subset I$, the consequent $Y \subset I$ and $X \cap Y = \phi$. The confidence of a rule denotes how likely it is to find item(s) of Y when item(s) of X occur(s), while the support of a rule is how frequent items of X and Y appear together in the dataset. Support and confidence, defined by Equations (4) and (5) respectively, are the most commonly used evaluation metrics that assess how strong the association rule is. The Apriori algorithm [34] is used to extract the association rules from transactional datasets. Minimum values for support and confidence have to be satisfied to avoid extracting meaningless rules. These minimum values need to be set by the designer. Lift is a metric used to confirm the dependency between the rule's antecedent and consequent as shown in Equation (6), a value of 1 indicates independency, while greater than 1 indicates dependency. The higher the lift value, the greater is the dependency.

$$\text{Sup}(X \rightarrow Y) = \frac{|\,\text{Transactions containing X\&Y}\,|}{|\,\text{Transactions}\,|} = P(X \cap Y) \tag{4}$$

$$\text{Conf}(X \rightarrow Y) = \frac{\mid \text{Transactions containing X\&Y} \mid}{\mid \text{Transactions containing X} \mid} = P(Y|X) \quad (5)$$

$$\text{Lift}(X \rightarrow Y) = \frac{\mid \text{Transactions containing X\&Y} \mid}{\mid \text{Transactions containing X} \mid * \mid \text{Transactions containing Y} \mid} = \frac{P(X \cap Y)}{P(X)P(Y)} \quad (6)$$

In the market basket analysis application, the items are the supermarket products, e.g., butter, bread, and a transaction contains the items that have been simultaneously bought by a customer in this transaction. In our AAL application, the items of the transactional database are the sensors installed in the AAL environment. However, a transaction contains the sensors that are ON simultaneously in an instant of time, as well as those sensors that are ON within its sliding window of size w seconds. This is because we are concerned to capture the temporal correlations between sensors within few seconds due to performing various activities by resident. The transactional database has been prepared in the preprocessing stage. Another concern in the AAL application is the uneven usage of the different areas of an apartment. A living room may be used by an older adult resident more often than the office room, leading to scarcity of the triggers of the office's sensors in the dataset. In such cases, the support of the rule that has the less often triggered sensors may not exceed the minimum support value that was preset in the Apriori algorithm, and thus will not appear in the extracted set of rules. To overcome this limitation, we define a metric as relative support to be used in the Apriori algorithm instead of the support for rules extraction. Support compares the number of transactions containing all items of X & items of Y to the total number of transactions present in the database as shown in Equation (4). While relative support is defined by Equation (7), it compares the number of transactions containing all items of X & items of Y to the minimum number of transactions that contain any of the individual items of X or Y.

$$\text{Rel. Sup}(X \rightarrow Y) = \frac{\mid \text{Transactions containing X\&Y} \mid}{\text{Min}(\mid \text{Transactions for each item in X or Y} \mid)} \quad (7)$$

3.1.3. Post-Pruning of Correlations

The mined set of rules that have already exceeded the minimum values for the relative support and confidence still needs further post-pruning to eliminate the redundant and/or less useful rules. Our proposed sensor failure detection method relies on the following hypothesis; if a rule has all of its antecedent sensors active during run-time, while its consequent sensors(s) did not become active within the specified sliding window size, then the sensors can be suspected to be faulty. Accordingly, we aim to have most of the sensors installed in the resident's home appear in consequent part of rules so that they could be checked for being faulty in the monitoring stage. Hence, the rules are grouped for each sensor in consequent, i.e., if there are 20 sensors that appear in the consequent parts of rules, then we will have 20 groups. From each group, the rule with highest confidence, the rule with highest support and the two top trade-off rules between confidence and support, are selected. In our opinion, the former would be the most interesting rules to our application. To obtain the trade-off rules, confidence and support of the rules within each group are normalised, then are summed with weights 1:1, and the rules with the top two highest sums, i.e., trade-off scores, are selected. For example, to prune the rules of sensor M012, the rules that have M012 as a consequent are grouped, and then those rules which have the highest confidence, highest support and the two top trade-off scores are selected to be on the final set of rules that will be used in the monitoring stage, while the rest of the rules that have M012 as a consequent are eliminated.

3.2. Sensor Correlations Monitoring

The pruned set of rules are the most interesting correlations that will be monitored online; they are stored using bitmap arrays [35]. The health status of each sensor, which is the probability that a sensor is healthy, will be computed according to the fulfilment of these correlations.

Every time a sensor trigger *event* occurs, the data is processed and the corresponding sliding window is prepared similar to Section 3.1.1, where the sensor signal value is updated and the sliding window logically OR the sensors' signals within the sliding window size of w seconds. A UML (Unified Modeling Language) diagram that describes the main workflow for the health status update is shown in Figure 3. The pseudocode in Algorithm A1 illustrates in details the health status update of sensors due to monitoring the pruned set of rules. Two satisfaction states of rules are possible: satisfaction and unsatisfaction. If the sliding window contains active sensors that satisfy a rule antecedent as well as its consequent, then this correlation is fully satisfied and the health status of these sensors are updated according to the satisfaction set of equations in Algorithm A2. It is assumed that only one sensor failure can occur at a time (single-sensor failure). Hence, if the sliding window contains active sensors that satisfy a rule antecedent but it fulfils the rule consequent except for one sensor, then this rule is unsatisfied. If this unsatisfied rule has already been satisfied in the previous sliding window or if it will be satisfied in the upcoming sliding window, then the health status will not be updated. In addition, if this rule has been unsatisfied in the previous sliding window then health will not be updated. Otherwise, the health status of this rule's sensors are going to be updated according to the unsatisfaction set of equations in Algorithm A3. The joint probabilities between sensors that are included in the equations can already be obtained from the intermediate calculations of the Apriori algorithm while scanning the training data for finding the frequent itemsets, hence no extra computation is needed. Whenever the health status of a sensor falls below the preset health threshold, failure of this sensor will then be flagged. Figure 4 shows a UML analysis object model of the online stage of our system.

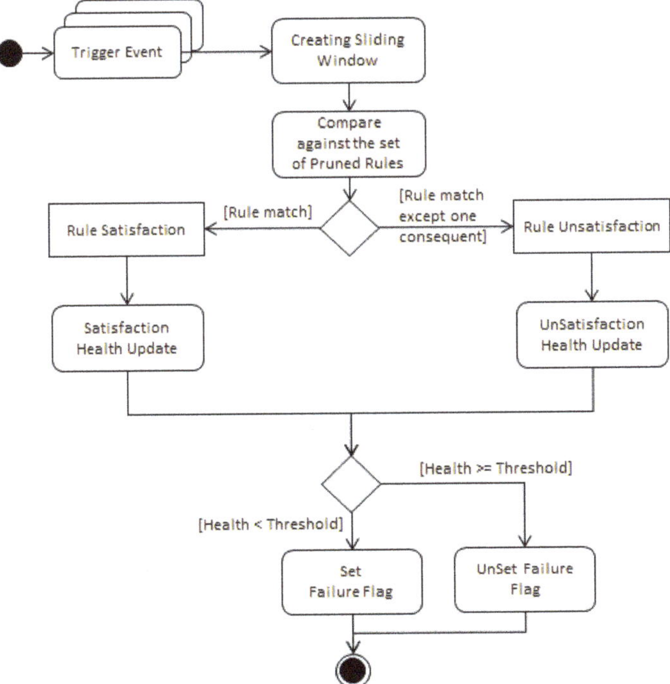

Figure 3. UML activity diagram of the health status update.

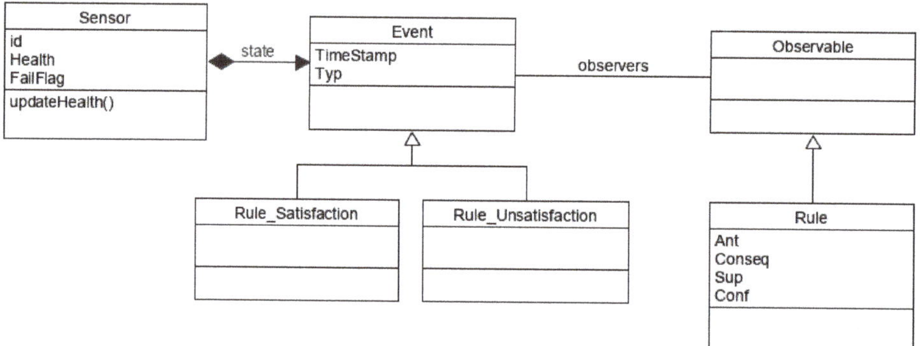

Figure 4. UML analysis object model of the online stage of the failure detection system.

4. Experimental Work and Results

Our proposed approach for sensor failure detection and isolation was evaluated using a publicly available dataset. In this section, the methodology of the experimental work and the results will be presented.

4.1. Dataset

The publicly available Aruba CASAS dataset [36] was used to evaluate the proposed approach for failure detection and isolation of non-intrusive sensors installed in AAL. The dataset was collected over a duration of 6 months from a single-resident elderly's home equipped with 31 motion sensors, 4 door contact sensors and 4 temperature sensors. As our approach is concerned with finding failure in event-driven binary sensors, temperature sensors were not included in the evaluation. In addition, the contact sensor D003, installed on a door located within the apartment as shown in Figure 5, does not have any triggers in the dataset. Thus in total, we have 34 sensors under investigation. The dataset was found to have some instances at which all of the sensors of the apartment get triggered at fractions of a second and all remain active for some time, thus filtering was done to remove such instances. To obtain the training and testing data, a split ratio of 50/50 was used. The training data was used for extracting the offline correlations, while the testing data was processed sequentially to simulate the run-time online processing using MATLAB 2019b software.

4.2. Evaluation Method

The following metrics are used for evaluating the sensor failure detection and isolation system: precision, recall and F1-measure. Precision is the percentage of true positives from the total number of sliding windows reported as positive, while recall is the percentage of true positives from the actual positive sliding windows. The testing dataset was divided into 6 segments, where the segment is approximately 2 weeks in length. Precision, recall and F1-measure are averaged over the segments.

In order to compute the true positives (TP) and false negatives (FN), the segments were duplicated and injected with failure. Failure is injected in each segment on each of the sensors that appear in the consequent parts of the extracted rules. Whenever a sliding window is reported to have a failure from our algorithm, the ground truth is compared with the report to determine whether it is a true positive or not. The start of sensor failure is chosen to be the first timestamp at which the sensor gets triggered in the segment. The faultless segments were used to count the false positives (FP) and true negatives (TN). Receiver Operating Characteristic (ROC) curve and the area under its curve (AUC) were also used to evaluate the performance of failure detection. The ROC curve shows the tradeoff between the true positive rate (TPR) and the false positive rate (FPR) as the health threshold value is varied from 0 to 1. The closer the curve to the left top corner of the plot, the better the performance of failure detection is, implying higher quality of rules that govern the failure detection. A diagonal ROC indicates that it is sort of random classification of failures.

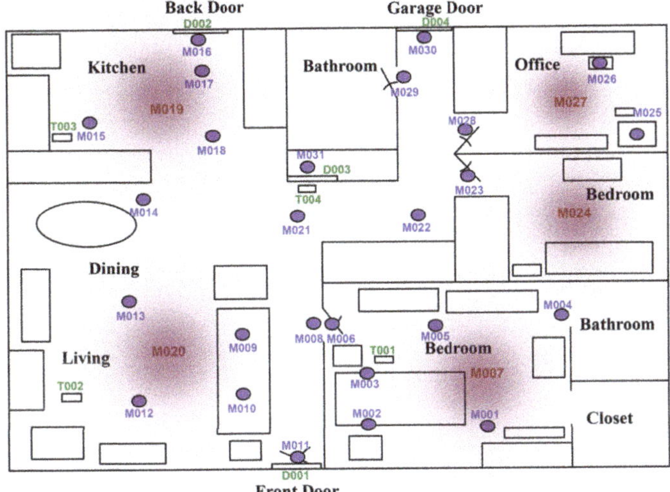

Figure 5. Aruba CASAS floor plan.

4.3. *Parameters of the Correlations Extraction*

To achieve high performance for the sensor failure detection and isolation system, optimum values for four parameters need to be selected. These parameters are the sliding window size, minimum relative support, minimum confidence and health threshold. The optimum parameters would output the best set of correlations and thus the best failure detection and isolation performance. During the selection of parameters, thresholds setting dataset is used. The thresholds setting dataset contains 4-week data (2 segments) of the testing dataset.

4.3.1. Parameter Effect

Before the selection phase, we wanted to study the effect of each parameter independently on the extracted rules and the performance of the system. Using the training dataset, we set the parameters and extract the correlations as described in Section 3.1. Then, the effect of the extracted rules on the performance of the failure detection system is evaluated on the threshold setting dataset that was injected with fail-stop failures. Fail-stop failure was injected for each of the sensors found in the consequent part of the extracted rules.

Increasing the size of the sliding window from 0 to 60 s, while keeping the minimum relative support at 45%, minimum confidence at 60% and health threshold at 0.4, was studied. It was observed that increasing the size of the sliding window increases the total number of sensors in the consequent parts of rules and increases the complexity of rules as well, i.e., more items/sensors per rule. Figure 6a,b plot the precision and recall of failure detection with the parameters set to the former values when the sensor ID of the x-axis is injected with fail-stop failure. For example, in Figure 6a the columns at sensor M007 show the values of precision and recall of failure detection when M007 was injected with fail-stop failure. High failure detection precision and recall can be observed in most of the cases of failed sensors. Note that the sensors with nonempty bar data in the figures are the consequent sensors of the extracted rules at the indicated values of parameters. Failure detection of only the consequent sensors were evaluated, i.e., in Figure 6a there are only 5 sensors that have bar data, denoting that only those sensors were present in the consequent parts of the rules extracted using 0 s sliding window, minimum support of 45% and minimum confidence of 60%, and failure was injected in each of those sensors and failure detection was evaluated then.

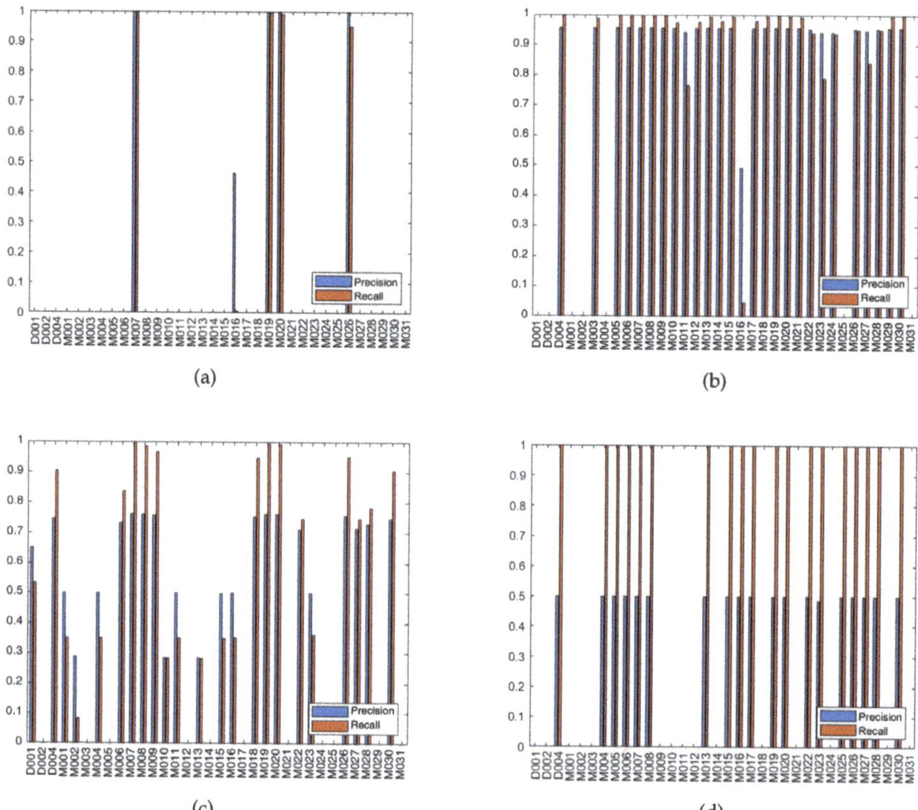

Figure 6. Precision and recall of failure detection when a sensor has fail-stop failure, at health threshold 0.4, (**a**) sliding window 0 s, minimum relative support 45%, and minimum confidence 60%. (**b**) sliding window 60 s, minimum relative support 45%, and minimum confidence 60%. (**c**) sliding window 0 s, minimum relative support 2%, and minimum confidence 60%. (**d**) sliding window 0 s, minimum relative support 45%, and minimum confidence 10%.

Figures 6a,c show the precision and recall of detecting failures with setting the minimum relative support at 45% and 2%, respectively, while maintaining the size of the sliding window at 0 s, minimum confidence at 60% and health threshold at 0.4. Observing the effect of decreasing the minimum relative support, it was found that the number of sensors in consequent part of rules increases but nearly half of them have low failure detection precision and recall. The low precision and recall are due to the low relative support of the rules that govern those sensors. Such sensors are the source of false positives, their governing rules seems to be spatially unrealistic, e.g., M001, M023 $\rightarrow$ M010, that was obtained using a sliding window of 0 s, implying that they are supposed to be ON simultaneously which cannot happen from a single resident even with the switch-off delays of motion sensors. The performance of the other sensors was also affected; the high false positives of the system have reduced their failure detection precision while maintaining their high recall. The complexity of the extracted rules has increased due to lowering the minimum relative support. Some sensors appeared in the consequent of rules when the sliding window has been increased but not when the relative support has been decreased, and vice versa. From Figure 6a–c, it is observed that D001, M001 and M002 have appeared in the consequent of rules, when relative support decreased from 45% to 2% and thus can be checked for being faulty, but they were not part of any rule's consequent when the sliding window was increased from 0 to 60 s.

Lowering the minimum confidence from 60% to 10%, while keeping the sliding window at 0 s and the minimum relative support at 45%, is presented in Figure 6a,d. More sensors appeared in the consequent part of rules, and the complexity of rules did not change when the minimum confidence was lowered. The low confidence rules imposed high number of false positives for its sensors, which has deteriorated the performance of the system. The false positives induced when the minimum confidence was decreased to 10% (average false positives of 84,178) are much greater than those induced when the minimum relative support was lowered to 2% (average false positives of 29,493). This is because some of the extracted low confidence rules have high support, thus their sensors will be triggered a lot by the user.

4.3.2. Setting Parameters

We aim to select the best combination of values for the sliding window size, minimum relative support, minimum confidence and health threshold, which would enable failure detection and isolation of as many sensors as possible with high precision and recall. The thresholds setting dataset is used to validate the selection. A set of guidelines that aids in the parameters selection process was formulated and is presented as follows:

1. First, extract the association rules for various combinations of values from wide range of sliding window size, minimum relative support and confidence > = 50%, while maintaining a single preliminary threshold value, using the training dataset.
2. Then, sort the combinations of parameters according to the total number of sensors in consequent part of their extracted rules in descending order.
3. Select the top-most set of parameters, which produces rules with the highest number of consequent sensors, then prune this set of rules as illustrated in Section 3.1.3.
4. Use the pruned rules to detect failure when each of the consequent sensors is injected with fail-stop failure in the thresholds setting dataset. Afterwards, plot the all-in-one ROC curve of failure detection, that is plotted with aggregating all the sensor failure cases. Furthermore, plot the individual ROC curves of failure detection when each sensor has failed to have more insights about the performance.
5. Find the optimal operating point and the AUC of the all-in-one ROC curve.
6. If the all-in-one ROC curve shows poor performance, i.e., optimal TPR is low (<0.8), optimal FPR is high (>0.02) and AUC is low (<0.9), then delete this set of parameters entry from the sorted combinations and repeat Steps 3–6 with the next highest number of consequent sensors. Otherwise, the selection process of parameters is done successfully, recording the corresponding sliding window size, minimum relative support and minimum confidence.
7. Record the health threshold value that corresponds to the optimal operating point of the all-in-one ROC curve.

The exclusion of the values of confidence that are below 50% in Step 1 is necessary, as when we experimented with below 50% confidence, its ROC curves had always showed poor performance with optimal TPR below 0.8 and/or optimal FPR above 0.02 and/or AUC below 0.9. In addition, the logic in Algorithm A2 which our calculations for failure detection rely upon in the case of rule satisfaction is sustained while using > = 50% confidence. If we used a low confidence rule, e.g., 10%, and it is satisfied then the probability that the sensors of the satisfied rule are faulty would be 90%, which would make rule satisfaction useless to confirm that its sensors are nonfaulty due to fulfilling the correlation.

To select the parameters for our case study, the proposed guidelines were followed. In Step 1, the set of values we used for the sliding window size was [0, 3, 5, 8, 10, 15, 20, 25, 30, 45, 60] s, the minimum relative support set was [2, 5, 10, 15, 20, 25, 30, 35, 45] %, and the minimum confidence set was [50, 60, 70, 80, 90, 100] %. Note that the number of sensor events of the dataset can be divided by its collection duration to get an estimate about the rate of sensors triggering and accordingly choose the range of set values of the sliding window size. The preliminary health threshold value was chosen to be 0.4. The highest number of consequent sensors that could be obtained using the various combinations of the sets was 31 sensors. However, the values of the parameters that yield 31 consequent sensors produce bad failure detection performance that is reflected on its ROC curves. Figure 7 shows the ROC curves that were plotted from setting the sliding window size to 60 s, minimum relative support to 5% and minimum confidence to 50%, this setting yields rules with 31 consequent sensors. The all-in-one ROC curve has an optimal TPR of 0.7169, optimal FPR of 0.06104 and AUC of 0.8903. Iterating back between Steps 3–6, until good ROC curves in Figure 8 are reached from setting the sliding window size to 30 s, minimum relative support to 15% and minimum confidence to 60%. These finally selected values of parameters could detect failures for 28 sensors. Its all-in-one ROC curve has an optimal TPR of 0.8773, optimal FPR of 0.01593 and AUC of 0.9419. The health threshold value that corresponds to the optimal operating point is 0.3591. Note that it may happen that multiple combinations of parameters for the same number of consequent sensors would produce similar overall performance but with one sensor performing better than the other, and vice versa. In our case study, the previously mentioned selected values for parameters produced close performance to that of using sliding window of 45 s, minimum relative support of 20% and minimum confidence of 60%. However, we favoured our selection because less computational effort during the monitoring stage is needed for the smaller sliding window size.

4.4. Experiments

Three types of failures were injected in the testing dataset; fail-stop, obstructed-view and moved-location failures. Each consequent sensor was injected with failure, and the failure detection as well as isolation was evaluated. The initial values of all health status of sensors were set to 1. The sliding window size, minimum relative support, minimum confidence and health threshold were set to 30 s, 15%, 60% and 0.3591, respectively, according to the selection of parameters conducted in Section 4.3.2. The following sensors, D001, D002, M002, M004, M025 and M031, were not checked for failure, as they did not appear in the consequent part of any rule.

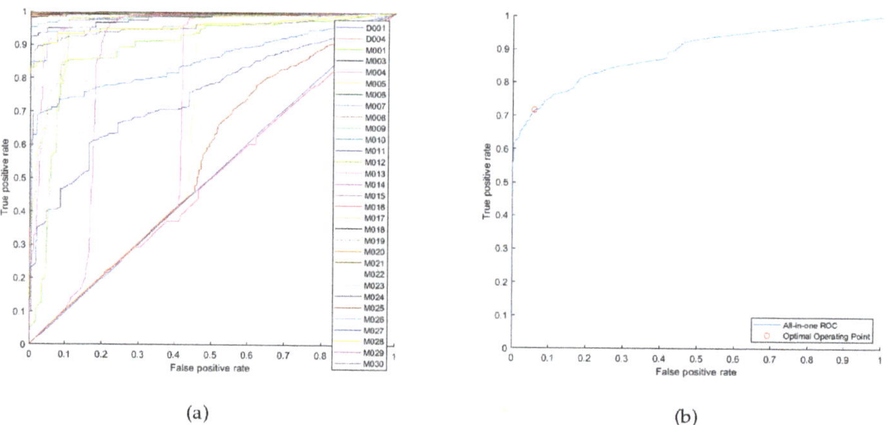

Figure 7. Using sliding window size of 60 s, minimum relative support 5% and minimum confidence of 50%: (**a**) ROC curves of failure detection when each consequent sensor has fail-stop failure. (**b**) All-in-one ROC curve.

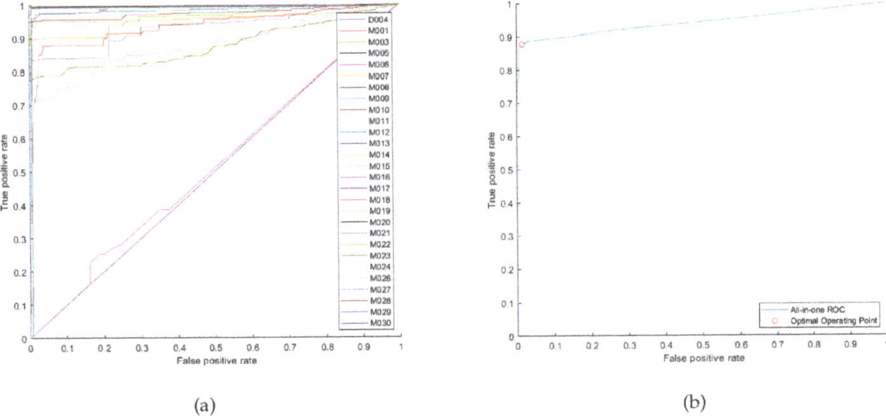

(a) (b)

Figure 8. Using sliding window size of 30 s, minimum relative support 15% and minimum confidence of 60%: (**a**) ROC curves of failure detection when each consequent sensor has fail-stop failure. (**b**) All-in-one ROC curve.

4.4.1. Fail-Stop Failure

Fail-stop failure was injected by replacing the readings of the sensor under test by zeros after its point of failure. Fail-stop failure was injected individually on each of the sensors that appeared in the consequent part of rules. The precision and recall of detecting fail-stop failure when failure is injected in each of those sensors is shown in Figure 9a. Meanwhile, the precision and recall for isolating the faulty sensor is shown in Figure 9b. The precision and recall metrics were computed as described in Section 4.2. On the x-axis of Figure 9 lie the IDs of all the event-driven sensors of the apartment shown in Figure 5. The figures are interpreted as follows, the bar columns at sensor D004 in Figure 9a are the precision and recall values of detecting that a failure has occurred when D004 was injected with fail-stop failure. While in Figure 9b, the columns at D004 show the precision and recall of identifying that D004 has failed. No columns were plotted at D001, D002, M002, M004, M025 and M031, as those sensors were not injected with failure nor evaluated as they did not appear as a consequent in the rules. Most of the consequent sensors have high precision and recall for its detection and isolation. There are 26 sensors that when injected with fail-stop failure cause failure detection precision ≥ 0.95, and 24 sensors that cause a recall ≥ 0.87. Isolation precision is ≥ 0.97 for 26 sensors, while the isolation recall is ≥ 0.87 for 24 sensors. The isolation latency was plotted in Figure 9c. The isolation latency is between 2 and 7 h in 13 sensors, 12 and 24 h in 6 sensors and 24 and 48 h in 5 sensors. There are 4 sensors (M001, M011, M016 and M017) that reported very high isolation latency ≥ 120 h. The higher the rate at which the sensor is triggered by the user, i.e., higher support, the shorter the time needed for isolation. It is observed that the sensors which have high isolation precision but along with low isolation recall and high latency, e.g., M001 and M011, are those governed by rules of low support. D002 appears as an antecedent in all the governing rules of M016 and M017, e.g., D002, M019 -> M016. In the first two segments of the testing data, D002 did not have any triggers. Thus, the rules that have M016 and M017 as consequent were never initiated in the first two segments. As a result, M016 and M017 have undefined isolation precision in Figure 9b because of the zero true positives of those two segments. Those segments that have undefined isolation precision were excluded when calculating the average isolation latency for each sensor plotted in Figure 9c. M016 and M017 have high trigger rates but their rules have low support, because one of its antecedent sensors, D002, has a low trigger rate. To calculate the average precision and recall of failure detection and isolation among the examined sensors of the experiment, the two segments of M016 and M017 that had undefined isolation precision were excluded. The average precision and recall of failure detection are 0.9493 and 0.9018, respectively, while the average failure isolation precision and recall are 0.9987 and 0.9116, respectively.

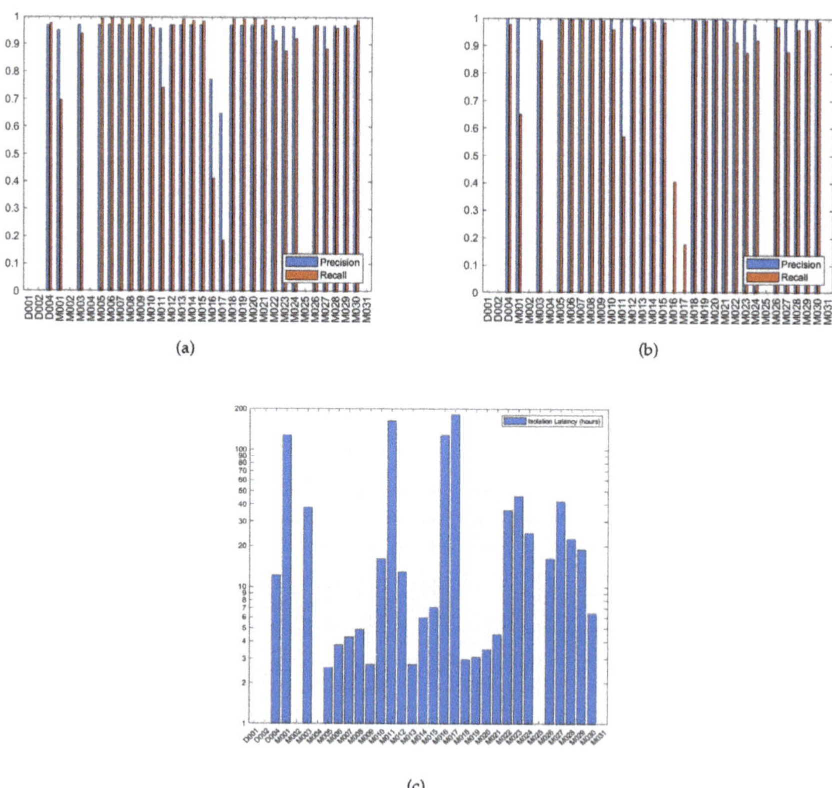

Figure 9. Fail-Stop Failure: (**a**) Precision and recall of failure detection. (**b**) Precision and recall of failure isolation. (**c**) Failure isolation latency.

4.4.2. Obstructed-View Failure

Obstructed-view failure is the failure at which the sensor view is obstructed, e.g., its view gets blocked by furniture. It was simulated by replacing the sensor readings by zeros along the duration at which the sensor view was obstructed. The obstruction duration was set to 5 days. Figure 10a shows the precision and recall of detecting 5 days of obstructed-view failure. The precision and recall for isolating the faulty sensor and its isolation latency are shown in Figure 10b,c, respectively. Similar to the fail-stop failures, detecting and isolating most consequent sensors show high detection and isolation performance except for M001, M011, M016 and M017. There are 20 sensors that when injected with obstructed-view failure cause failure detection precision ≥ 0.9, and 4 sensors between 0.8 and 0.9. Meanwhile, 24 sensors can be isolated with precision ≥ 0.92, and 19 sensors can be isolated with recall ≥ 0.87. The average failure detection precision and recall among examined sensors are 0.8563 and 0.8089, respectively. The average failure isolation precision and recall are 0.9954 and 0.8285, respectively. The isolation latency for the sensors injected with obstructed-view failure is almost the same as when injected with fail-stop failure.

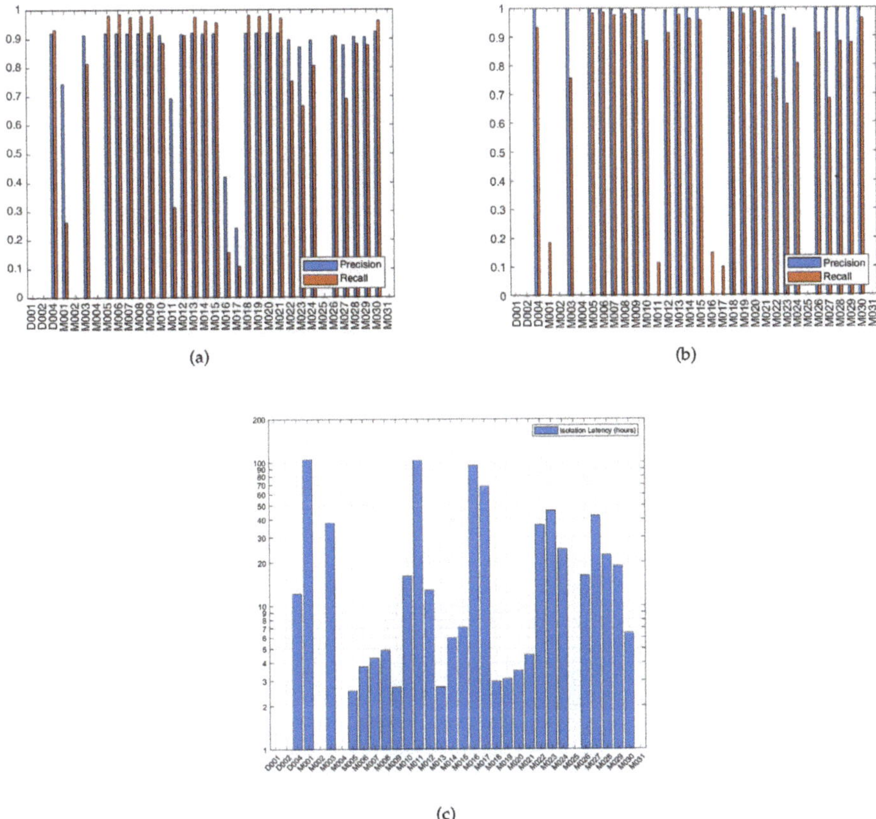

Figure 10. Obstructed-View (5 days) Failure: (**a**) Precision and recall of failure detection. (**b**) Precision and recall of failure isolation. (**c**) Failure isolation latency.

4.4.3. Moved-Location Failure

Moved-location failure means that a sensor's location has changed, this may happen when a sensor gets remounted by the user in the wrong location or when it is mounted on a piece of furniture that has been moved to another location. This type of failure was simulated by changing the readings of the sensor after its point of failure by readings of its newly moved location. Figure 11 shows the performance of detecting and isolating the moved-location of some of the consequent sensors. The x-axis of Figure 11 describes the moved-location case, e.g., D004 -> D002, means that the sensor D004 has moved to the location of sensor D002. Figure 11a plots the precision and recall of detecting failure, and Figure 11b shows the precision and recall of identifying that the moved sensor has failed, i.e., the failed sensor is D004 in our previous example. The precision of failure detection in the presented 13 moved-location cases are $\geq$0.9, and the precision of the failure isolation is $\geq$0.99 in the presented cases except for M010 -> M013 is 0.83. On the other hand, the recall of failure detection is $\geq$0.82 for 6 cases, between 0.7 and 0.8 for 5 cases, and $\leq$ 0.6 for 2 cases. Meanwhile, the recall of failure isolation is $\geq$0.8 for 5 cases, between 0.68 and 0.8 for 5 cases, and $\leq$0.6 for 3 cases. The average failure detection precision and recall among the presented cases are 0.9580 and 0.74, respectively, while the average failure isolation precision and recall are 0.9863 and 0.6839, respectively. The isolation latency is $\leq$7 h in 8 cases, between 16 and 19 h in 2 cases, and $\geq$42 h in 3 cases. The distance of the new location from the old one is not what dominates the precision or recall of detecting the moved-location failure. Moving a sensor within the same room could be detected with higher recall when M005 was moved to the location of M001 within the bedroom than that of moving M010 to M013 within the living room. Similarly, moving a sensor to another room could be detected with higher recall when D004 was moved from the garage door to replace the D002 at the kitchen back door than that of moving M005 from the bedroom to M009 in living room.

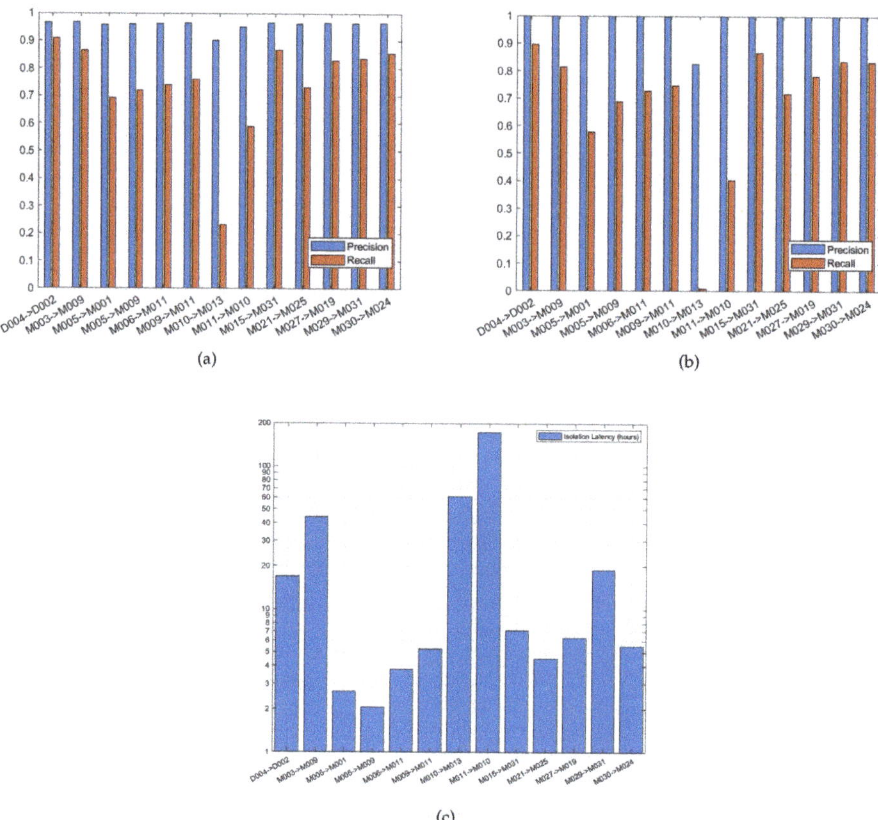

Figure 11. Moved-Location Failure: (**a**) Precision and recall of failure detection. (**b**) Precision and recall of failure isolation. (**c**) Failure isolation latency.

5. Discussion

Our proposed failure detection and isolation system is distinguished by its low computational effort and high interpretability, in addition to its use of unlabelled datasets. The results show that the consequent sensors that were injected with fail-stop and obstructed-view failures could be detected and isolated with high precision and recall. The isolation latency is highly dependable on the behaviour of the resident as well as the start time of failure with respect to his behaviour. The more frequent the usage of the area of an apartment that has the failed sensor is, the shorter the time to isolate this sensor failure. In addition, the start time of the failure affects the isolation latency, i.e., if the sensor failure has occurred just before the resident goes to bed at night, then the failure will not be isolated before the next morning by any means. Detecting moving a sensor to another place can be achieved with high precision and recall only when this newly moved location has minimal correlation to the old location. This is on contrary to the fail-stop and obstructed-view failures, where the sensor failure detection performance is proportional to its correlation to other sensors.

A summary table of the related work was presented in our survey paper [11]. Although the results are not directly comparable due to the use of different datasets, design of experiments and evaluation methodology, the benefits of our proposed system over the other relevant state of the art was presented in Section 2. The limitation of our approach is that the sensors that do not appear as consequent to the activation of other sensor(s) in the apartment cannot be checked for failure. However, our approach can be used to determine these sensors, and thus can help to highlight the needed reconfiguration of sensors' positioning in the apartment to obtain a fully functional sensor failure detection and isolation system.

As for future work, the use of variable size sliding window for detecting failures may further improve the system performance, especially for the moved-location failures. Rules will be extracted for the consequent sensors that have strong rules using shorter duration sliding window during the correlations extraction stage, and only those sensors that did not appear will be extracted over a longer duration sliding window. However, this should be weighed against its computational complexity during the real-time correlations monitoring stage. Furthermore, the use of an auxiliary system to detect failure for those sensors that did not appear as consequent could be investigated. This auxiliary system may exploit the following features for those sensors; its trigger day, trigger time and duration of activation.

6. Conclusions

This paper proposed a failure detection and isolation system for binary event-driven sensors deployed in the AAL environment. Correlations between sensors were extracted with no prior knowledge of the sensor placement on the floor plan and using unlabelled datasets. Guidelines for the selection of the user defined parameters for correlations extraction were presented. The correlations are monitored during run-time to detect sensor failures. The proposed approach was evaluated using publicly available dataset injected with fail-stop, obstructed-view and moved-location failures. The system was able to detect and isolate the various types of failures. The results show that fail-stop failures could be detected with an average precision and recall of 0.9493 and 0.9018, and isolated with average precision and recall of 0.9987 and 0.9116, respectively. Obstructed-view failures were detected with average precision of 0.8563 and recall of 0.8089, and isolated with average precision of 0.9954 and recall of 0.8285. Meanwhile, the moved-location failures were detected at 0.9580 average precision and at 0.74 average recall and isolated at 0.9863 average precision and 0.6839 average recall.

Author Contributions: Conceptualization, N.E.E., S.J., J.P. and V.S.; Methodology, N.E.E.; Investigation, N.E.E.; Validation, N.E.E.; Writing—original draft preparation, N.E.E.; Writing—review and editing, S.J. and J.P.; Supervision, S.J. and V.S. All authors have read and agreed to the published version of the manuscript.

Funding: This research paper is part of a Ph.D. Thesis granted by the Ministry of Higher Education of Egypt.

Acknowledgments: The authors acknowledge the Technical University of Munich for supporting the publication in the framework of the Open Access Publishing Program. Furthermore, the authors would like to thank Maximilian Kapsecker, Jens Klinker and Lara Marie Reimer for their technical assistance during the revision stage of the manuscript.

Conflicts of Interest: The authors declare no conflict of interest.

Abbreviations

The following abbreviations are used in this manuscript:

AAL	Ambient assisted living
ICT	Information and communication technologies
AmI	Ambient intelligence
ADL	Activities of daily living
PIR	Passive infrared sensor
TP	True positives
FN	False negatives
FP	False positives

TN	True negatives
ROC	Receiver operating characteristic
AUC	Area under curve
TPR	True positive rate
FPR	False positive rate

Appendix A

Algorithm A1 Failure detection.

Input:

DataStream: the stream of the AAL sensors *events*

Sen: the set of sensors represented by tuples {(*id*, *Health*, *FailFlag*)}, where *id* is the sensor's id number, *Health* is the health status of sensor, and *FailFlag* is the failure flag of sensor

R: the set of rules represented by tuples {(*Ant*, *Conseq*, *Sup*, *Conf*)}, where *Ant* contains the sensors in the rule antecedent, *Conseq* contains the sensors in the rule consequent, *Sup* is the support of rule, and *Conf* is the rule's confidence

SatRulHist: the set of rules that were satisfied in the previous sliding window $SwNum - 1$, where *SwNum* is the sliding window's running number

UnSatRulHist: the set of rules that were unsatisfied, i.e., has one missing sensor in the rule consequent, in the previous sliding window $SwNum - 1$

FutSw: the set of sensors that are active in the next sliding window $SwNum + 1$

HealthThresh: the threshold value for the health status of sensors

Output:

Sen: updated *Health* and *FailFlag* of the set of sensors

```
while CurrSW = ProcessSW(DataStream) do
  // CurrSw is the set of active sensors in the current sliding window SwNum
  for each Rul ∈ R ∧ Rul.Ant ⊆ CurrSw do
    if Rul.Conseq ⊂ CurrSw then
      Sen.Health ← SatisfHealthUpdate(Rul, CurrSw, Sen)
      SatRulHist ← Rul
    else if |Rul.Conseq − CurrSw| = 1 then
      if Rul ∉ SatRulHist ∧ |(Rul.Ant ∪ Rul.Conseq) − FutSw| ≠ ϕ then
        if Rul ∉ UnSatRulHist then
          Sen.Health ← UnSatisfHealthUpdate(Rul,CurrSw,Sen)
        end if
        UnSatRulHist ← Rul
      end if
    end if
  end for
  for each s ∈ Sen do
    if s.Health < HealthThresh then
      s.FailFlag ← 1
    else
      s.FailFlag ← 0
    end if
  end for
end while
```

Algorithm A2 Health Status Update due to Rule Satisfaction.

SatisfHealthUpdate $\{Rul, CurrSw, Sen\}$
for each $s \in Sen, s.id \in (Rul.Ant \cup Rul.Conseq)$ **do**
  $PrF \leftarrow 1 - Rul.Conf$ // PrF is the probability that the sensor is faulty
  $s.Health \leftarrow 0.1 \times (1 - PrF) + 0.9 \times s.Health$
end for
return $Sen.Health$

Algorithm A3 Health Status Update due to Rule UnSatisfaction.

UnSatisfHealthUpdate $\{Rul,CurrSw,Sen\}$
for each $s \in Sen, s.id \in Rul.Ant$ **do**
  if $|Rul.Conseq| = 1 \land |Rul.Ant| = 1$ **then**
    $PrF \leftarrow 1 - P(s)$
  else
    $PrF \leftarrow 1 - \frac{P(\bigcap_{\{x \in (CurrSW \cap (Rul.Ant \cup Rul.Conseq)\}} x)}{P(\bigcap_{\{x \in (CurrSW \cap (Rul.Ant \cup Rul.Conseq))|x \neq s\}} x)}$
  end if
  $s.Health \leftarrow 0.1 \times (1 - PrF) + 0.9 \times s.Health$
end for
for each $s \in Sen, s.id \in Rul.Conseq$ **do**
  if $s.id \in CurrSW$ **then**
    $PrF \leftarrow 1 + \frac{Rul.Sup - P(\bigcap_{\{x \in (CurrSW \cap (Rul.Ant \cup Rul.Conseq))\}} x)}{P(\bigcap_{\{x \in (CurrSW \cap (Rul.Ant \cup Rul.Conseq))|x \neq s\}} x) - P(\bigcap_{\{x \in (Rul.Ant \cup Rul.Conseq)|x \neq s\}} x)}$
  else
    $PrF \leftarrow \frac{Rul.Sup}{P(\bigcap_{\{x \in (CurrSW \cap (Rul.Ant \cup Rul.Conseq))\}} x)}$
  end if
  $s.Health \leftarrow 0.1 \times (1 - PrF) + 0.9 \times s.Health$
end for
return $Sen.Health$

References

1. Nations, U. *World Population Ageing*; Technical Report; Department of Economic and Social Affairs, Population Division: New York City, NY, USA, 2019.
2. Rashidi, P.; Mihailidis, A. A Survey on Ambient-Assisted Living Tools for Older Adults. *IEEE J. Biomed. Health Inform.* **2013**, *17*, 579–590. [CrossRef] [PubMed]
3. Monekosso, D.; Florez-Revuelta, F.; Remagnino, P. Ambient Assisted Living [Guest editors' introduction]. *IEEE Intell. Syst.* **2015**, *30*, 2–6. [CrossRef]
4. Dobre, C.; Mavromoustakis, C.X.; Garcia, N.M.; Mastorakis, G.; Goleva, R.I. Introduction to the AAL and ELE Systems. In *Ambient Assisted Living and Enhanced Living Environments*; Butterworth-heinemann: Oxford, UK, 2017; pp. 1–16.
5. Viard, K.; Fanti, M.P.; Faraut, G.; Lesage, J.J. An event-based approach for discovering activities of daily living by hidden Markov models. In Proceedings of the 2016 15th International Conference on Ubiquitous Computing and Communications and 2016 International Symposium on Cyberspace and Security (IUCC-CSS), Granada, Spain, 14–16 December 2016; pp. 85–92.

6. Kapitanova, K.; Hoque, E.; Stankovic, J.A.; Whitehouse, K.; Son, S.H. Being SMART about failures: Assessing repairs in SMART homes. In Proceedings of the 2012 ACM Conference on Ubiquitous Computing, Pittsburgh, PA, USA, 5–8 September 2012; pp. 51–60.
7. Rahal, Y.; Pigot, H.; Mabilleau, P. Location Estimation in a Smart Home: System Implementation and Evaluation Using Experimental Data. *Int. J. Telemed. Appl.* **2008**, *2008*, 142803. [CrossRef] [PubMed]
8. Feng, Z.; Fu, J.Q.; Wang, Y. Weighted distributed fault detection for wireless sensor networks Based on the distance. In Proceedings of the 33rd Chinese Control Conference, Nanjing, China, 28–30 July 2014; pp. 322–326.
9. Fan, C.; Tan, J. A majority voting scheme in wireless sensor networks for detecting suspicious node. In Proceedings of the 2009 Second International Symposium on Electronic Commerce and Security, Nanchang, China, 22–24 May 2009; Volume 2, pp. 495–498.
10. Nguyen, T.A.; Bucur, D.; Aiello, M.; Tei, K. Applying time series analysis and neighbourhood voting in a decentralised approach for fault detection and classification in WSNs. In Proceedings of the Fourth Symposium on Information and Communication Technology, Da Nang, Vietnam, 5–6 December 2013; pp. 234–241.
11. ElHady, N.E.; Provost, J. A Systematic Survey on Sensor Failure Detection and Fault-Tolerance in Ambient Assisted Living. *Sensors* **2018**, *18*, 1991. [CrossRef]
12. Ballardini, A.L.; Ferretti, L.; Fontana, S.; Furlan, A.; Sorrenti, D.G. An indoor localization system for telehomecare applications. *IEEE Trans. Syst. Man Cybern. Syst.* **2016**, *46*, 1445–1455. [CrossRef]
13. Ahvar, E.; Lee, G.M.; Han, S.N.; Crespi, N.; Khan, I. Sensor network-based and user-friendly user location discovery for future smart homes. *Sensors* **2016**, *16*, 969. [CrossRef] [PubMed]
14. Mckeever, S.; Ye, J.; Coyle, L.; Bleakley, C.; Dobson, S. Activity recognition using temporal evidence theory. *J. Ambient. Intell. Smart Environ.* **2010**, *2*, 253–269. [CrossRef]
15. Javadi, E.; Moshiri, B.; Yazdi, H.S. Activity Recognition In Smart Home Using Weighted Dempster-Shafer Theory. *Int. J. Smart Home* **2013**, *7*, 23–34. [CrossRef]
16. Amri, M.H.; Aubry, D.; Becis, Y.; Ramdani, N. Robust fault detection and isolation applied to indoor localization. *IFAC-PapersOnLine* **2015**, *48*, 440–445. [CrossRef]
17. Danancher, M. A Discrete Event Approach for Model-Based Location Tracking of Inhabitants in Smart Homes. Ph.D. Thesis, École Normale Supérieure de Cachan-ENS Cachan, Cachan, France, 2013.
18. Veronese, F.; Pour, D.S.; Comai, S.; Matteucci, M.; Salice, F. Method, Design and Implementation of a Self-checking Indoor Localization System. In *International Workshop on Ambient Assisted Living*; Springer: Berlin, Germany, 2014; pp. 187–194.
19. Veronese, F.; Comai, S.; Matteucci, M.; Salice, F. Method, design and implementation of a multiuser indoor localization system with concurrent fault detection. In Proceedings of the 11th International Conference on Mobile and Ubiquitous Systems: Computing, Networking and Services, ICST (Institute for Computer Sciences, Social-Informatics and Telecommunications Engineering), London, UK, 2–5 December 2014; pp. 100–109.
20. Mohamed, A.; Jacquet, C.; Bellik, Y. A fault detection and diagnosis framework for ambient intelligent systems. In Proceedings of the Ubiquitous Intelligence & Computing and 9th International Conference on Autonomic & Trusted Computing (UIC/ATC), Fukuoka, Japan, 4–7 September 2012; pp. 394–401.
21. Jacquet, C.; Mohamed, A.; Bellik, Y. An ambient assisted living framework with automatic self-diagnosis. *Int. J. Adv. Life Sci.* **2013**, *5*, 10.
22. Oliveira, C.H.S.; Giroux, S.; Ngankam, H.; Pigot, H. Generating Bayesian Network Structures for Self-diagnosis of Sensor Networks in the Context of Ambient Assisted Living for Aging Well. In Proceedings of the International Conference on Smart Homes and Health Telematics, Paris, France, 29–31 August 2017; pp. 198–210.
23. Munir, S.; Stankovic, J.A. Failuresense: Detecting sensor failure using electrical appliances in the home. In Proceedings of the Mobile Ad Hoc and Sensor Systems (MASS), Philadelphia, PA, USA, 28–30 October 2014; pp. 73–81.
24. Kodeswaran, P.A.; Kokku, R.; Sen, S.; Srivatsa, M. Idea: A system for efficient failure management in smart iot environments. In Proceedings of the 14th Annual International Conference on Mobile Systems, Applications, and Services, Singapore, 25–30 June 2016; pp. 43–56.

25. Ye, J.; Stevenson, G.; Dobson, S. Using temporal correlation and time series to detect missing activity-driven sensor events. In Proceedings of the Pervasive Computing and Communication Workshops (PerCom Workshops), St. Louis, MO, USA, 23–27 March 2015; pp. 44–49.
26. Ye, J.; Stevenson, G.; Dobson, S. Fault detection for binary sensors in smart home environments. In Proceedings of the Pervasive Computing and Communications (PerCom), St. Louis, MO, USA, 23–27 March 2015; pp. 20–28.
27. Ye, J.; Stevenson, G.; Dobson, S. Detecting abnormal events on binary sensors in smart home environments. *Pervasive Mob. Comput.* **2016**, *33*, 32–49. [CrossRef]
28. Kapitanova, K.; Hoque, E.; Stankovic, J.A.; Son, S.H.; Whitehouse, K.; Alessandrelli, D. Being SMART About Failures: Assessing Repairs in Activity Detection. 2011. Available online: https://www.semanticscholar.org/paper/Being-SMART-About-Failures-%3A-Assessing-Repairs-in-Kapitanova-Hoque/f05968403b88738e869a360ca3910bebad5218b4#citing-papers (accessed on 19 August 2020).
29. Choi, J.; Jeoung, H.; Kim, J.; Ko, Y.; Jung, W.; Kim, H.; Kim, J. Detecting and identifying faulty IoT devices in smart home with context extraction. In Proceedings of the 2018 48th Annual IEEE/IFIP International Conference on Dependable Systems and Networks (DSN), Luxembourg, 25–28 June 2018; pp. 610–621.
30. Agrawal, R.; Imieliński, T.; Swami, A. *Mining Association Rules between Sets of Items in Large Databases;* Acm Sigmod Record; ACM: New York, NY, USA, 1993; Volume 22, pp. 207–216.
31. Yairi, T.; Kato, Y.; Hori, K. Fault detection by mining association rules from house-keeping data. In Proceedings of the International Symposium on Artificial Intelligence, Robotics and Automation in Space, Montréal, QC, Canada, 18–22 June 2001; Volume 3.
32. Hou, Z.; Lian, Z.; Yao, Y.; Yuan, X. Data mining based sensor fault diagnosis and validation for building air conditioning system. *Energy Convers. Manag.* **2006**, *47*, 2479–2490. [CrossRef]
33. Liu, J.; Shi, D.; Li, G.; Xie, Y.; Li, K.; Liu, B.; Ru, Z. Data-driven and association rule mining-based fault diagnosis and action mechanism analysis for building chillers. *Energy Build.* **2020**, *216*, 109957. [CrossRef]
34. Agarwal, R.; Srikant, R. Fast algorithms for mining association rules. In Proceedings of the 20th VLDB Conference, Santiago de Chile, Chile, 12–15 September 1994; pp. 487–499.
35. Jacquenet, F.; Largeron, C.; Udréa, C. Efficient management of non redundant rules in large pattern bases: Bitmap approach. In Proceedings of the Eighth International Conference on Enterprise Information Systems: Databases and Information Systems Integration, Paphos, Cyprus, 23–27 May 2006; pp. 208–215.
36. CASAS Datasets. Available online: http://ailab.wsu.edu/casas/datasets/ (accessed on 19 August 2020).

Publisher's Note: MDPI stays neutral with regard to jurisdictional claims in published maps and institutional affiliations.

MDPI
St. Alban-Anlage 66
4052 Basel
Switzerland
Tel. +41 61 683 77 34
Fax +41 61 302 89 18
www.mdpi.com

Sensors Editorial Office
E-mail: sensors@mdpi.com
www.mdpi.com/journal/sensors

www.ingramcontent.com/pod-product-compliance
Lightning Source LLC
LaVergne TN
LVHW071246170726
843515LV00006BA/1346

* 9 7 8 3 0 3 6 5 0 2 4 8 9 *